Selected Titles in This Series

152 **R. S. Ismagilov,** Representations of infinite-dimensional groups, 1996

151 **S. Yu. Slavyanov,** Asymptotic solutions of the one-dimensional Schrödinger equation, 1996

150 **B. Ya. Levin,** Lectures on entire functions, 1996

149 **Takashi Sakai,** Riemannian geometry, 1996

148 **Vladimir I. Piterbarg,** Asymptotic methods in the theory of Gaussian processes and fields, 1996

147 **S. G. Gindikin and L. R. Volevich,** Mixed problem for partial differential equations with quasihomogeneous principal part, 1996

146 **L. Ya. Adrianova,** Introduction to linear systems of differential equations, 1995

145 **A. N. Andrianov and V. G. Zhuravlev,** Modular forms and Hecke operators, 1995

144 **O. V. Troshkin,** Nontraditional methods in mathematical hydrodynamics, 1995

143 **V. A. Malyshev and R. A. Minlos,** Linear infinite-particle operators, 1995

142 **N. V. Krylov,** Introduction to the theory of diffusion processes, 1995

141 **A. A. Davydov,** Qualitative theory of control systems, 1994

140 **Aizik I. Volpert, Vitaly A. Volpert, and Vladimir A. Volpert,** Traveling wave solutions of parabolic systems, 1994

139 **I. V. Skrypnik,** Methods for analysis of nonlinear elliptic boundary value problems, 1994

138 **Yu. P. Razmyslov,** Identities of algebras and their representations, 1994

137 **F. I. Karpelevich and A. Ya. Kreinin,** Heavy traffic limits for multiphase queues, 1994

136 **Masayoshi Miyanishi,** Algebraic geometry, 1994

135 **Masaru Takeuchi,** Modern spherical functions, 1994

134 **V. V. Prasolov,** Problems and theorems in linear algebra, 1994

133 **P. I. Naumkin and I. A. Shishmarev,** Nonlinear nonlocal equations in the theory of waves, 1994

132 **Hajime Urakawa,** Calculus of variations and harmonic maps, 1993

131 **V. V. Sharko,** Functions on manifolds: Algebraic and topological aspects, 1993

130 **V. V. Vershinin,** Cobordisms and spectral sequences, 1993

129 **Mitsuo Morimoto,** An introduction to Sato's hyperfunctions, 1993

128 **V. P. Orevkov,** Complexity of proofs and their transformations in axiomatic theories, 1993

127 **F. L. Zak,** Tangents and secants of algebraic varieties, 1993

126 **M. L. Agranovskiĭ,** Invariant function spaces on homogeneous manifolds of Lie groups and applications, 1993

125 **Masayoshi Nagata,** Theory of commutative fields, 1993

124 **Masahisa Adachi,** Embeddings and immersions, 1993

123 **M. A. Akivis and B. A. Rosenfeld,** Élie Cartan (1869–1951), 1993

122 **Zhang Guan-Hou,** Theory of entire and meromorphic functions: Deficient and asymptotic values and singular directions, 1993

121 **I. B. Fesenko and S. V. Vostokov,** Local fields and their extensions: A constructive approach, 1993

120 **Takeyuki Hida and Masuyuki Hitsuda,** Gaussian processes, 1993

119 **M. V. Karasev and V. P. Maslov,** Nonlinear Poisson brackets. Geometry and quantization, 1993

118 **Kenkichi Iwasawa,** Algebraic functions, 1993

117 **Boris Zilber,** Uncountably categorical theories, 1993

116 **G. M. Fel′dman,** Arithmetic of probability distributions, and characterization problems on abelian groups, 1993

(Continued in the back of this publication)

Representations of Infinite-Dimensional Groups

Translations of
MATHEMATICAL MONOGRAPHS

Volume 152

Representations of Infinite-Dimensional Groups

R. S. Ismagilov

American Mathematical Society
Providence, Rhode Island

Р. С. Исмагилов

ПРЕДСТАВЛЕНИЯ БЕСКОНЕЧНОМЕРНЫХ ГРУПП

Translated by D. Deart from an original Russian manuscript

1991 *Mathematics Subject Classification.* Primary 22E65, 22E67;
Secondary 17B65, 58B25, 58D05, 60G55, 81R10.

ABSTRACT. This book is devoted to representations of two classes of infinite-dimensional groups: current groups and diffeomorphism groups. The author presents a complete treatment of the subject, including general methods for constructing irreducible representations of infinite-dimensional groups and general results about such representations. He also exhibits deep relations between representations of infinite-dimensional groups and the theory of Fock spaces, the theory of point random processes, and other branches of mathematics.

The book will be useful for researchers and graduate students working in representation theory, functional analysis, and mathematical physics.

Library of Congress Cataloging-in-Publication Data
Ismagilov, R. S. (Rais Salmanovich)
Representations of infinite-dimensional groups / R. S. Ismagilov; [translated by D. Deart from an original Russian manuscript].
p. cm.—(Translations of mathematical monographs, ISSN 0065-9282; v. 152)
Includes bibliographical references.
ISBN 0-8218-0418-9 (acid-free)
1. Lie groups. 2. Representations of groups. 3. Loops (Group theory) I. Title. II. Series.
QA387.I76 1996
512′.55—dc20 96-15810
CIP

∞ The paper used in this book is acid-free and falls within the guidelines
established to ensure permanence and durability.
Printed on recycled paper.

10 9 8 7 6 5 4 3 2 1 01 00 99 98 97 96

To my parents

Contents

Introduction

This book is mainly devoted to representations of two classes of infinite-dimensional groups that appear in problems close to mathematical analysis. The first class consists of current groups, i.e., groups of mappings (measurable, smooth, etc.) of a given set to a given group. The second class consists of diffeomorphism groups of smooth manifolds and their "large" (infinite-dimensional) subgroups.

The literature devoted to the loop groups (smooth currents on the circle) and groups of diffeomorphisms of the circle is ample; a lot of deep results in mathematics in the last ten years are related to these groups. In the present book, talking about groups of smooth currents and diffeomorphism groups, we will mainly consider manifolds of dimension greater than one (although in the last, eighth chapter, this aspect of the theory will appear, though in a rather nonstandard setup; more about this later). Of course, in such an approach a lot of deep relations and properties that are inherent in the case of a circle are lost. On the other hand, in such "dimension increase" some new problems appear, and the present book is devoted to these problems.

A few words about the content of the book. The first, introductory, chapter is devoted to certain general techniques that are used later in the construction of representations. We consider Fock spaces and representations of infinite-dimensional symplectic and orthogonal groups. Our exposition differs from the one used in, say, Berezin's classical book "The method of second quantization" [**2**] in that we devote our main attention to computations with quadratic exponents. In particular, representations of the above infinite-dimensional groups are defined by their action on quadratic exponents. This method immediately leads to exact formulas; here we partly follow Segal [**64**].

After that we consider a general scheme for the construction of representations of a given group by imbedding the group into the group of motions of a Hilbert spaces and taking the canonical projective representation of the latter group in the Fock space; the imbedding is defined by a unitary representation of the original group and a cocycle of this representation. Here we present results of Araki [**41**] and Vershik, Gelfand, Graev [**6, 7**]. Also, we present an irreducibility condition using the notion of "singularly related measures" on an abelian group; later this condition is essentially used in the analysis of groups of smooth currents and of diffeomorphism groups.

Finally, in the first chapter we consider generalized determinant bundles introduced by Mickelsson and Rajeev [**60**]. More precisely, we present a simple general construction that leads to representations of abelian extensions of groups acting on manifolds. One of the realizations of this scheme gives the generalized determinant bundle.

The second chapter is devoted to representation of groups of measurable currents. It starts with "factorizable operators" in the Fock space introduced in Araki, Woods [**43**] and Araki [**42**]. The main result here is the "convexity theorem", which gives a simple approach to the analysis of irreducibility of factorizable representations for groups and semigroups of measurable currents. This approach essentially solves the irreducibility problem.

Another subject of the second chapter is the construction of representations of groups of measurable currents in continuous tensor products of Hilbert spaces.

In the third chapter we consider representations of groups of smooth currents with values in Lie groups. The main results are related to the case when this Lie group is compact and semisimple. There exists essentially unique "interesting" series of such representations. The corresponding representations were discovered around 1976 simultaneously by several authors, and were later called "energy representations." This is one of the realizations of the above-mentioned "Araki scheme" based on the use of the Maurer–Cartan cocycle. The analysis of the irreducibility of these representations reduces to a nontrivial problem about translations of a Gaussian measure in a function space.

The solution of this problem depends on the dimension n of the corresponding space X. It was obtained by several authors (although the case $n = 2$ is not completely studied yet). We present a simple and uniform proof of the corresponding results. This proof is a combination of methods of Kahane [**55**], Kusuoka [**56**], and the author [**17**]. The second tool in the proof is the irreducibility criterion mentioned earlier, which uses the notion of singularly related measures in function spaces.

Also in the third chapter we consider representations of certain extensions of the group of smooth currents. First, a certain modification of the energy representations provides representations of the universal covering of the group of smooth current. Then we describe unitary representations of some abelian extension. Finally, in this chapter we construct representations of the Mickelsson–Rajeev extension using the above generalized determinant bundle. Here we follow the above-mentioned paper by Mickelsson and Rajeev.

Chapter 4 contains the survey of properties of "large" diffeomorphism groups, including, in particular, groups that preserve a volume form, or a symplectic or a contact structure. We consider the following problems: subgroups of diffeomorphisms that can be factored into the product of diffeomorphisms with arbitrarily small supports, "abelianization", stabilizers of finite configurations of points on the manifold. The results are either rather simple or well known; they are used later in the construction of representations. Among the new results here we mention the content of §21 (theorem on adjoint elements) and of §25, where an example of an infinite-dimensional symplectic manifold and the corresponding "geometric quantization" are considered.

In Chapter 5 we consider representations of diffeomorphism groups. Of course, it is impossible to describe all representations (say, unitary and irreducible). A reasonable formulation of the problem would be to find an explicit realization of representations determined by some natural property. The first result of this nature in Chapter 5 is the description of those representations of the diffeomorphism group (more explicitly, of the connected component in the group of compactly supported diffeomorphisms) that contain a nonzero vector invariant with respect to the subgroup of diffeomorphisms supported in a given open subset of the manifold. The result claims that any such representation can be realized by a dynamical system with the diffeomorphism group as "time" and the set of all closed subsets (configurations) as the set with measure.

This general result sets the style of the subsequent exposition: all further considerations are related to the space of configurations. For representations related to finite configurations we obtain conditions for irreducibility and nonequivalence, and prove an analog of the Mackey theorem on induced representations. (Let us note that the last result is obtained for "smooth" representations only.) We present examples of quasi-invariant measures supported on the space of countable configurations. A special role is played by the Poisson measure on the space of locally finite configurations in $\mathbb{R}^n$; this measure appears in the classification of representations of the diffeomorphism group of $\mathbb{R}^n$ that are spherical with respect to the subgroup of volume preserving diffeomorphisms.

In Chapter 6 representations of certain group extensions are considered. In particular, we consider the semidirect product of the diffeomorphism group of a space with measure to the group of measurable currents. Also, we consider the semidirect product of the group of diffeomorphisms (with compact support) of the space $\mathbb{R}^n$ to the group of smooth functions. We reproduce (with certain modifications) the result of Goldin–Grodnik–Powers–Sharp that describes a class of representations of this group.

Chapter 7 is devoted to the group of diffeomorphisms preserving a volume form or a symplectic form. Once again, we use the Araki scheme to construct a representation in the Fock space. For the first of these groups the required cocycle is obtained from an arbitrary closed nonexact 1-form on the manifold; for the second group it is obtained using mappings of the group of symplectomorphisms studied by Calabi [**45**]. The irreducibility theorem is proved. Finally, for the first of these groups irreducible representations spherical under the subgroup supported on a fixed maximal cell of the manifold are considered.

In Chapter 8 we consider the simplest example of the group of those diffeomorphisms of a manifold that preserve a certain foliation, namely, the foliation of the two-dimensional torus with the irrational rotation number. Let us briefly describe the main result.

It is known that the group of orientation preserving diffeomorphisms of the circle can be embedded into the infinite-dimensional symplectic or into the infinite-dimensional orthogonal group. Taking canonical representations of these two groups in the Fock spaces (mentioned earlier), we obtain highest weight projective representations of the group of diffeomorphisms of the circle. Chapter 8 contains a variation on this theme related to the group of diffeomorphisms of the torus preserving the foliation with the irrational rotation number and type II factors. More precisely, for each factor of type II_∞ we construct an analog of the symplectic group and an analog of the corresponding classical homogeneous domain. Then we construct a projective (nonunitary) representation of the group in the space of holomorphic functions on this domain. Finally, the group of diffeomorphisms of the torus preserving the foliation with the irrational rotation number is imbedded into the above "symplectic" group. In this way we obtain a projective representation of our diffeomorphism group. Passing to the Lie algebra (an analog of the Virasoro algebra), we prove the existence of an analog of the highest weight.

We could also construct an analog of the orthogonal group related to a II_∞ factor and embed it into the group of smooth currents on the torus. However, this does not yield a projective representation. Also, we indicate the construction of unitary (ordinary and projective) representations without highest weight.

Prerequisites for the book are standard university courses (including elements of representation theory, results about manifolds, differential forms, homology, Gaussian measures). Some of the required results are contained in the first chapter of the book.

CHAPTER 1

Some General Constructions Related to Representations

The results of this chapter will be used later in the book to construct representations of current groups and of diffeomorphism groups. However, our intention is to make this chapter not just an auxiliary one. In it, we briefly (but with complete proofs) present some topics of general interest (Fock spaces, infinite-dimensional symplectic and orthogonal group, generalized determinant bundle, etc.).

§1. Preliminary results (group representations and cohomology, measurable spaces, dynamical systems)

Topics discussed in this section are covered in several books (see, e.g., [**9, 10, 24**]). Our goal is to fix the notation and to outline some simple facts to be used later.

1.1. Group representations. We will consider linear representations of a topological group G in a complex (sometimes real) linear topological space V; a representation will be written as $g \mapsto T(g)$. We assume that the mapping $(g, v) \mapsto T(g)v$ is continuous. Most interesting for our purposes are unitary representations in Hilbert spaces.

If T_i is a representation of the group G in the space V_i, $i = 1, 2$, then by $L(T_1, T_2)$ we denote the space of intertwining operators, i.e., operators $A\colon V_1 \to V_2$ satisfying the condition $AT_1(g) = T_2(g)A$ for any $g \in G$.

1.2. Cohomology of Lie groups and Lie algebras. Let a group G act by automorphisms of the group A. We will write this action as ga, $g \in G$, $a \in A$. First, let the group A be abelian, with the group operation $a + b$ and the unit element 0 (such a group A is called a G-module). Then an 1-cocycle φ and a 2-cocycle ψ are functions $\varphi\colon G \to A$, $\psi\colon G \times G \to A$ satisfying the conditions

$$(\partial_1\varphi)(g_1, g_2) \equiv g_1\varphi(g_2) - \varphi(g_1g_2) + \varphi(g_1) = 0,$$
$$(\partial_2\psi)(g_1, g_2, g_3) \equiv g_1\psi(g_2, g_3) - \psi(g_1g_2, g_3) + \psi(g_1, g_2g_3) - \psi(g_1, g_2) = 0$$

for all $g_i \in G$ (the operators ∂_i are called the coboundary operators). Trivial cocycles (or coboundaries) are the cocycles of the form $\varphi(g) = ga_0 - a_0$, $\psi(g_1, g_2) = (\partial_1 b_0)(g_1, g_2)$ for some element $a_0 \in A$ and some function $b_0\colon G \to A$. The quotient groups of all cocycles with respect to the subgroups formed by coboundaries are the cohomology groups $H^i(G, A)$, $i = 1, 2$.

Now, let A be a not necessarily abelian group, with the group operation written in the form $a \cdot b$ and with the unit element e. The group G is still assumed to act by automorphisms of A. A function $\varphi : G \to A$ satisfying the condition

$$\varphi(g_1, g_2) = \varphi(g_1) \cdot (g_1 \varphi(g_2))$$

is called a cocycle (or a 1-cocycle). Cocycles φ_1, φ_2 are called equivalent (cohomologous) if $\varphi_2(g) = c \cdot \varphi_1(g) \cdot (gc)^{-1}$ for some $c \in A$. The set of equivalence classes of 1-cocycles is denoted by $H^1(G, A)$.

If the above groups are assumed to be topological and the action of G on A is continuous, we come to the continuous cohomology groups $H^1_c(G, A)$ (if A is abelian) and to the set $H^1_c(G, A)$ (if A is not assumed to be abelian).

Let there be given a representation $g \mapsto T(g)$, $g \in G$, in a linear space V. Then the groups $H^1_c(G, V)$ defined for the additive group of the space V become linear spaces.

Finally, let $\mathfrak{g}$ be a Lie algebra, V a linear space, $\rho\colon \mathfrak{g} \to L(V)$ a linear representation. A 1-cocycle is a linear map $\varphi\colon \mathfrak{g} \to V$ satisfying the condition

$$(\partial_1 \varphi)(l_1, l_2) \equiv \varphi([l_1, l_2]) - \rho(l_1)\varphi(l_2) + \rho(l_2)\varphi(l_1) = 0,$$

and a 2-cocycle is a linear map $\psi\colon \mathfrak{g} \times \mathfrak{g} \to V$ satisfying the condition

$$\begin{aligned}(\partial_2)(\psi(l_1, l_2, l_3) \equiv \psi([l_1, l_2], l_3) - \psi([l_1, l_3], l_2) + \psi([l_2, l_3], l_1) \\ - \rho(l_1)\psi(l_2, l_3) + \rho(l_2)\psi(l_1, l_3) - \rho(l_3)\psi(l_1, l_2) = 0.\end{aligned}$$

Trivial cocycles (coboundaries) are given by the formulas $\varphi(l) = \rho(l)v_0$ and $\psi(l_1, l_2) = \partial_1 \varphi$, where $v_0 \in V$ and φ is a linear map from $\mathfrak{g}$ to V. Quotient spaces (cocycles/coboundaries) are the cohomology spaces $H^i(\mathfrak{g}, V)$, $i = 1, 2$. The spaces $H^i_c(\mathfrak{g}, V)$ are defined in an obvious way.

1.3. Spaces with measure. The Gaussian measure. The spaces with measure (Ω, μ) that will occur below are assumed to be equivalent to the interval of the real line endowed with the Lebesgue measure defined in Borel sets, with possible addition of a finite or countable set of atoms (points of positive measure). The space (Ω, μ) is called the Lebesgue space and the underlying σ-algebra is called standard. We will often use the notation $\omega \in \Omega \bmod 0$ for a point ω running over the entire set Ω with a removed subset of measure 0.

Let us recall certain facts about the Gaussian measure. Let V_0 be a real linear space with a positive definite inner product, V the Hilbert space obtained as the completion of V_0. The inner product and the length of a vector are denoted by $(\cdot, \cdot)$ and $|\cdot|$ respectively. Let us assume that V_0 is contained in the range of a positive Hilbert-Schmidt operator $A\colon V \to V$. Denote by Ω the space of all linear functionals $\omega\colon V_0 \to \mathbb{R}$ that are continuous with respect to the norm $v \mapsto |A^{-1}v|$. The value of a functional ω on a vector v is denoted by $\langle \omega, v\rangle$. Identifying each vector $u \in V$ with the functional $v \mapsto (v, u)$, we obtain an embedding $V \hookrightarrow \Omega$. Under these conditions there is a measure μ on Ω such that

$$\int_\Omega e^{\langle v, \omega\rangle}\, d\mu(\omega) = e^{|v|^2/2}, \qquad v \in V_0. \tag{1.1}$$

Let us define linear and quadratic functionals on Ω. Choose a basis $\{e_k,\ k \geq 1\}$ of the space V such that $e_k \in V_0$ for all k. For a vector $v \in V$ define

$$(v, \omega) = \sum_{k=1}^{\infty} (v, e_k)\omega_k, \qquad \omega_k = \langle e_k, \omega\rangle. \tag{1.2}$$

For a trace-class operator $T\colon V \to V$ define

$$(1.3)\qquad (T\omega, \omega) = \sum a_{jk}\omega_j\omega_k, \qquad \omega_k = \langle e_k, \omega\rangle, \quad \omega \in \Omega,$$

where $a_{jk} = (Te_k, e_j)$. The series (1.2) and (1.3) converge in the topology defined by the L_2-norm and their sums do not depend on the choice of the basis $\{e_k\}$.

Let us mention the following generalization of formula (1.2). Suppose we are given yet another Hilbert space H and a Hilbert–Schmidt operator $s\colon V \to H$. It can be extended to $\Omega \bmod 0$ by the formula

$$s(\omega) = \sum_{k=1}^{\infty} (s^* h_k, \omega) h_k,$$

where $\{h_k\}$ is an orthonormal basis in H.

Let $V^{\mathbb{C}}$ be the complexification of the space V. For $v = v_1 + iv_2$ we set $\overline{v} = v_1 - iv_2$. Definitions (1.2) and (1.3) can be extended in an obvious way to vectors $v \in V^{\mathbb{C}}$ and trace-class operators $T\colon V^{\mathbb{C}} \to V^{\mathbb{C}}$. For $\|T\| \le 1$ we have

$$(1.4)\quad \int_\Omega e^{(v,\omega)+(1/2)(T\omega,\omega)}\, d\mu(\omega) = \det(I-T)^{-1/2} e^{(1/2)((I-T)^{-1}v,\overline{v})}, \qquad v \in V^{\mathbb{C}}.$$

Here det is the Fredholm determinant of an operator (see [**35**, Chap. XIII]). In particular, we get the following generalization of formula (1.1):

$$(1.5)\qquad \int_\Omega e^{(v,\omega)}\, d\mu(\omega) = e^{(1/2)(v,\overline{v})}, \qquad v \in V^{\mathbb{C}}.$$

Let us consider some transformations of the space Ω. The shift $\omega \mapsto \omega + \omega_0$ maps the measure μ into an equivalent measure if and only if $\omega_0 \in V$ (we use the embedding $V \subset \Omega$ mentioned earlier). The following formula gives the corresponding Radon-Nikodým derivative:

$$(1.6)\qquad \frac{d\mu(\omega+\sigma)}{d\mu(\omega)} = e^{-(1/2)|v|^2-(v,\omega)}, \qquad \omega \in \Omega \bmod 0, \quad v \in V.$$

Let $T\colon V \to V$ be an orthogonal invertible operator. Then there exists a one-to-one measure preserving transformation $T^*\colon \Omega \bmod 0 \to \Omega \bmod 0$ such that $(v, T^*\omega) = (Tv, \omega)$, $\omega \in \Omega \bmod 0$, for any $v \in V$.

1.4. Dynamical systems, cocycles, representations. A dynamical system is a triple (Ω, μ, G), where (Ω, μ) is a space with measure and G is a topological group that acts on Ω, with the measure being quasi-invariant under this action (i.e., each element from G maps μ into an equivalent measure). We will write the action as ωg. Define the Radon-Nikodým derivative

$$j(g, \omega) = \frac{d\mu(\omega g)}{d\mu(\omega)}.$$

Let us require the following continuity conditions to be satisfied:

(D_1) $\mu(E \triangle Eg) \to 0$,

(D_2) $\int_{E\triangle Eg^{-1}} \left|\sqrt{j(g,\omega)-1}\right|^2 d\mu(\omega) \to 0$,

as $g \to e$; here E is any measurable set such that $\mu(E) < \infty$.

Let H^0 be a Hilbert space and $U(H^0)$ the group of unitary operators in H^0. Let $a(g, \omega)$ be a function that is defined almost everywhere on Ω for each $g \in G$, takes values in $U(H^0)$, and is weakly measurable in ω for any g (in other words, for any $h_1, h_2 \in H^0$ the function $\omega \mapsto (a(g, \omega h_1, h_2)$ is measurable). Following 1.2, such a function is called a *cocycle* if the following conditions are satisfied:

(D_3) $a(g_1 g_2, \omega) = a(g_1, \omega)a(g_2, \omega g_1)$, $a(e, \omega) = I$ for all $g_1, g_2 \in G$, $\omega \in \Omega \bmod 0$.

(D_4) $\int_E (a(g, \omega)h, h)d\mu(\omega) \to (h, h)\mu(E)$ as $g \to e$ for any $h \in H^0$ and any measurable set E such that $\mu(E) < \infty$.

Condition (D_3) means that the mapping $g \to a(g, \cdot)$ is a 1-cocycle of the action of the group G on the group of measurable mappings $\Omega \to U(H^0)$; as usual, we identify mappings that coincide almost everywhere. The equivalence of cocycles $a_i(g, \omega)$, $i = 1, 2$, is expressed by the formula

$$a_2(g, \omega) = c(\omega)a_1(g, \omega)(c(\omega g))^{-1},$$

where $c \colon \Omega \to U(H^0)$ is a weakly measurable mapping.

Using such a cocycle, let us introduce the operators $T(g)$, $g \in G$, in the space $H = L_2(\Omega, \mu, H^0)$ by the formula

$$T(g)f(\omega) = \sqrt{j(g, \omega)}a(g, \omega)f(\omega g), \qquad f \in H, \quad g \in G. \tag{1.7}$$

PROPOSITION 1.1. *Conditions* (D_1)–(D_4) *are equivalent to the condition that the operators* (1.7) *define a strongly continuous unitary representation of the group* G.

The proof is easy, and we do not present it here.

Let us consider cocycles $a_i(\omega, g)$, $i = 1, 2$, of a dynamical system (Ω, μ, G), taking values in the groups $U(H_i^0)$ of unitary operators in Hilbert spaces H_i^0. Let T_i be the corresponding unitary representations of the group G. Consider a weakly measurable family $\omega \mapsto A(\omega)$, $\omega \in \Omega \bmod 0$, of mappings from H_1^0 to H_2^0 satisfying the conditions $\operatorname{ess\,sup} \|A\omega\| < \infty$ and

$$a_2(g, \omega)A(\omega g) = A(\omega)a_1(g, \omega), \qquad g \in G, \quad \omega \in \Omega \bmod 0.$$

To the family $A(\omega)$ there corresponds the operator $A \colon H_1 \to H_2$, $H_i = L_2(\Omega, \mu, H_i^0)$, acting by the formula $(Af)(\omega) = A(\omega)f(\omega)$, $f \in H$. We say that the operator A (or the family $A(\omega)$) intertwines a_1 and a_2. Denote the set of all such operators by $L(a_1, a_2)$. It is clear that A intertwines also the representations T_1 and T_2. Therefore, we get the embedding

$$L(a_1, a_2) \hookrightarrow L(T_1, T_2). \tag{1.8}$$

DEFINITION 1.1. The cocycle $(g, \omega) \mapsto a(g, \omega)$ is called *irreducible* if $L(a, a) = \mathbb{C}I$, where I is the identity operator.

PROPOSITION 1.2. *If the representation* (1.7) *is irreducible, then the dynamical systems* (Ω, μ, G) *is ergodic and the cocycle* $a(g, \omega)$ *is irreducible.*

Let us recall that the ergodicity is defined as follows: if $E \subset \Omega$ is a measurable set and $\mu(E \triangle Eg) = 0$ for all $g \in G$, then either $\mu(E) = 0$ or $\mu(\Omega \setminus E) = 0$.

The proof of the proposition is clear.

1.5. Induced representations. Let a dynamical system (Ω, μ, G) be transitive. It would be convenient to assume that Ω is a topological space with a continuous action of G and μ a Borel measure on Ω. Take a point $\omega_0 \in \Omega$, its stabilizer $G^{\omega_0} = \{g \in G \mid \omega_0 g = \omega_0\}$, and, finally, a unitary representation $g \mapsto U(g)$, $g \in G^{\omega_0}$, in a Hilbert space H^0. Let us recall how to construct the induced representation $T = \operatorname{Ind} U$ of the group G.

Let us consider the space $G \times H^0$ and the action of the group G^{ω_0} on this space given by the formula $g_0(g, h) = (g_0 g, U(g_0)h)$, $g_0 \in G^{\omega_0}$, $(g, h) \in G \times H^0$. Let W be the space of orbits of this action. There is a mapping $W \to \Omega$ that sends the orbit of $(g, h) \in G \times H^0$ to the point $\omega_0 g$. This mapping defines a vector bundle on Ω, whose fibers are Hilbert spaces. The right action of G on $G \times H^0$ given by the formula $(g, h)g_1 = (gg_1, h)$, $g_1 \in G$, defines the right action of the group G on W by automorphisms of the bundle $W \to \Omega$. Let H be the Hilbert space of sections $s \colon \Omega \to W$ with the norm

$$|s|^2 = \int_\Omega |s(\omega)|^2_{H(\omega)}\, d\mu(\omega),$$

where $H(\omega)$ is the fiber over $\omega \in \Omega$. The operators

$$(T(g)s)(\omega) = \sqrt{j(g, \omega)}\, s(\omega g) g^{-1}$$

form the induced representation $T = \operatorname{Ind} U$.

Let us give another realization of the induced representation. Let us assume that we are given a Borel mapping $\theta \colon \Omega \to G$ such that $\omega_0 \theta(\omega) = \omega$ for all $\omega \in \Omega$. Then the mapping $(g, \omega) \mapsto a(g, \omega)$ of the form

$$a(g, \omega) = U(\theta(\omega) g (\theta(\omega g))^{-1}), \qquad g \in G, \quad \omega \in \Omega \bmod 0, \tag{1.9}$$

is a cocycle (with values in $U(H^0)$) and formula (1.7) defines a unitary representation of the group G. This is just the induced representation $T = \operatorname{Ind} U$.

For two representations U_i, $i = 1, 2$, of the stabilizer G^{ω_0} there is a natural embedding

$$L(U_1, U_2) \to L(T_1, T_2), \qquad T_i = \operatorname{Ind} U_i, \quad i = 1, 2.$$

Hence, any unitary representation of the subgroup G^{ω_0} yields a cocycle by formula (1.9). If the group G is locally compact and the subgroup G^{ω_0} is closed, then any cocycle is equivalent to a cocycle of the form (1.9) (this is a well-known theorem of Mackey [**58**]).

For a group A and a homomorphism $G^{\omega_0} \to A$ a similar formula defines a cocycle with values in A.

§2. Some generalizations of induced representations. Generalized determinant bundle and an extension of the group GL_p^0

In this section we consider constructions related to an action of a group G on a manifold X. This yields linear representations of certain extensions of the group G. An important application of this construction is the generalized determinant bundle on the complexified Grassmannian, which was constructed by Mickelsson and Rajeev [**60**]. We consider also a modification of the main construction that yields central extensions of the group G.

2.1. Construction yielding a representation of an abelian extension of a group. In this subsection all groups, manifolds, actions, etc., are assumed to be Banach and complex analytic. This restriction is stipulated by further applications of the main construction to the generalized determinant bundle. The case when all objects are, say, real analytic can be considered similarly. All necessary information about infinite-dimensional manifolds can be found, for example, in [**4**, §5].

Let a group G act on a manifold transitively and effectively. Choose a point $x_0 \in X$ and define

$$G_0 = \{g \mid g \in G,\ x_0 g = x_0\}.$$

Let $H \subset G$ be a closed normal subgroup such that its action on X is also transitive. The last condition is equivalent to the equality $G = G_0 \cdot H$. Denote $H_0 = H \cap G_0$, so that $X = G_0 \setminus G = H_0 \setminus H$. (We recall that the groups G, G_0, H and the manifold X are complex analytic and the action $x \mapsto xg$ is holomorphic in x and g).

Assume that $w : H_0 \times H \to \mathbb{C}^*$ is a holomorphic function satisfying the condition

$$w(ab, h) = w(a, bh)w(b, h), \qquad a, b \in H_0, \quad h \in H; \tag{2.1}$$

in other words, w is a holomorphic 1-cocycle of the left action of the group H_0 on H. Using this cocycle, we construct a linear bundle $\xi\colon E \to X$ as follows.

Define the action of the group H_0 on $H \times \mathbb{C}$ by the formula

$$a(h, \lambda) = (a \cdot h, w(a, h) \cdot \lambda), \qquad a \in H_0, \quad (h, \lambda) \in H \times \mathbb{C}.$$

Denote by $[h, \lambda]$ the orbit of an element (h, λ) and by E the set of all orbits. The mapping $\xi\colon E \to X$ given by the formula $\xi([h, \lambda]) = x_0 h$ defines a desired holomorphic linear bundle on X.

Now we "lift" elements of the subgroups H and G_0 to automorphisms of the bundle $\xi\colon E \to X$. This will be done under the following assumptions:

a) We have

$$w(a^g, h^g) = w(a, h), \qquad g \in G_0, \quad a \in H_0, \quad h \in H \tag{2.2}$$

(we use the notation $g_1^{g_2} = g_2^{-1} g_1 g_2$).

b) There exists a holomorphic function $u\colon H \times H \to \mathbb{C}$ such that

$$\begin{aligned} \frac{w(a, h_1 h_2)}{w(a, h_1)} &= \frac{u(ah_1, h_2)}{u(h_1, h_2)}, \qquad h_1, h_2 \in H, \quad a \in H_0, \\ u(a, e) &= 1, \qquad a \in H_0 \end{aligned} \tag{2.3}$$

(hence, for any $h' \in H$ the cocycles $(a, h) \mapsto w(a, h)$ and $(a, h) \mapsto w(a, hh')$, $(a, h) \in H_0 \times H$, are equivalent).

Under these assumptions, we associate to any elements $h \in H$, $g \in G_0$ automorphisms $\alpha(h)$, $\beta(g)$ of the bundle $\xi\colon E \to X$ by the formulas

$$\begin{aligned} &[h_1, \lambda]\alpha(h) = [h_1 \cdot h, u(h_1, h) \cdot \lambda], \\ &[h_1, \lambda]\beta(g) = [g^{-1} h_1 g, \lambda], \qquad (h_1, \lambda) \in H \times \mathbb{C}. \end{aligned}$$

Using conditions (2.2) and (2.3) one can easily verify that these formulas define automorphisms of the bundle and that the corresponding actions on the base X coincide with the initial actions $x \mapsto xh$, $x \mapsto xg$. Denote by $\widehat{G}$ the group of all holomorphic automorphisms of the bundle $\xi\colon E \to X$ that generate on the base X

a transformation by an element of the group G. The above construction shows that the natural mapping $\widehat{G} \to G$ is a surjection and that there exists an exact sequence

$$1 \to \mathcal{H}^*(X) \to \widehat{G} \to G \to 1, \tag{2.4}$$

where $\mathcal{H}^*(X)$ is the group of holomorphic functions $f\colon X \to \mathbb{C}^*$.

Now we construct a natural linear representation of the group $\widehat{G}$. For that, we use the dual bundle $\xi'\colon E' \to X$ and the space S of its holomorphic sections. Any section $s \in S$ is a holomorphic function on E which is complex linear at each fiber $\xi^{-1}(x)$. The representation of the group $\widehat{G}$ is given by the formula

$$(T(g)s)(v) = s(vg), \qquad v \in E, \quad g \in \widehat{G}, \quad s \in S.$$

Let us describe another construction of the representation T. For that, we note that the space of sections S can be identified with the space F of holomorphic functions $\varphi\colon H \to \mathbb{C}$ such that

$$\varphi(h) = w(a,h)\varphi(ah), \qquad a \in H_0, \quad h \in H. \tag{2.5}$$

Carrying over the representation T to the space F (and keeping the notation T for this new representation) we obtain the formulas

$$\begin{aligned} (T(\alpha(h))\varphi)(h_1) &= u(h_1,h)\varphi(h_1h) \qquad h_1 \in H, \quad h \in H,\\ (T(\beta(g))\varphi)(h_1) &= \varphi(g^{-1}hg), \qquad g \in G_0, \quad h \in H,\\ (T(\psi)\varphi)(h) &= \psi(x_0h)\varphi(h), \qquad \psi \in \mathcal{H}^*(X), \quad h \in H \end{aligned}$$

Concluding this subsection, let us mention the following special case of the above construction. Let the cocycle $w(a,h)$, $(a,h) \in H_0 \times H$, not depend on h, so that it is reduced to a homomorphism $H_0 \to \mathbb{C}^*$. In this case the construction of the bundle $\xi\colon E \to X$ is the holomorphic version of the construction from 1.5. Instead of the group $\widehat{G}$ here it is reasonable to consider a smaller group $\widetilde{G}$ generated by automorphisms $\alpha(h)$, $\beta(h)$, $h \in H$, $g \in G_0$. We obtain a surjection $p\colon \widetilde{G} \to G$. Simple computations show that an element from $\ker p$ is determined by a number $z \in \mathbb{C}^*$ and is reduced to multiplication by z in each fiber. We do not present these computations, since in 2.4 they will appear in a more general situation.

2.2. Passing to the Lie algebra. We continue to study the construction from 2.1. Let $\mathfrak{g}$ be the Lie algebra of the group G; note that $\mathfrak{g}$ is a complex Banach space which was used in the definition of the Banach analytic group G, together with the Lie bracket on this space. Denote by $\mathfrak{g}_0$, $\mathfrak{h}$, $\mathfrak{h}_0$ the subalgebras corresponding to the subgroups G_0, H, H_0. The adjoint representation of the group G in $\mathfrak{g}$ will be written as $g \mapsto \mathrm{Ad}(g)$. Let us assume that we are given a complex Banach subspace $\mathfrak{g}_1 \subset \mathfrak{g}$ such that

$$\begin{gathered} \mathrm{Ad}(g)(\mathfrak{g}_1) = \mathfrak{g}_1 \quad \text{for any } g \in G_0,\\ \mathfrak{g} = \mathfrak{g}_0 \oplus \mathfrak{g}_1, \quad \mathfrak{h} = \mathfrak{h}_0 \oplus \mathfrak{g}_1, \quad [\mathfrak{g}_1, \mathfrak{g}_1] \subset \mathfrak{h}_0. \end{gathered}$$

Recall also that H is a normal subgroup in G, so that $\mathfrak{h}$ is an ideal in $\mathfrak{g}$. Also, we have $\mathfrak{h}_0 = \mathfrak{h} \cap \mathfrak{g}_0$.

Let $\widehat{\mathfrak{g}}$ be the Lie algebra of the group $\widehat{G}$. From (2.4), we get the Lie algebra extension

$$0 \to \mathcal{H}(X) \to \widehat{\mathfrak{g}} \to \mathfrak{g} \to 0, \tag{2.6}$$

where $\mathcal{H}(X)$ is the algebra of holomorphic functions on X. Inclusions $\alpha\colon H \to \widehat{G}$ and $\beta\colon G_0 \to \widehat{G}$ yield inclusions of linear spaces $\alpha_0\colon \mathfrak{h} \to \widehat{\mathfrak{g}}$ and $\beta_0\colon \mathfrak{g}_0 \to \widehat{\mathfrak{g}}$. Using these inclusions, let us define the linear map (inclusion) $\gamma_0\colon \mathfrak{g} \to \widehat{\mathfrak{g}}$ by the formulas $\gamma_0(Y) = \alpha_0(Y)$ for $Y \in \mathfrak{g}_1$, $\gamma_0(Y) = \beta_0(Y)$ for $Y \in \mathfrak{g}_0$. This inclusion defines a 2-cocycle on $\mathfrak{g}$ with values in $\mathcal{H}(X)$ by the formula

$$c(Y_1, Y_2) = [\gamma_0(Y_1), \gamma_0(Y_2)] - \gamma_0([Y_1, Y_2]). \tag{2.7}$$

This cocycle completely determines the extension (2.6). Our goal is to obtain an explicit formula for the function $x \mapsto c(Y_1, Y_2)(x)$, $x \in X$. To do this, let us introduce the functions $u_0\colon \mathfrak{g}_1 \times \mathfrak{g}_1 \times H \to \mathbb{C}$, $w_0\colon \mathfrak{h}_0 \times H \to \mathbb{C}$ by the formulas

$$\begin{aligned} u_0(Y_1, Y_2, h) &= \frac{\partial^2}{\partial t\, \partial s} u(h \exp tY_1, \exp sY_2)\big|_{t=0,\, s=0}, \\ w_0(Y, h) &= \frac{d}{dt} w(\exp tY, h)\big|_{t=0}, \\ Y_1, Y_2 &\in \mathfrak{g}_1, \quad Y \in \mathfrak{h}_0, \quad h \in H. \end{aligned}$$

LEMMA 2.1. (1) *If* $Y_1, Y_2 \in \mathfrak{g}_1$, *then*

$$\begin{gathered} c(Y_1, Y_2)(x) = u_0(Y_1, Y_2, h) - u_0(Y_2, Y_1, h) - w_0([Y_1, Y_2], h), \\ x \in X, \quad h \in H, \quad x_0 h = x. \end{gathered}$$

(2) *If* $Y_1 \in \mathfrak{g}_0$, $Y_2 \in \mathfrak{g}$, *then* $c(Y_1, Y_2) = 0$.

The proof of this lemma is not too difficult, although it requires cumbersome computations.

Concluding this subsection we consider the (already mentioned) case when the cocycle $w\colon H_0 \times H \to \mathbb{C}$ is of the form $w(a, h) = \chi(a)$, where $\chi\colon H_0 \to \mathbb{C}^*$ is a holomorphic homomorphism. This homomorphism defines a one-dimensional Lie algebra representation $\chi_0\colon \mathfrak{h}_0 \to \mathbb{C}$. Lemma 2.1 shows that in this case the 2-cocycle (2.7) is given by the following formulas:

$$\begin{aligned} c(Y_1, Y_2) &= -\chi_0([Y_1, Y_2]), \qquad Y_1, Y_2 \in \mathfrak{g}_1, \\ c(Y_1, Y_2) &= 0, \qquad Y_1 \in \mathfrak{g}_0, \quad Y_2 \in \mathfrak{g}. \end{aligned}$$

If χ_0 does not vanish identically on $[\mathfrak{g}_1, \mathfrak{g}_1]$, this cocycle generates a one-dimensional central extension of the Lie algebra $\mathfrak{g}$; this extension corresponds to the group extension $1 \to \mathbb{C}^* \to \widetilde{G} \to G \to 1$ mentioned in 2.1.

2.3. Example: generalized determinant bundle and a representation of the group GL_p^0. The determinant bundle on the Grassmannian of an infinite-dimensional Hilbert space is described in [**33**, Chap. 7]. This notion was generalized in the paper by Mickelsson and Rajeev [**60**]. We reproduce here only one result from [**60**] that deals with the complexification of the above bundle. It will be an important example of the realization of the general scheme presented in 2.1 and 2.2. The required holomorphic cocycle will be obtained from the generalized Fredholm determinant $\det_p$. We start with the definition, and then review some properties of this notion. Our exposition here will be brief (for details, see [**35**, Chap. XIII]).

Let H be a complex Hilbert space of countable dimension, $L(H)$ the algebra of bounded operators in H, $J_p(H)$ the ideal defined by the condition $\|A\|_p =$

$(\operatorname{tr}(A^*A)^{p/2})^{1/p} < \infty$, $p \geq 1$. For an operator of the form $A = I + K$, $K \in J_1(H)$ we can define the Fredholm determinant $\det A$ as follows. If $\|K\| < 1$, then

$$\det A = e^{\operatorname{tr}\log A}, \qquad \log A = \log(1+K) = \sum_{n=1}^{\infty} \frac{(-1)^{n-1}}{n} K^n.$$

Now let us take an integer $p > 0$. For an operator A of the form $A = I + K$ define

$$\det_p(I+K) = \det\left((I+K)\exp\left\{\sum_{n=1}^{p-1} \frac{(-1)^n}{n} K^n\right\}\right), \qquad K \in J_p(H).$$

One can easily prove that this definition makes sense, i.e., that the operator under the det sign on the right-hand side belongs to $I + J_1(H)$. Of course, $\det_1 A = \det A$ for $K \in J_1(H)$. Properties of $\det_p$ that we will need later are summarized in the following two lemmas.

LEMMA 2.2. *We have*

$$\det_p(AB) = \det_p A \cdot \det_p B \cdot e^{\gamma_p(A,B)},$$

where $\gamma_p(A,B)$ is a polynomial in A, B of the form

$$\gamma_p(A,B) = \operatorname{tr}\left(\sum_{n=1}^{p-1} \frac{(-1)^n}{n}((AB-I)^n - (A-I)^n - (B-I)^n)\right).$$

LEMMA 2.3. *If $A \in I + J_p(H)$, then the inequality $\det_p(A) \neq 0$ is equivalent to the invertibility of A.*

Both lemmas follow from the definition.

This concludes the review of the properties of $\det_p$, and we start the construction of the generalized determinant bundle.

Let H_+ and H_- be Hilbert spaces of countable dimension. Let $J_p(H_+, H_-)$ denote the set of linear maps $D\colon H_+ \to H_-$ such that $\|D\|_p = (\operatorname{tr}(D^*D)^{p/2})^{1/p} < \infty$. Define the set $J_p(H_-, H_+)$ similarly. Let GL_p be the group of all invertible operators in the Hilbert space $H_+ \oplus H_-$ given by the matrices of the form

$$g = \begin{pmatrix} a & b \\ c & d \end{pmatrix}, \qquad b \in J_{2p}(H_+, H_-), \quad c \in J_{2p}(H_-, H_+).$$

The conditions on the elements b, c of the matrix g can be written in the form

$$[g, J] \in J_{2p}, \qquad J = \begin{pmatrix} I_+ & 0 \\ 0 & -I_- \end{pmatrix}. \tag{2.8}$$

Define the topology in GL_p by the metric $\|a_1 - a_2\| + \|b_1 - b_2\|_{2p} + \|c_1 - c_2\|_{2p} + \|d_1 - d_2\|$. Denote by GL_p^0 the connected component of the identity in the group GL_p. Elements of the group GL_p^0 are characterized by the identity $\operatorname{ind}(a) = \operatorname{ind}(d) = 0$, where ind is the Fredholm index of an operator.

Now we apply the construction from 2.1, taking (in the notation of 2.1)

$$G = GL_p^0, \qquad H = \{g \in G \mid a \in I_+ + J_p(H_+),\ d \in I_- + J_p(H_-)\},$$

$$G_0 = \left\{\begin{pmatrix} a & 0 \\ 0 & d \end{pmatrix}\right\}, \qquad H_0 = G_0 \cap H.$$

Finally, the space $X = G_0 \setminus G$ can be identified with the space of all operators in $H_+ \oplus H_-$ of the form

$$F = \begin{pmatrix} F_{11} & F_{12} \\ F_{21} & F_{22} \end{pmatrix}$$

such that $F_{11} \in I_+ + J_p(H_+)$, $F_{22} \in -I_- + J_p(H_-)$, $F^2 = I$, $F_{12} \in J_{2p}(H_-, H_+)$, $F_{21} \in J_{2p}(H_+, H_-)$. The group $G = GL_p^0$ acts on X by the formula $Fg = g^{-1}Fg$; it is clear that the stationary subgroup of the element J from (2.8) is G_0.

Let us mention another, somewhat more geometric, description of the homogeneous space X. Denote by P_+ and P_- the projections from $H_+ \oplus H_-$ to H_+ and H_- respectively. For any subspace $L \subset H_+ \oplus H_-$ denote by P_L the projection to L. Then the space X can be identified with the set of pairs of subspaces (L_+, L_-) in $H_+ \oplus H_-$ such $H_+ \oplus H_-$ is the direct sum of L_+ and L_- and all four operators $P_+ - P_+ P_{L_+} P_+$, $P_{L_+} P_-$, $P_- - P_- P_{L_-} P_-$, and $P_{L_-} P_+$ belong to the class J_{2p}. This identification can be done as follows: to $F \in X$ we associate the pair (L_+, L_-), where $L_\pm = \{x \in H_+ \oplus H_- \mid Fx = \pm x\}$.

According to 2.1, now we need a holomorphic cocycle $w\colon H_0 \times H \to \mathbb{C}^*$ and a holomorphic function $u\colon H \times H \to \mathbb{C}^*$ that satisfy the conditions (2.2) and (2.3). This can be done for $p = 1, 2, \dots$. Namely, the required conditions are satisfied if we take

$$w(a,h) = \frac{\det_p h_{11}}{\det_p(a_{11}h_{11})}, \qquad a \in H_0, \quad h \in H,$$

$$u(h_1,h_2) = \frac{\det_p(h_1)_{11} \det_p((h_1^{-1})_{11}(h_1h_2)_{11})}{\det_p(h_1h_2)_{11} \det_p((h_1^{-1})_{11}(h_1)_{11})}, \qquad h_1, h_2 \in H$$

(here g_{11} denotes the upper right element of the matrix $g \in GL_p$). Using Lemma 2.2, we can rewrite these functions in the form

$$w(a,h) = (\det_p a_{11})^{-1} e^{-\gamma_p(a_{11},h_{11})}, \qquad a \in H_0, \quad h \in H,$$

$$u(h_1,h_2) = \exp\left(\gamma_p((h_1^{-1})_{11},(h_1h_2)_{11}) - \gamma_p((h_1^{-1})_{11},(h_1)_{11})\right), \qquad h_1, h_2 \in H$$

These formula imply that w and u do not vanish. Formulas (2.2) and (2.3) easily follows from Lemma 2.2.

Now we have all we need to apply the construction from 2.1. This construction provides us with a holomorphic linear bundle $\xi\colon E \to X$ and an extension $\widehat{G}$ of the group G by the multiplicative group of nowhere vanishing holomorphic functions on X. The action of G on X can be lifted to an embedding of the group $\widehat{G}$ into the automorphism group of the bundle $\xi\colon E \to X$.

Finally, let us take the dual bundle $\xi'\colon E' \to X$ and the space S of its holomorphic sections. One can easily see that the function $h \mapsto \det_p h_{11}$, $h \in H$, belongs to S (here we identify sections from S with functions $H \to \mathbb{C}$ satisfying the condition (2.5); this identification was described in 2.1). Hence, $S \neq \{0\}$ and the general construction from 2.1 yields an exact linear representation of the group $\widehat{G}$ in the space S.

The last thing we need to do is to pass to the Lie algebras of the groups G and $\widehat{G}$; we denote these Lie algebras by $\mathfrak{g}$ and $\widehat{\mathfrak{g}}$. We have the extension

$$0 \to \mathcal{H}(X) \to \widehat{\mathfrak{g}} \to \mathfrak{g} \to 0, \tag{2.9}$$

where $\mathcal{H}(X)$ is the group of all holomorphic functions on X. Our goal is to find an explicit expression for the 2-cocycle that defined this extension.

The Lie algebras of the groups G and $\widehat{G}$ are of the form

$$\mathfrak{g} = \left\{ \begin{pmatrix} a & b \\ c & d \end{pmatrix}, \ a \in L(H_+),\ d \in L(H_-),\ b \in J_{2p}(H_-, H_+),\ c \in J_{2p}(H_+, H_-) \right\},$$

$$\mathfrak{g}_0 = \left\{ \begin{pmatrix} a & 0 \\ 0 & d \end{pmatrix}, \ a \in L(H_+),\ d \in L(H_-) \right\}.$$

Let us introduce also the subspace

$$\mathfrak{g}_1 = \left\{ \begin{pmatrix} 0 & b \\ c & 0 \end{pmatrix}, \ b \in J_{2p}(H_-, H_+),\ c \in J_{2p}(H_+, H_-) \right\}.$$

It is clear that $\mathfrak{g}$, $\mathfrak{g}_0$, $\mathfrak{g}_1$ satisfy the conditions from 2.2. After all these preparations the required cocycle can be obtained from Lemma 2.1. To avoid cumbersome formulas, we consider only the case $p = 2$.

LEMMA 2.4. *For $p = 2$ the extension* (2.9) *is given by the 2-cocycle $c\colon \mathfrak{g} \times \mathfrak{g} \to \mathcal{H}(X)$ of the form*

$$c(Y, Z)(F) = \frac{1}{2} \operatorname{tr}((I_+ - F_{11})(Y_{12}Z_{21} - Z_{12}Y_{21}) + (I_- + F_{22})(Z_{21}Y_{12} - Y_{21}Z_{12})),$$

where $Y, Z \in \mathfrak{g}$, $F \in X$.

This formula can be rewritten in the form

$$c(Y, Z)(F) = \frac{1}{8} \operatorname{tr}(P_+KP_+ - P_-KP_-),$$

$$K = [[J, Y], [J, Z]](F - J), \quad P_+ = \begin{pmatrix} I_+ & 0 \\ 0 & 0 \end{pmatrix}, \quad P_- = \begin{pmatrix} 0 & 0 \\ 0 & I_- \end{pmatrix}$$

The proof follows from computations similar to those indicated in the proof of Lemma 2.1.

For $p = 1$ the above construction is reduced to the construction described in general form in Lemma 2.1. This is the case of the ordinary determinant bundle, when we get a one-dimensional central extension of the group GL_1^0. The corresponding 2-cocycle on the Lie algebra $\mathfrak{g}$ takes the form

$$c(Y, Z) = \operatorname{tr}(Y_{12}Z_{21} - Z_{12}Y_{21}).$$

The reader can find the details for the case $p = 1$ in [**33**, Chap. 6].

2.4. A construction leading to central extensions. As we have already said, if we take in the main construction from 2.1 a 2-cocycle of the form $w(a, h) = \chi(a)$, where χ is a homomorphism $H_0 \to \mathbb{C}^*$, then we obtain a central extension of the group G. Here we describe a somewhat more complicated version of this construction, when we pass from the group H to a central extension. The group G, the manifold X, etc. are assumed to be complex-analytic.

As in 2.1, let G act on X transitively and effectively. Let G_0 be the stabilizer of a point x_0, H a normal subgroup of G that also acts on X transitively and effectively, and $H_0 = G_0 \cap H$. Suppose we are given a central extension

$$1 \to \mathbb{C}^* \xrightarrow{i} \widetilde{H} \xrightarrow{p} H \to 1$$

that splits over the subgroup H_0. Therefore, denoting $\widetilde{H}_0 = p^{-1}(H_0)$, we have a homomorphism $\chi\colon \widetilde{H}_0 \to \mathbb{C}^*$ such that $\chi \circ i$ is the identity on $\mathbb{C}^*$. Assume also that for any $g \in G_0$ the homomorphism $h \mapsto ghg^{-1}$ can be lifted to an automorphism A_g of the group $\widetilde{H}$, so that

$$p(A_g(\widetilde{a})) = gp(\widetilde{a})g^{-1}, \qquad \widetilde{a} \in \widetilde{H}, \quad g \in G_0.$$

Furthermore, we assume that

1. The mapping $g \mapsto A_g$, $g \in G_0$, is a homomorphism from G_0 to the group $\operatorname{Aut} \widetilde{H}$.

2. If $g \in H_0$, then A_g coincides with the automorphism $\widetilde{a} \mapsto \widetilde{g}\widetilde{a}\widetilde{g}^{-1}$, $\widetilde{a} \in \widetilde{H}$, $\widetilde{g} \in p^{-1}(g)$.

3. We have $\chi(\widetilde{a}) = \chi(A_g(\widetilde{a}))$ for $\widetilde{a} \in \widetilde{H}_0$, $g \in G_0$.

Having prepared all this, let us construct the promised central extension of the group G.

First of all, using the obvious transitive action of the group $\widetilde{H}$ on X (by the formula $x\widetilde{a} = xp(\widetilde{a})$, $\widetilde{a} \in \widetilde{H}$) and the homomorphism $\chi\colon \widetilde{H}_0 \to \mathbb{C}^*$, we construct, in the usual way, a linear bundle on X. Let us recall this construction. The group $\widetilde{H}_0$ acts on $\widetilde{H} \times \mathbb{C}^*$ by the formula $\widetilde{a}_1(\widetilde{a}, \lambda) = (\widetilde{a}_1\widetilde{a}, \chi(\widetilde{a}_1)\lambda)$. Denote by $[\widetilde{a}, \lambda]$ the orbit of the element $(\widetilde{a}, \lambda) \in \widetilde{H} \times \mathbb{C}^*$ and by E the set of all orbits. We have the mapping $\xi\colon E \to X$ of the form $\xi([\widetilde{a}, \lambda]) = x_0\widetilde{a}$, which obviously is a linear bundle on X.

Now we "lift" each transformation $x \mapsto xg$, $x \in X$, $g \in G$, to an automorphism of the bundle $\xi\colon E \to X$. First, for any $h \in H$ we choose an element $\widetilde{h} \in \widetilde{H}$ such that $p(\widetilde{h}) = h$. Then to elements $h \in H$, $g \in G_0$ we associate transformations $\alpha(h)$, $\beta(g)$ of the space E by the formulas

$$[\widetilde{a}, \lambda]\alpha(h) = [\widetilde{a}\widetilde{h}, \lambda],$$
$$[\widetilde{a}, \lambda]\beta(g) = [A_{g^{-1}}(\widetilde{a}), \lambda].$$

One can easily see that $\alpha(h)$, $\beta(g)$ are automorphisms of the bundle $\xi\colon E \to X$, and the corresponding mappings of the base X coincide with h and g respectively. Denote by $\widetilde{G}$ the group generated by all automorphisms $\alpha(h)$, $\beta(g)$, $h \in H$, $g \in G_0$. We have an obvious surjection $\pi\colon \widetilde{G} \to G$.

LEMMA 2.5. *Each automorphism of the bundle $\xi\colon E \to X$ that belongs to* $\ker \pi$ *is determined by a number $c \in \mathbb{C}^*$ and is given by the multiplication by c in each fiber $\xi^{-1}(x)$. We have the following central extension:*

$$1 \to \chi(\widetilde{H}_0) \to \widetilde{G} \xrightarrow{\pi} G \to 1. \tag{2.10}$$

PROOF. An arbitrary automorphism $t \in \ker \pi$ can be represented in the form $t = \alpha(h_1)\beta(g_1) \cdots \alpha(h_n)\beta(g_n)$, where

$$h_i \in H, \quad g_i \in G_0, \quad h_1g_1 \dots h_ng_n = e \tag{2.11}$$

(here e is the identity of the group G). If $[\widetilde{a}, \lambda] \in E$, then

$$\begin{aligned}[\widetilde{a}, \lambda]t &= [\widetilde{a}, \lambda]\alpha(h_1)\beta(g_1) \dots \alpha(h_n)\beta(g_n) \\ &= [A_{g_n^{-1}\cdots g_1^{-1}}(\widetilde{a}) \cdot A_{g_n^{-1}\cdots g_1^{-1}}(\widetilde{h}_1) \cdot A_{g_n^{-1}\cdots g_1^{-2}}(\widetilde{h}_2) \cdots A_{g_n^{-1}}(\widetilde{h}_n), \lambda] \\ &= [A_{g_n^{-1}\cdots g_1^{-1}}(\widetilde{a}) \cdot \widetilde{u}, \lambda] = [A_r(\widetilde{a}) \cdot \widetilde{u}, \lambda],\end{aligned} \tag{2.12}$$

where

$$\widetilde{u} = A_{g_n^{-1}\cdots g_1^{-1}}(\widetilde{h}_1)A_{g_n^{-1}\cdots g_2^{-1}}(\widetilde{h}_2)\cdots A_{g_n^{-1}}(\widetilde{h}_n), \quad r = g_n^{-1}\cdots g_1^{-1}.$$

Let us consider elements $\widetilde{u} \in \widetilde{H}$, $r \in G$. First of all, from (2.11) and from the fact that H is a normal subgroup it follows that $r \in H \cap G_0 = H_0$. But then the element $A_r(\widetilde{a})$ from (2.12) can be written in the form $A_r(\widetilde{a}) = \widetilde{r}\widetilde{a}\widetilde{r}^{-1}$ (recall that for each $h \in H$ we have selected the element $\widetilde{h} \in p^{-1}(h)$). Next, using the definition of the automorphism A_g, we see that $p(\widetilde{u}) = r$. Hence, $\widetilde{u} = \widetilde{r}k$ for some element $k \in \ker p$, which also belongs to the center of the group $\widetilde{H}$. Hence, $A_r(\widetilde{a})\widetilde{u} = \widetilde{r}\widetilde{a}\widetilde{r}^{-1}\widetilde{r}k = \widetilde{r}\widetilde{a}k = k\widetilde{r}\widetilde{a}$. Now we can continue the chain of equalities in (2.12) as follows:

$$[\widetilde{a}, \lambda]t = [k\widetilde{r}\widetilde{a}, \lambda] = [\widetilde{a}, \chi^{-1}(k\widetilde{r})\lambda].$$

We see that the transformation t is reduced to multiplication by $c = \chi^{-1}(k\widetilde{r})$ in each fiber. This easily implies the exact sequence (2.11). The lemma is proved.

Now we construct the representation of the group $\widetilde{G}$ associated to the dual bundle $\xi' \colon E' \to X$. Denote by S the space of smooth sections of ξ'. By definition, each section $s \in S$ is a smooth function on E that is complex-linear along the fibers $\xi^{-1}(x)$, $x \in X$. The group $\widetilde{G}$ acts on S in an obvious way: to an element $g \in \widetilde{G}$ there corresponds the operator of the form

$$(T(g)s)(v) = s(vg), \qquad g \in \widetilde{G}, \quad s \in S. \tag{2.13}$$

The required representation is constructed. It can be rewritten in a different form as follows. A section $s \in S$ can be identified with a function $\widetilde{H} \times \mathbb{C} \to \mathbb{C}$ which is linear in the second variable and invariant under the transformation $(\widetilde{a}, \lambda) \mapsto (\widetilde{a}_0 a, \chi(\widetilde{a}_0)\lambda)$, where $(\widetilde{a}, \lambda) \in \widetilde{H} \times \mathbb{C}$, $\widetilde{a}_0 \in \widetilde{H}_0$. Such a function has the form $(\widetilde{a}, \lambda) \mapsto u(\widetilde{a})\lambda$, where

$$u(\widetilde{a}_0\widetilde{a}) = \chi^{-1}(\widetilde{a}_0)u(\widetilde{a}), \qquad \widetilde{a}_0 \in \widetilde{H}_0, \quad \widetilde{a} \in \widetilde{H}. \tag{2.14}$$

Hence, the space S is identified with the space F of smooth functions $\widetilde{H} \to \mathbb{C}$ satisfying the condition (2.14). Under this identification, the representation (2.13) becomes the representation (denoted by the same letter T) given by the formulas

$$(T(\alpha(h)u)(\widetilde{a}) = u(\widetilde{a}\widetilde{h}), \qquad h \in H, \quad \widetilde{a} \in \widetilde{H},$$
$$(T(\beta(g)u)(\widetilde{a}) = u(A_{g^{-1}}(\widetilde{a})), \qquad g \in G_0.$$

2.5. Remarks. As we have already mentioned, the generalized determinant bundle was constructed by Mickelsson and Rajeev [**60**]. The general approach from 2.1 based on the application of holomorphic cocycles allowed us to simplify the exposition somewhat. We omitted several important details from [**60**]. First of all, we restricted ourselves to the linear bundle from [**60**] over the complexification of the Grassmannian (more precisely, over the part of this complexification that is denoted by $\mathbb{C}\,\mathrm{Gr}$ in [**60**]). This resulted in the representation of the group GL_p^0 in the space of holomorphic sections. As is explained in [**60**], this cannot be done using the original bundle over Gr_p. Moreover, we consider only the connected component GL_p^0 of the group GL_p.

The importance of the above constructions is explained by the fact that some infinite-dimensional groups (current groups, groups of diffeomorphisms, etc.) can be imbedded into GL_p^0. The case $p = 1$ (and the corresponding current groups) is studied in the book by Segal and Pressley [**33**]. The imbedding of smooth current groups into GL_p^0, $p > 1$, is considered in the same paper [**60**]; one simple case of such an embedding will be discussed below in §19. Mickelsson's book [**59**] contains an exposition of all these problems with applications to mathematical physics.

§3. The space $\overline{S}(H)$ (bosonic Fock space) and a representation of the group $H \cdot Sp(H)$

The space indicated in the title of this section is often denoted by $\operatorname{Exp} H$. We describe this space and show how it is related with Gaussian measures; for a more detailed exposition of all these topics see Guichardet [**54**]. Then we describe a representation of the infinite-dimensional symplectic group $Sp(H)$ in the space $\overline{S}(H)$ (Shale–Weil representation). The classical exposition of this subject can be found in Berezin's book [**2**]. Our exposition is based on computations with quadratic exponents; here we follow Segal (see [**64**]).

3.1. The algebra $\widehat{S}(H)$ and the space $\overline{S}(H)$. Let H be a complex Hilbert space, $\overline{S}^n(H)$ its nth symmetric power, $n \geq 0$. More precisely, $\overline{S}^0(H) = \mathbb{C}$, and for $n > 0$ the space $\overline{S}^n(H)$ is obtained by the completion of the algebraic symmetric nth power of the space H with the respect to the norm generated by the following inner product: on decomposable tensors, it has the form

$$(h_1 \cdots h_n, h_1' \cdots h_n') = \sum_s \prod (h_k, h_{ks}'),$$

where s runs over the group of all permutations of the set $\{1, 2, \dots, n\}$. Denote by $\widehat{S}(H)$ the space of all formal series $r = \sum_{m \geq 0} r_m$, $r_m \in \overline{S}^m(H)$. The obvious extension of the multiplication in the symmetric algebra of the space H to the space $\widehat{S}(H)$ makes $\widehat{S}(H)$ an algebra. The product of elements r, r' is denoted by $r * r'$; also, $r^{*n} = r * \cdots * r$ (n times). Finally, we endow $\widehat{S}(H)$ with the topology of the coordinatewise strong convergence, i.e., a sequence $r^{(k)} = \sum_{m \geq 0} r_{mk}$ of elements from $\widehat{S}(H)$ converges to $r = \sum_{m \geq 0} r_m$ if and only if for any $m \geq 0$ we have $|r_{mk} - r_m| \to 0$ as $k \to \infty$.

Now we define the bosonic Fock space $\overline{S}(H)$ as the Hilbert space of all elements $r = \sum_{m \geq 0} r_m$ of the algebra $\widehat{S}(H)$ such that $\sum_{m \geq 0} |r_m^2| < \infty$. The inner product in $\overline{S}(H)$ is defined by the formula $(r, r') = \sum_{m \geq 0} (r_m, r_m')$. Another notation for $\overline{S}(H)$ is $\operatorname{Exp} H$.

3.2. Exponents in the algebra $\widehat{S}(H)$ and in the space $\overline{S}(H)$. For an element $r = \sum_{m \geq 0} r_m$ of the algebra $\widehat{S}(H)$, define the element $\operatorname{Exp} r$ of the same algebra by the formula

$$\operatorname{Exp} r = \sum_{k=0}^{\infty} \frac{r^{*k}}{k!}. \tag{3.1}$$

One can easily see that this series converges in $\widehat{S}(H)$. In particular, for any $h \in H$ there exists an element $\operatorname{Exp} h$; clearly, $\operatorname{Exp} h \in \overline{S}(H)$. One can easily prove that

$$(\operatorname{Exp} h_1, \operatorname{Exp} h_2) = e^{(h_1, h_2)}, \qquad h_1, h_2 \in H. \tag{3.2}$$

LEMMA 3.1. *Let L be an $\mathbb{R}$-linear subset in H such that $L+iL$ be dense in H. Then the set $\{\operatorname{Exp} h,\ h \in L\}$ is total in $\overline{S}(H)$.*

PROOF. Let a vector $f = \sum_{m\geq 0} f_m$ from $\overline{S}(H)$ be orthogonal to all vectors $\operatorname{Exp} h$, $h \in L$. For any $h_1, h_2 \in L$ consider the function $F(z) = (\operatorname{Exp}(h_1 + zh_2), f)$, $z \in \mathbb{C}$. It is holomorphic everywhere in $\mathbb{C}$, and $F(x) = 0$ for $x \in \mathbb{R}$. Therefore, $F(z) = 0$ for all z. Hence, $(\operatorname{Exp} h, f) = 0$ for all $h \in L + iL$ and $(\operatorname{Exp} zh, f) = 0$ for $z \in \mathbb{C}$, $h \in L + iL$. Using (3.1), we get that $(h^n, f_n) = 0$ for all $n \geq 0$, $h \in L + iL$. Since $L + iL$ is dense in H, this implies $f_n = 0$, $n = 0, 1, \ldots$. Hence $f = 0$, and Lemma 3.1 is proved.

Lemma 3.1 and formula (3.2) allow us to give another definition of the space $\overline{S}(H)$ as follows. Consider the complex linear space with the basis consisting of formal symbols $\operatorname{Exp} h$, $h \in H$, and introduce in this space the inner product defined on basic elements by formula (3.2). This inner product is positive-definite, and the completion of the above linear space with respect to the corresponding Hilbert norm is the space $\overline{S}(H)$. In this definition one can take only the vectors $\operatorname{Exp} h$, $h \in L$, where L is a set satisfying the conditions of Lemma 3.1

An orthogonal decomposition $H = H_0 \oplus H_1$ results in the natural isomorphism of Hilbert spaces $\overline{S}(H_0) \otimes \overline{S}(H_1) \simeq \overline{S}(H)$. This isomorphism associates to the vector $\operatorname{Exp} h_0 \otimes \operatorname{Exp} h_1$ the vector $\operatorname{Exp}(h_0 \oplus h_1)$, $h_i \in H_i$, $i = 0, 1$.

3.3. Relations with the Gaussian measure. From now on we will assume that H is a complex Hilbert space of countable or finite dimension, with the involution $h \mapsto \overline{h}$. Recall that the involution is antilinear, $(\overline{h}_1, \overline{h}_2) = (h_2, h_1)$, and $\overline{\overline{h}} = h$. Let us introduce the real Hilbert space $V = \{h \in H \mid h = \overline{h}\}$. Then H is the complexification of the space V, i.e., $H = V + iV = V_{\mathbb{C}}$.

Let us take a dense $\mathbb{R}$-linear subset $V_0 \subset V$ belonging to the range of a positive Hilbert–Schmidt operator $A\colon V \to V$. Following 1.3, let us take the space Ω consisting of all linear functionals on V_0 that are continuous in the norm $h \mapsto |A^{-1}h|$, and define the Gaussian measure μ on Ω by formula (1.1). Our goal is to construct a natural isomorphism of the Hilbert spaces $\overline{S}(H)$ and $L_2(\Omega, \mu)$. To this end, we choose $c \in \mathbb{C}$, $|c| = 1$, and to each vector $\operatorname{Exp} h$ associate the function $f_h \in L_2(\Omega, \mu)$ of the form

$$f_h(\omega) = e^{-(1/2)c^2(h,\overline{h})+(ch,\omega)}, \qquad \omega \in \Omega \bmod 0. \tag{3.3}$$

Formulas (3.2) and (1.5) imply that the correspondence $\operatorname{Exp} h \mapsto f_h$ preserves the inner product. Since the sets $\{\operatorname{Exp} h\}$ and $\{f_h\}$ are total in the spaces $\overline{S}(H)$ and $L_2(\Omega, \mu)$ respectively, this correspondence can be extended to an isomorphism of Hilbert spaces $\Phi_c\colon \overline{S}(H) \to L_2(\Omega, \mu)$. Let us recall that here $c \in \mathbb{C}$, $|c| = 1$. The isomorphism Φ_1 will be important below.

3.4. The isomorphism of the spaces $\overline{S}^2(H)$ and $J_2^0(H)$. We continue to consider a space H with an involution $h \mapsto \overline{h}$. This involution can be extended to the space of linear operators $s \in L(H)$ by the formula $\overline{s(h)} = \overline{s(\overline{h})}$. Let us introduce also the transpose operator $s' = \overline{s^*}$.

As usual, denote by $J_2(H)$ the space of all Hilbert-Schmidt operators in H. It is a Hilbert space with respect to the inner product $(s_1, s_2) = 2\operatorname{tr}(s_1 s_2^*)$ (the factor 2 will be convenient later). Consider the subspace

$$J_2^0(H) = \{s \in J_2(H) \mid s = s'\}.$$

To each tensor $h_1 * h_2 \in \overline{S}^2(H)$ we associate the operator

$$h \mapsto \tfrac{1}{2}((h, \overline{h}_1)h_2 + (h, \overline{h}_2)h_1), \qquad h \in H.$$

This gives an isometric isomorphism of the Hilbert spaces $\overline{S}^2(H)$ and $J_2^0(H)$. Let us recall that for two elements $s_1, s_2 \in \overline{S}^2(H)$ their product in the algebra $\widehat{S}(H)$ is denoted by $s_1 * s_2$. The composition of elements s_1 and s_2 considered as operators is denoted by $s_1 s_2$. For an element $s \in \overline{S}^2(H)$ we have the operator norm $\|s\|$ and the Hilbert-Schmidt norm $|s| = \sqrt{(s,s)} = \sqrt{2\,\mathrm{tr}(ss^*)}$.

The next lemma describes the canonical form of a vector from $\overline{S}^2(H)$.

LEMMA 3.2. *Any vector $s \in \overline{S}^2(H)$ can be represented in the form*

$$s = \sum_{k \geq 1} \lambda_k e_k * e_k, \tag{3.4}$$

where $\lambda_1 \geq \lambda_2 \geq \cdots$, $\sum \lambda_k^2 < \infty$, and $\{e_k\}$ is an orthonormal set of elements from H.

The proof follows easily from the polar decomposition of the operator s.

3.5. Quadratic exponents. For any elements $s \in \overline{S}^2(H)$ and $h \in H$ consider the element $\mathrm{Exp}(h + s/2)$ (see (3.1)). The necessary properties of these element are collected in the following theorem.

THEOREM 3.1. (1) *If $\|s\| < 1$, then* $\mathrm{Exp}(h + s/2) \in \overline{S}(H)$ *for any $h \in H$.*

(2) *The isomorphism $\Phi_1 \colon \overline{S}(H) \to L_2(\Omega, \mu)$ (see* 3.3) *maps* $\mathrm{Exp}(h+s/2)$ *to the function*

$$\begin{aligned} f_{h,s}(\omega) = \det(I+s)^{-1/2} \exp\{\tfrac{1}{2}(s(I+s)^{-1}\omega, \omega) + ((I+s)^{-1}h, \omega) \\ - \tfrac{1}{2}((I+s)^{-1}h, \overline{h})\}. \end{aligned} \tag{3.5}$$

(3) *If $\|s_i\| < 1$, $h_i \in H$, $i = 1, 2$, then*

$$\begin{aligned} &\det(I - s_1\overline{s}_2)^{1/2}(\mathrm{Exp}(h_1 + s_1/2), \mathrm{Exp}(h_2 + s_2/2)) \\ &\quad = \exp\{\tfrac{1}{2}(\overline{s}_2(I - s_1\overline{s}_2)^{-1}h_1, \overline{h}_1) + \tfrac{1}{2}(s_1(I - \overline{s}_2 s_1)^{-1}\overline{h}_2, h_2) \\ &\qquad + ((I - s_1\overline{s}_2)^{-1}h_1, h_2)\}. \end{aligned} \tag{3.6}$$

PROOF. We start with two simple lemmas.

LEMMA 3.3. *For $h \in H$, $f \in \overline{S}^n(H)$ we have*

$$|h * f| \leq \sqrt{n+1}\,|h|\,|f|. \tag{3.7}$$

To prove the lemma it suffices to assume that $|h| = 1$. Take an orthonormal basis $\{e_k,\ k = 1, 2, \dots\}$ in H with $e_1 = h$. Then the vectors $e_1^{m_1} e_2^{m_2} \dots$, $m_i \in \mathbb{Z}$, $m_i \geq 0$, $\sum m_i = n$, $m_i = 0$ for sufficiently large i, form an orthonormal basis in $\overline{S}^n(H)$. Multiplying elements of this basis by h we get a set of orthogonal vectors. Computing the lengths of these vectors, we obtain (3.7).

LEMMA 3.4. *Fix $a > 0$ and set*

$$\psi(n,a) = \sum_{k=0}^{\infty} \frac{a^k}{k!} \sqrt{\frac{(n+k)!}{n!}}, \qquad n = 1, 2, \ldots .$$

Then $\psi(a,n) < B^{\sqrt{n}}$ for all n with a constant B depending only on a.

To prove this lemma, denote by d_k the elements of the above series. Then $d_k/d_{k-1} = ak^{-1}\sqrt{n+k}$. This easily implies that the maximum term of the series does not exceed $A^{\sqrt{n}}$ and $d_k/d_{k-1} < 1/2$ for $k > A\sqrt{n}$, where A depends only on a. This implies the desired estimate for $\psi(n,a)$.

Now we prove Theorem 3.1. (1) First let $h = 0$. Represent an element $s \in \overline{S}^n(H)$ in the form (3.4). Then $0 < \lambda_i < 1$ for all j. Therefore,

$$(s^{*k}, s^{*k}) = \sum \left(\frac{k!}{k_1! k_2! \ldots} \right)^2 (2k_1)!(2k_2)! \cdots \lambda_1^{2k_1} \lambda_2^{2k_2} \cdots ,$$

where $k = k_1 + k_2 + \ldots$. Hence,

$$|\mathrm{Exp}(s/2)|^2 = \sum_{k=0}^{\infty} \frac{4^{-k}}{(k!)^2} (s^{*k}, s^{*k}) = \prod_{k=1}^{\infty} \left(1 - \lambda_k^2\right)^{-1/2}.$$

Now let $h \neq 0$. Using Lemmas 3.3 and 3.4 we get (using also the Cauchy-Schwarz-Bunyakovskiĭ inequality at the last step)

$$\begin{aligned}
\sum_{m=1}^{\infty} \frac{1}{m!} |(h + s/2)^m| &\leq \sum_{m,k} \frac{2^{-k}|s^{*k}|\,|h|^{m-k}}{k!(m-k)!} \sqrt{(2k+1)\ldots(m+k)} \\
&\leq \sum_{k,p} \frac{2^{-k}|s^{*k}|\,|h|^{p}}{k!p!} \sqrt{\frac{(2k+p)!}{(2k)!}} = \sum_k \frac{2^{-k}|s^{*k}|}{k!} \psi(2k, |h|) \\
&\leq \sum_k \frac{2^{-k}|s^{*k}|}{k!} B^{\sqrt{2k}} \leq \mathrm{const} \cdot \sqrt{\sum_k \frac{4^{-k}}{(k!)^2} |((1+\varepsilon)s)^{*k}|^2} \\
&= \left| \mathrm{Exp}\left(\frac{1+\varepsilon}{2} s \right) \right| < \infty,
\end{aligned}$$

where ε is positive and so small that $\|(1+\varepsilon)s\| < 1$. This implies the first statement of the theorem.

(2) Before proving the second statement, let us prove the equality

$$(3.8) \qquad (\mathrm{Exp}(h + s/2), \mathrm{Exp}\, h_0) = \exp\{\tfrac{1}{2}(s\overline{h}_0, h_0) + (h, h_0)\},$$

where $s \in \overline{S}^2(H)$, $\|s\| < 1$, $h, h_0 \in H$. We can assume that $h_0 \neq 0$. Take $H_1 = H \ominus \mathbb{C}h_0$. Then $h = \alpha h_0 + h_1$, $s = \lambda h_0^{*2} + h_0 * h_2 + t$, where $h_1, h_2 \in H_1$, $t \in \overline{S}^2(H_1)$. It is clear that $\alpha = (h, h_0)(h_0, h_0)^{-1}$, $\lambda = (s\overline{h}_0, h_0)(h_0, h_0)^{-2}$. It is clear also that the vector h_0^{*2n}, $n \geq 1$, is orthogonal to any product of vectors from

$\overline{S}(H)$ that has a factor equal to either h_1, or h_2, or t. Therefore,

$$(\operatorname{Exp}(h+s/2), \operatorname{Exp} h_0) = \left(\operatorname{Exp}\left(\frac{\lambda}{2}h_0^{*2} + \alpha h_0\right), \operatorname{Exp} h_0\right)$$
$$= \sum_{m,n} \frac{1}{n!\,m!}\left(\left(\frac{\lambda}{2}h_0^{*2} + \alpha h_0\right)\right)^{*m}, h_0^{*n}\right) = \sum \frac{(\lambda/2)^k \alpha^{m-k}}{n!\,k!\,(m-k)!}(h_0^{*(k+m)}, h_0^{*n})$$
$$= \sum \frac{1}{k!\,(m-k)!}\left(\frac{\lambda}{2}\right)^k \alpha^{m-k}(h_0, h_0)^{m+k} = \sum \frac{1}{k!\,p!}\left(\frac{\lambda}{2}\right)^k \alpha^p (h_0, h_0)^{2k+p}$$
$$= \exp\left\{\frac{1}{2}(s_0, h_0) + (h, h_0)\right\},$$

and formula (3.8) is proved.

Now we prove the second statement of the theorem. Let us recall (see 3.3) that the isomorphism $\Phi_1 : \overline{S}(H) \to L_2(\Omega, \mu)$ maps the vector $\operatorname{Exp} h$ to the function

$$f_h(\omega) = e^{-(h,h)/2+(h,\omega)}, \qquad \omega \in \Omega \bmod 0.$$

Formulas (1.4) and (3.8) imply that

$$(f_{h,s}, f_{h_0}) = (\operatorname{Exp}(h+s/2), \operatorname{Exp} h_0).$$

Since the sets $\{f_h\}$ and $\{\operatorname{Exp} h\}$ are total in the spaces $L_2(\Omega, \mu)$ and $\overline{S}(H)$ respectively, the isomorphism Φ_1 maps $\operatorname{Exp}(s+h/2)$ to the function $f_{h,s}$, as required.

(3) The statements we have already proved imply the formula

$$(\operatorname{Exp}(h_1 + s_1/2), \operatorname{Exp}(h_2 + s_2/2)) = (f_{h_1,s_1}, f_{h_2,s_2}).$$

Computing the right-hand side of this formula using (3.5) and (1.4) we get (3.6), and the third statement of the theorem is proved.

3.6. The groups $Sp(H)$ and $H \cdot Sp(H)$. As before, by H we denote a complex Hilbert space of countable or finite dimension, endowed with an involution $h \mapsto \overline{h}$. Any $\mathbb{R}$-linear continuous map of the space H into itself is of the form $h \mapsto Ah + B\overline{h}$, where A and B are bounded complex linear operators. The symplectic group $Sp(H)$ is defined as the group of all invertible $\mathbb{R}$-linear operators $h \mapsto Ah + B\overline{h}$ satisfying the following conditions: (1) they preserve the form $\operatorname{Im}(h_1, h_2)$, $h_1, h_2 \in H$; (2) B is a Hilbert-Schmidt operator. It is clear that the group $Sp(H)$ is isomorphic to the group of matrices

$$g = \begin{pmatrix} A & B \\ \overline{B} & \overline{A} \end{pmatrix}$$

satisfying the conditions

$$g^* J g = g J g^* = J, \quad B \in J_2(H), \quad J = \begin{pmatrix} I_H & 0 \\ 0 & -I_H \end{pmatrix};$$

here I_H is the identity operator in H. The topology in $Sp(H)$ is determined by the metric $\|A_1 - A_2\| + |B_1 - B_2|$, where the first summand is the operator norm and second summand is the Hilbert-Schmidt norm. The identity element of the group $Sp(H)$ is denoted by I.

Finally, by $H \cdot Sp(H)$ we denote the group of all transformations of the space H that have the form $h \mapsto Ah + B\overline{h} + h_0$, where A, B are as above and $h_0 \in H$. This group is the semi-direct product of the group $Sp(H)$ to the additive group of the space H. The above element of the group $H \cdot Sp(H)$ is denoted by (h_0, g), $h_0 \in H$, $g \in Sp(H)$. The elements (h, I) and $(0, g)$ of the group $H \cdot Sp(H)$ will be denoted simply by h and g.

Our goal is to construct a unitary representation of the group $H \cdot Sp(H)$ in the space $\overline{S}(H)$.

3.7. The action of the group $H \cdot Sp(H)$ on the space $H \times \mathcal{D}(H)$. Introduce the set

$$\mathcal{D}(H) = \{s \in J_2(H) \mid s = s', \ \|s\| < 1\}. \tag{3.9}$$

This set is an infinite-dimensional analog of one of the classical domains ([**34**, Chap. II]. Let us recall (see 3.4) that an operator $s \in J_2(H)$ with $s = s'$ is identified with an element of the space $\overline{S}^2(H)$.

Let us construct the action mentioned in the title of this subsection. First of all, the formula

$$gs = (B + As)(\overline{A} + \overline{B}s)^{-1}, \qquad s \in \mathcal{D}(H), \quad g \in Sp(H), \tag{3.10}$$

defines a transitive action of the group $Sp(H)$ on the space $\mathcal{D}(H)$. Next, define the function $\varphi \colon Sp(H) \times \mathcal{D}(H) \to GL(H)$ (here $GL(H)$ is the group of invertible operators in the space H) by the formula

$$\varphi(g, s) = (A^* + sB^*)^{-1}, \qquad g \in Sp(H), \quad s \in \mathcal{D}(H). \tag{3.11}$$

This function satisfies the condition $\varphi(g_1g_2, s) = \varphi(g_1, g_2s)\varphi(g_2, s)$. In the next lemma we use this function to construct an action of the group $H \cdot Sp(H)$ on the space $H \times \mathcal{D}(H)$.

LEMMA 3.5. *Formulas*

$$h(a, s) = (a + h - s\overline{h}, s), \qquad h \in H, \tag{3.12}$$

$$g(a, s) = (\varphi(g, s)a, gs), \qquad g \in Sp(H), \tag{3.13}$$

where $(a, s) \in H \times \mathcal{D}(H)$, $(h, g) \in H \cdot Sp(H)$, *define the left action of the group* $H \cdot Sp(H)$ *on* $H \times \mathcal{D}(H)$.

The proof follows from simple computations. The action of an element $(h, g) \in H \cdot Sp(H)$ on an element $(a, s) \in H \times \mathcal{D}(H)$ will be written in the form $(h, g)(a, s)$. Let us mention the equality

(3.14)

$$\begin{aligned} &\det(I_H - gs_1\overline{gs_2}) \\ &\quad = \det(I_H - s_1\overline{s}_2)\det(AA^*)^{-1}\det(I_H + s_1BA^{-1})\det(I_H + A^{-1}Bs_2)^{-1}. \end{aligned}$$

3.8. Shale–Weil representation. We will use the action of the group $H \cdot Sp(H)$ on $H \times \mathcal{D}(H)$ to construct representations of the above group. To do that, to each element $(a, s) \in H \times \mathcal{D}(H)$ we associate the vector

$$\mathrm{Exp}(a + s/2) \tag{3.15}$$

of the space $\overline{S}(H)$ (see 3.5). Next, for each element (h, g) of the group $H \cdot Sp(H)$ we construct a unitary operator $W(h, g)$ in the space $\overline{S}(H)$ that acts on a vector of the form (3.15) by the formula

$$W(h, g)\,\mathrm{Exp}(a+s/2) = f(h, g, s, a)\,\mathrm{Exp}(a'+s'/2), \qquad (a', s') = (h, g)(a, s); \tag{3.16}$$

here the last formula is the action of the element (h, g) of the group $H \cdot Sp(H)$ on the element $(a, s) \in H \times \mathcal{D}(H)$, and $f(h, d, a, s)$ is a nonzero complex number. Our goal is to select $f(h, g, a, s)$ is such a way that formula (3.16) defines a unitary projective representation of the group $H \cdot Sp(H)$. For that, it suffices to find the restriction of the required representation W to each of the subgroups H and $Sp(H)$.

Let us begin with the group H. Its action on $H \times \mathcal{D}(H)$ is given by (3.12). In the next lemma we construct the representation generated by this action.

LEMMA 3.6. *For any element $h \in H$ there exists a unique unitary operator $K(h)$ in the space $\overline{S}(H)$ such that*

(3.17)
$$K(h)\,\mathrm{Exp}(a + s/2) = \exp\left\{-\frac{|h|^2}{2} + \frac{(s\bar{h}, h)}{2} - (a, h)\right\} \mathrm{Exp}((s/2) + a + h - \bar{s}h),$$
$$(a, s) \in H \times \mathcal{D}(H).$$

These operators satisfy the condition

$$K(h_1)K(h_2) = e^{\mathrm{Im}(h_1, h_2)} K(h_1 + h_2) \tag{3.18}$$

(hence, form a projective representation of the additive group of the space H).

Let us note that for $s = 0$ formula (3.17) becomes

$$K(h)\,\mathrm{Exp}\,a = \exp\left\{-\frac{|h|^2}{2} - (a, h)\right\} \mathrm{Exp}(a + h) \qquad a, h \in H. \tag{3.19}$$

PROOF OF THE LEMMA. The mapping sending the vector $\mathrm{Exp}\,a$, $a \in H$, to the right-hand side of (3.19) preserves inner product (this follows from 3.2). Since the set $\{\mathrm{Exp}\,a,\ a \in H\}$ is total in $\overline{S}(H)$, there exists a unique unitary operator $K(h)$ in the space $\overline{S}(H)$ satisfying (3.19). Formula (3.19) easily implies that the operators $K(h)$ satisfy (3.18). It remains to show that the action of the operator $K(h)$ on elements of the form $\mathrm{Exp}(a + s/2)$ is given by the formula (3.17). This follows from the following computations based on formulas (3.6) and (3.19):

$$\begin{aligned}
(K(h)\,\mathrm{Exp}(a + s/2), \mathrm{Exp}\,b) &= (\mathrm{Exp}(a + s/2), K(-h)\,\mathrm{Exp}\,b) \\
&= \exp\left\{-\frac{|h|^2}{2} + (h, b)\right\} (\mathrm{Exp}(a + s/2), \mathrm{Exp}(b - h) \\
&= \exp\left\{-\frac{|h|^2}{2} + (h, b) + \frac{1}{2}s(\bar{b} - \bar{h}, b - h) + (a, b - h)\right\} \\
&= \exp\left\{-\frac{|h|^2}{2} + \frac{(s\bar{b}, b)}{2} - (a, b)\right\} (\mathrm{Exp}(a + h - s\bar{h} + s/2), \mathrm{Exp}\,b).
\end{aligned}$$

Fixing s, h, a and taking an arbitrary $b \in H$, we obtain the desired formula (3.17).

Now we turn to the group $Sp(H)$. Its action on $H \times \mathcal{D}(H)$ is given by formula (3.13). The next lemma provides the corresponding representation.

LEMMA 3.7. *For any $g \in Sp(H)$ there exists a unique unitary operator $V(g)$ in the space $\overline{S}(H)$ such that*

$$V(g)\,\mathrm{Exp}(a + s/2) = f(g, s, a)\,\mathrm{Exp}(\varphi(g, s)a + (1/2)gs), \tag{3.20}$$

where

$$f(g,s,a) = (\det AA^*)^{-1/4} \det(I_H + sB^*(A^*)^{-1})^{-1/2} \times \exp\{-\tfrac{1}{2}((\overline{A} + \overline{B}s)^{-1}\overline{B}a, \overline{a}))\}. \tag{3.21}$$

The operators $V(g)$ *satisfy the conditions*

$$V(g_1)V(g_2) = \lambda(g_1,g_2)V(g_1g_2), \qquad \lambda(g_1,g_2) = \arg\left(\det(g_1g_2)_{11}(g_2)_{11}^{-1}(g_1)_{11}^{-1}\right)^{1/4},$$

where g_{11} *is the upper left element of the matrix* g *and* $\arg z$, $z \in \mathbb{C}$, *is defined by* $\arg z = z|z|^{-1}$.

The proof is reduced to computations using the relation (3.14).

Finally, one more computation yields the following equality relating the representations that we have constructed:

$$V(g)K(h)V(g^{-1}) = K(ghg^{-1}), \qquad g \in Sp(H), \quad h \in H.$$

Now the representation W of the group $H \cdot Sp(H)$ from (3.16) is given by the formula

$$W(h, g = K(h)V(g), \qquad (h,g) \in H \cdot Sp(H). \tag{3.22}$$

We have proved the following result.

THEOREM 3.2. *The operators* (3.22) *form a unitary projective representation of the group* $H \cdot Sp(H)$ (*the Shale–Weil representation*).

3.9. Even and odd part of the representation $g \mapsto V(g)$, $g \in Sp(H)$. Consider subspaces $\overline{S}_0(H) = \bigoplus \overline{S^{2n}(H)}$ and $\overline{S}_1(H) = \bigoplus \overline{S^{2n+1}(H)}$ of the space $\overline{S}(H)$ (even and odd subspaces). One can easily see that the sets of vectors $\operatorname{Exp}(s/2)$ and $a * \operatorname{Exp}(s/2)$, where $a \in H$, $s \in \mathcal{D}(H)$, are total in $\overline{S}_0(H)$ and $\overline{S}_1(H)$ respectively. Each of these subspaces is invariant under the representation $g \mapsto V(g)$, $g \in Sp(H)$. Therefore, we obtain representations V_i of the group $Sp(H)$ in $\overline{S}_i(H)$, $i = 0,1$. They are given by the formulas

$$V_0(g)\operatorname{Exp}(s/2) = \det(AA^*)^{-1/4}\det(I_H + sB^*(A^*)^{-1})^{-1/2}\operatorname{Exp}(gs/2), \tag{3.23}$$

$$V_1(g)(a * \operatorname{Exp}(s/2)) = \det(AA^*)^{-1/4}\det(I_H + sB^*(A^*)^{-1})^{-1/2}\varphi(g,s)a * \operatorname{Exp}(gs/2). \tag{3.24}$$

The first formula immediately follows from (3.20) and (3.21). To obtain the second formula, one must replace the element $a \in H$ by ta in (3.20) and take the derivative of both sides with respect to t at $t = 0$. The representations V_i, $i = 0,1$, are irreducible (we do not give the proof here).

3.10. Realization of representations V_0, V_1 in the space of holomorphic functions. Denote by F_0 the space of holomorphic functions $f\colon \mathcal{D} \to \mathbb{C}$. Define the right action of the group $Sp(H)$ in the space $\mathcal{D}(H)$ by the formula

$$sg = (sB + \overline{A})^{-1}(sA + \overline{B}), \qquad g = \begin{pmatrix} A & B \\ \overline{B} & \overline{A} \end{pmatrix} \in Sp(H), \quad s \in \mathcal{D}(H).$$

For any $\gamma \in \mathbb{C}$ define the operators $T_\gamma(g)$, $g \in Sp(H)$, in the space F_0 by the formula

$$(T_\gamma(g)f)(s) = \det(I_H + sB\overline{A}^{-1})^\gamma f(sg), \qquad f \in F_0. \tag{3.25}$$

One can easily show that these operators form a projective representation of the group $Sp(H)$. We will be most interested with the case $\gamma = -1/2$. As we shall now see, in this case it is reasonable to multiply the operator (3.25) by the factor $(\det AA^*)^{1/4}$ (depending only on g), i.e., to consider the projective representation S_0 given by

$$(S_0(g)f)(s) = (\det AA^*)^{-1/4} \det(I_H + sB\overline{A}^{-1})^{-1/2} f(sg).$$

Let us show that the representation V_0 constructed in 3.9 can be "imbedded" into S_0. Indeed, consider the imbedding $j_0 \colon \overline{S}_0(H) \to F_0$ that associates to $v \in \overline{S}_0(H)$ the following function $f_v \in F_0$:

$$f_v(s) = (\mathrm{Exp}(-s/2), \overline{v}), \qquad s \in \mathcal{D}(H).$$

The imbedding j_0 intertwines the representations V_0 and S_0, i.e., $j_0 V_0(g) = S_0(g) j_0$ for any $g \in Sp(H)$. The proof follows from simple calculations.

Hence, we have realized the representation V_0 in a space of holomorphic functions.

Now we consider the representation $g \mapsto V_1(g)$, $g \in Sp(H)$, in the space $\overline{S}_1(H)$. Let F_1 be the space of all holomorphic functions $f \colon H \times \mathcal{D}(H) \to \mathbb{C}$ that are linear in the first variable $a \in H$. In this space we have a projective representation $g \mapsto S_1(g)$, $g \in Sp(H)$, given by the formula

$$(S_1(g)f)(a,s) = (\det AA^*)^{-1/4} \det(I_H + sB\overline{A}^{-1})^{-1/2} f((\overline{A} + sB)^{-1}a, sg), \quad f \in F_1.$$

Consider the imbedding $j_1 \colon \overline{S}_1(H) \to F_1$ that associates to $v \in \overline{S}_1(H)$ the following function $f_v \in F_1$:

$$f_v(a,s) = (a * \mathrm{Exp}(-s/2), \overline{v}), \qquad (a,s) \in H \times \mathcal{D}(H)$$

(here $a * \mathrm{Exp}(-s/2)$ is the product in the algebra $\widehat{S}(H)$; see 3.1). This imbedding intertwines the representations V_1 and S_1.

To complete this section, we return to the representation T_γ and show that it can be obtained by the construction described in 2.4.

Let us apply the above construction with (in the notation of 2.4)

$$\begin{gathered} G = Sp(H), \quad X = \mathcal{D}(H), \quad x_0 = o \in \mathcal{D}(H), \\ G_0 = \left\{ \begin{pmatrix} A & 0 \\ 0 & \overline{A} \end{pmatrix}, \ A^*A = AA^* = I_H \right\}, \\ H = \left\{ g = \begin{pmatrix} A & B \\ \overline{B} & \overline{A} \end{pmatrix}, \ I_H - A \in J_1(H) \right\}, \\ H_0 = G_0 \cap H = G_0 = \left\{ g = \begin{pmatrix} A & 0 \\ 0 & \overline{A} \end{pmatrix}, \ I_H - A \in J_1(H),\ A^*A = AA^* = I_H \right\} \end{gathered}$$

(here $J_1(H)$ is the space of trace-class operators in H). As a central extension $\widetilde{H}$ of the group H we take the set $H \times \mathbb{C}^*$ with the composition

$$(h_1, z_1)(h_2, z_2) = (h_1 h_2, z_1 z_2 \det(I + A_1^{-1} B_1 \overline{B}_2 A_2^{-1})^{\gamma}),$$
$$h_i = \begin{pmatrix} A_i & B_i \\ \overline{B}_i & \overline{A}_i \end{pmatrix} \in H, \quad z_i \in \mathbb{C}^*, \quad i = 1, 2.$$

We have the projection $p\colon \widetilde{H} \to H$ given by $p(h, x) = h$. The group $p^{-1}(H_0) = \widetilde{H}_0$ is the direct product $H_0 \times \mathbb{C}^*$. Finally, following 2.4, we construct a projective representation of the group $Sp(H)$. In the case under consideration the linear bundles described in 2.4 are trivial, so that our representation is realized in the space of smooth functions on $\mathcal{D}(H)$. Direct computations show that this representation is equivalent to the above representation T_γ.

§4. Representations of the Lie algebra $H\dot{+}sp(H)$; representations of the canonical commutation relations (CCR)

4.1. The Lie algebra $H\dot{+}sp(H)$ and its Fock representation. The Lie algebra of the Banach analytical group $Sp(H)$ consists of matrices

$$l = \begin{pmatrix} u & r \\ \overline{r} & \overline{u} \end{pmatrix}, \qquad u = -u^* \in L(H), \quad r = r' \in J_2(H). \tag{4.1}$$

We denote this Lie algebra by $sp(H)$. The Lie algebra of the group $G = H \cdot Sp(H)$ is the semidirect sum $\mathfrak{g} = H\dot{+}sp(H)$, where H is an ideal and an element of the form (4.1) acts on H by the formula $l(h) = u(h) + \overline{r}(h)$. Construct the projective representation π of the Lie algebra $\mathfrak{g}$ corresponding to the representation of the group $H \cdot Sp(H)$ from the previous section. Consider a dense linear subset $\overline{S}_*(H)$ of the space $\overline{S}(H)$ consisting of all vectors $f = \sum_{k \geq 0} f_k$ such that $f_k = 0$ for sufficiently large k. For $f \in \overline{S}_*(H)$ define the operator $a^+(f)$ in the space $\overline{S}_*(H)$ by the formula $a^+(f)\varphi = f * \varphi$ (here we use the multiplication in the algebra $\widehat{S}(H)$). By $a(f)$ we denote the operator in $\overline{S}_*(H)$ satisfying the condition $(a(f)\varphi, \psi) = (\varphi, a^+(f)\psi)$ for all $\varphi, \psi \in \overline{S}_*(H)$; one can easily verify that such an operator exists. The operators $a^+(f)$ and $a(f)$, $f \in H$, are called creation and annihilation operators respectively. The following relations hold:

$$a^+(f) = (a(\overline{f}))^*, \qquad f \in H, \tag{4.2}$$

$$\begin{gathered}[a(f_1), a^+(f_2)] = (f_1, \overline{f}_2) I, \\ [a(f_1), a(f_2)] = [a^+(f_1), a^+(f_2)] = 0, \quad f_1, f_2 \in H.\end{gathered} \tag{4.3}$$

Let us remark that the equality (4.2) must be viewed as the formal adjointness of the operators $a^+(f)$ and $a(\overline{f})$ in the space $\overline{S}_*(H)$. One can prove that these operators admit closures, which are adjoint to each other in the space $\overline{S}(H)$.

Any bounded operator $u\colon H \to H$ can be extended to a derivation Γ_u of the algebra $\overline{S}_*(H)$. The operator Γ_u leaves invariant each subspace $\overline{S^n}(H)$ and acts in this subspace by the formula

$$\Gamma_u(f_1 * \cdots * f_n) = u(f_1) * f_2 * \cdots * f_n + \cdots + f_1 * \cdots * f_{n-1} * u(f_n).$$

After all these preparations we describe the representation π of the Lie algebra $\mathfrak{g} = H\dot{+}sp(H)$. Operators of this representation are defined in the space $\overline{S}_*(H)$ by the formulas

$$\pi(l) = \Gamma_u + \frac{1}{2}(a^+(r) - a(\overline{r})), \qquad l = \begin{pmatrix} u & r \\ \overline{r} & \overline{u} \end{pmatrix} \in sp(H), \tag{4.4}$$

$$\pi(h) = a^+(h) - a(\overline{h}), \qquad h \in H. \tag{4.5}$$

If $\xi_i = h_i \dot{+} l_i \in \mathfrak{g}$, then

$$[\pi(\xi_1), \pi(\xi_2)] = \pi([\xi_1, \xi_2]) + \tfrac{1}{2}\operatorname{tr}(u_1\overline{u}_2 - \overline{u}_1 u_2) + 2i \operatorname{Im}(h_1, h_2); \tag{4.6}$$

here I is the identity operator in the space $\overline{S}(H)$. On the right-hand side of the last equality we have the cocycle corresponding to our projective representation. We skip the easy proof of (4.4)–(4.6).

Formulas (4.2)–(4.5) imply the following relations:

$$\begin{gathered} (\pi(h))^* = -\pi(h), \\ [\pi_1(h), \pi_2(h)] = 2i \operatorname{Im}(h_1, h_2). \end{gathered} \tag{4.7}$$

4.2. Representations of the canonical commutation relations (CCR). Let us consider relations of the form (4.7) in a more general situation.

Let V be a real linear topological space endowed with a bilinear skew-symmetric form τ. Let K be a complex Hilbert space with a dense linear subset $\mathcal{D}$. Assume that to any $v \in V$ one associates a skew-Hermitian operator $\pi(v) \colon \mathcal{D} \to \mathcal{D}$ in such a way that the correspondence $v \mapsto \pi(v)$ is real-linear and

$$[\pi(v_1), \pi(v_2)] = \tau(v_1, v_2)I.$$

Then the correspondence $v \mapsto \pi(v)$ is called a representation of CCR.

In particular, let V be a complex linear topological space with an involution $v \mapsto \overline{v}$ and with an Hermitian nonnegative definite inner product (v_1, v_2). Considering V as a real space, define the form τ by the formula $\tau(v_1, v_2) = 2\operatorname{Im}(v_1, v_2)$. For a representation $v \mapsto \pi(v)$ of CCR, define the operators $a^+(v)$, $a(v)$ in $\mathcal{D}$ by the formulas

$$\begin{aligned} a^+(v) &= -\frac{1}{2}(i\pi(v) + \pi(iv)), \\ a(v) &= \frac{1}{2}(\pi(i\overline{v}) - i\pi(\overline{v})). \end{aligned}$$

Then the maps $v \mapsto a(v)$, $v \mapsto a^+(v)$ are complex linear and

$$\begin{gathered} (a^+(v))^* = a(\overline{v}), \\ [a(v_1), a^+(v_2)] = (v_1, \overline{v}_2)I, \quad [a(v_1), a(v_2)] = [a^+(v_1), a^+(v_2)] = 0. \end{gathered} \tag{4.8}$$

Relations (4.8) are also called representations of CCR. The representations of CCR constructed in 4.1 are called Fock representations.

4.3. The representation of the Lie algebra of operators generated by a representation of CCR. Let there be given a representation of CCR in the form $v \mapsto (a^+(v), a(v))$, $v \in V$. Denote by $L_0(V)$ the Lie algebra of all operators $s\colon V \to V$ given by

$$s(v) = \sum_{k=1}^{n} (v, \overline{u}_k) v_k, \qquad v \in V, \quad u_k, v_k \in V. \tag{4.9}$$

To such an operator s we associate the operator $T(s)\colon \mathcal{D} \to \mathcal{D}$ of the form

$$T(s) = \sum_{k=1}^{n} a^+(v_k) a(u_k).$$

One can easily see that $T(s)$ does not depend on the presentation of the operator s in the form (4.9). The correspondence $s \mapsto T(s)$ is a representation of the Lie algebra $L_0(V)$. For the Fock representation the operator $T(s)$ becomes the operator Γ_s from 4.1

It may happen that the correspondence $s \mapsto T(s)$ can be extended to a larger Lie algebra $L \supset L_0(V)$, forming a representation of this larger algebra. (For the Fock representation L consists of all Hilbert-Schmidt operators.) We omit rigorous discussion of this topic, concluding the subsection with an example containing convenient formal expressions for $T(s)$.

Let $V = C^\infty(\mathbb{R}^n, \mathbb{C})$ with the standard inner product. A representation $v \mapsto (a^+(v), a(v))$ can be written (formally) in the form

$$a^+(v) = \int_{\mathbb{R}^n} v(x) a^+(x)\, dx,$$
$$a(v) = \int_{\mathbb{R}^n} v(x) a(x)\, dx,$$

where $a^+(x)$, $a(x)$ are operator-valued distributions satisfying the conditions

$$(a^+(x))^* = a(x),$$
$$[a(x), a^+(y)] = \delta(x-y), \quad [a^+(x), a^+(y)] = [a(x), a(y)] = 0.$$

Let us write an operator s in the space $C^\infty(\mathbb{R}^n, \mathbb{C})$ using the Schwartz kernel

$$s(v)(x) = \int_{\mathbb{R}^n} s(x,y) v(y)\, dy.$$

Then

$$T(s) = \int_{\mathbb{R}^n}\int_{\mathbb{R}^n} s(x,y) a^+(x) a(y)\, dx\, dy.$$

§5. The space $\overline{\Lambda}(H)$ (the fermionic Fock space) and a representation of the group $O(H)$

Berezin's book [**2**] contains a remarkable analogy between the space $\overline{S}(H)$ and the fermionic space $\overline{\Lambda}(H)$, which will be defined in 5.1. In particular, in [**2**] an analog of the Gaussian measure with respect to anticommuting variables is constructed. We will not touch this last topic, restricting ourselves to a short description of the space $\overline{\Lambda}(H)$ and of a certain representation of the infinite-dimensional orthogonal group $O(H)$. Similarly to §3, the exposition is based on quadratic exponents.

However, here the proofs are omitted. Formally, the results of this section will not be used later in the book. Other expositions can be found in [**2**] and [**3**]. As before, H is a complex Hilbert space of finite or countable dimension with the involution $h \mapsto \overline{h}$.

5.1. The algebra $\widehat{\Lambda}(H)$ and the Hilbert space $\overline{\Lambda}(H)$. Let $\overline{\Lambda^n}(H)$ denote the nth exterior power of the space H. More precisely, $\overline{\Lambda^0}(H) = \mathbb{C}$, and $\overline{\Lambda^n}(H)$ for $n > 0$ is the completion of the nth exterior power of the space H with respect to the Hilbert norm corresponding to the inner product

$$(x_1 \wedge \cdots \wedge x_n, y_1 \wedge \cdots \wedge y_n) = \sum_s \varepsilon(s)(x_1, y_{1s}) \cdots (x_n, y_{ns}),$$

where $s \in S_n$, and $\varepsilon(s)$ is the sign of the permutation s. Denote by $\widehat{\Lambda}(H)$ the algebra of formal power series

$$r = \sum_{k \geq 0} r_k, \qquad r_k \in \overline{\Lambda^n}(H),$$

with the multiplication, denoted $\wedge$, obtained by extending the multiplication in the exterior algebra. Finally, let

$$\overline{\Lambda}(H) = \left\{ r = \sum_{k \geq 0} r_k \,\middle|\, \sum_{k \geq 0} |r_k|^2 < \infty \right\}, \qquad \text{where } |r_k| = \sqrt{(r_k, r_k)}.$$

The inner product $(r, r') = \sum_{k \geq 0} (r_k, r'_k)$ makes $\overline{\Lambda}(H)$ a Hilbert space. Its even and odd parts are defined by the formulas $\overline{\Lambda}_0(H) = \bigoplus \overline{\Lambda^{2k}}(H)$ and $\overline{\Lambda}_1(H) = \bigoplus \overline{\Lambda^{2k+1}}(H)$ respectively.

5.2. An isomorphism between the spaces $\widehat{\Lambda}^2(H)$ and $J_2^1(H)$. Denote

$$J_2^1(H) = \{ s \in J_2(H) \mid s = -s' \}.$$

The inner product $(s_1, s_2) = 2 \operatorname{tr}(s_1 s_2^*)$ makes $J_2^1(H)$ a Hilbert space. To each element $h_1 \wedge h_2 \in \widehat{\Lambda}^2(H)$ we associate an operator in H by the formula

$$h \mapsto \frac{1}{2}((h, \overline{h}_1) h_2 - (h, \overline{h}_2) h_1).$$

This correspondence can be extended to an isomorphism of Hilbert spaces $\widehat{\Lambda}^2(H) \simeq J_2^1(H)$. Below we identify the corresponding elements of these two spaces. The operator norm of an element s is denoted by $\|s\|$, and its Hilbert-Schmidt norm by $|s|$.

5.3. Quadratic exponents in $\overline{\Lambda}(H)$. In the algebra $\widehat{\Lambda}(H)$ consider the elements of the form

$$\operatorname{Exp}(s/2), \quad a \wedge \operatorname{Exp}(s/2), \qquad a \in H, \quad s \in \overline{\Lambda^2}(H). \tag{5.1}$$

(Recall that for each $w \in \widehat{\Lambda}(H)$ the element $\operatorname{Exp} w$ is defined by the formula

$$\operatorname{Exp} w = \sum_{n=0}^{\infty} \frac{1}{n!} w^{\wedge n},$$

where $w^{\wedge n} = w \wedge \cdots \wedge w$, $w^{\wedge 0} = 1$.)

The next theorem summarizes the properties of elements (5.1).

THEOREM 5.1. (1) $\operatorname{Exp}(s/2) \in \overline{\Lambda}(H)$ *and* $a \wedge \operatorname{Exp}(s/2) \in \overline{\Lambda}(H)$, *for all* $a \in H$, $s \in \overline{\Lambda^2}(H)$.

(2) *The following equalities hold:*

$$(\operatorname{Exp}(u/2), \operatorname{Exp}(v/2)) = \sqrt{\det(I - u\overline{v})}, \tag{5.2}$$

$$(a \wedge \operatorname{Exp}(u/2), b \wedge \operatorname{Exp}(v/2)) = ((I - u\overline{v})^{-1} a, b)\sqrt{\det(I - u\overline{v})}, \tag{5.3}$$

$$u, v \in \overline{\Lambda^2}(H), \quad a, b \in H.$$

Let us clarify the meaning of (5.2). For $\|u\| < 1$, $\|v\| < 1$ the right-hand side is defined as a single-valued branch of the two-valued function, which is holomorphic in u, antiholomorphic in v, and takes the value 1 at $u = v = 0$. Formula (5.2) claims that this branch can be extended (as a function which is holomorphic in u and antiholomorphic in v) to all u, v in $\overline{\Lambda^2}(H)$, and this extension coincides with the left-hand side of (5.2). Formula (5.3) has a similar meaning.

The proof of this theorem can be extracted from computations in [**33**, Chapter 12].

5.4. The group $O(H)$ and the Shale–Weil representation. Denote by $O(H)$ the group of all invertible $\mathbb{R}$-linear transformations $h \mapsto Ah + B\overline{h}$, $h \in H$ (here A, B are complex linear bounded operators with $B \in J_2(H)$), that preserve the form $\operatorname{Re}(h_1, h_2)$. The group $O(H)$ can be identified with the group of matrices

$$g = \begin{pmatrix} A & B \\ \overline{B} & \overline{A} \end{pmatrix}, \qquad B \in J_2(H), \quad gg^* = g^*g = I.$$

The group $O(H)$ is not connected. We consider its connected component $O_+(H)$ and construct a unitary projective representation of $O_+(H)$ (the Shale–Weil representation). For any $g \in O_+(H)$ consider the set

$$E_g = \{s \in \overline{\Lambda^2}(H) \mid A + B\overline{s} \text{ is invertible}\}.$$

(In the expression $A + B\overline{s}$ we identified the element $s \in \overline{\Lambda^2}(H)$ with the corresponding operator from $J_2^{\frac{1}{2}}(H)$.) Define the action of an element $g \in O_+(H)$ on an element $s \in E_g$ by the formula

$$gs = (As + B)(\overline{A} + \overline{B}s)^{-1}.$$

Next, introduce the set

$$O^1(H) = \left\{ g = \begin{pmatrix} A & B \\ \overline{B} & \overline{A} \end{pmatrix} \middle| \, A \text{ is invertible.} \right\}$$

Theorem 5.1 easily implies that there exists a unique unitary operator $T_0(g)\colon \widehat{\Lambda}_0(H) \to \widehat{\Lambda}_0(H)$, $g \in O^1(H)$, satisfying the condition

$$T_0(g)\operatorname{Exp}(s/2) = (\det AA^*)^{1/4} (\det(I + \overline{A}^{-1}\overline{B}s))^{1/2} \operatorname{Exp}(gs/2), \qquad s \in E_g.$$

Let us note that $\overline{A}^{-1}\overline{B} \in \widehat{\Lambda}^2(H)$, so that $(\det(I + \overline{A}^{-1}\overline{B}s))^{1/2}$ in the above formula is defined uniquely using (5.2). The correspondence $g \mapsto T_0(g)$ can be extended to a projective unitary representation of the group $O_+(H)$ in the space $\overline{\Lambda}_0(H)$.

We can also construct a projective unitary representation T_1 in the space $\widehat{\Lambda}_1(H)$ by the formula

$$T_1(g)(a\wedge \operatorname{Exp}(s/2)) = (\det AA^*)^{1/4}(\det(I+\overline{A}^{-1}\overline{B}s))^{1/2}(A^*-sB^*)^{-1}a\wedge \operatorname{Exp}(gs/2).$$

In conclusion, let us note that the representation T_0 can be obtained using the general procedure described in 2.4. One must take the group $G = O_+(H)$ and set

$$H = \left\{\begin{pmatrix} A & B \\ \overline{B} & \overline{A} \end{pmatrix} \in O_+(H),\ I - A \in J_1(H)\right\},$$

$$G_0 = \left\{\begin{pmatrix} A & 0 \\ 0 & \overline{A} \end{pmatrix} \in O_+(H),\ AA^* = A^*A = I\right\},$$

$$H_0 = H \cap G_0 = \left\{\begin{pmatrix} A & 0 \\ 0 & \overline{A} \end{pmatrix} \in O_+(H),\ AA^* = A^*A = I,\ I - A \in J_1(H)\right\}.$$

We have a nontrivial $\mathbb{Z}_2$-covering $\widetilde{H} \to H$, which splits into the direct product $H_0 \times \{-1, 1\}$ over the subgroup H_0. As was described in 2.1, these data allow us to construct a projective representation of the group $O_+(H)$ in the space of holomorphic sections of the corresponding bundle. This representation is equivalent to the representation T_0.

§6. Representations of groups in the space $\overline{S}(H)$; the Araki scheme

In this section we describe one of the most frequently used methods for constructing representations of a "large" group; it is based on an embedding of the given group into the motion group of the Hilbert space and on using the canonical projective representation of this motion group in the Fock space (see §3). As applied to groups of measurable currents, this method was developed by Araki in [**42**]; these groups are discussed in more detail in Chapter 2. Important results have been obtained by Vershik, Gelfand, and Graev. In this and the following sections we describe some of these results.

6.1. Representations of the motion group of the Hilbert space. The group $H \cdot Sp(H)$ introduced in §3 contains the subgroup $H \cdot U(H)$, the semidirect product of the additive group of the space H to the group of unitary operators in H. Elements $(h, U) \in H \cdot U(H)$ are identified with motions $x \mapsto Ux + h$ of the space H. Therefore, the Shale–Weil representation constructed in §4 yields a projective representation of the group $\operatorname{Mot} H \simeq H \cdot U(H)$. Let us note that formulas for this representation do not involve the involution in H. Hence, for any complex Hilbert space H we have a unitary projective representation W of the group $\operatorname{Mot} H$ in the space $\overline{S}(H)$ defined by the formula

$$W(h, U)\operatorname{Exp} x = e^{-(1/2)|h|^2 - (Ux, h)}\operatorname{Exp}(Ux + h), \quad x \in H, \quad (h, U) \in \operatorname{Mot} H. \tag{6.0}$$

We have

$$W(h_1, U_1)W(h_2, U_2) = e^{i\operatorname{Im}(U_1^{-1}h_1, h_2)}W(h_1 + U_1h_2, U_1U_2), \quad (h_1, U_1), (h_2, U_2) \in \operatorname{Mot} H.$$

The operators $h \mapsto K(h) = W(h, I)$, $h \in H$, form a projective representation of the additive group of the space H (the Fock representation of CCR in the Weil form).

6.2. Group representations using embeddings into Mot H**; the Araki scheme.** Let $g \mapsto U(g)$, $g \in G$, be a unitary representation of a group G in the Hilbert space H, and let r be a 1-cocycle of this representation, i.e., a weakly continuous mapping $r\colon G \to H$ satisfying the conditions

$$r(g_1g_2) = r(g_1) + U(g_1)r(g_2), \quad r(e) = 0. \tag{6.1}$$

Then the mapping $g \mapsto (r(g), U(g))$, $g \in G$, is a homomorphism from G to Mot H. Its composition with the above representation of the group Mot H is a projective representation $g \mapsto T(g)$, $g \in G$, in the space $\overline{S}(H)$. We will denote this composition (whenever necessary) by $T^{U,r}$. Hence, the representation $T^{U,r}$ satisfies the following conditions:

(6.2)

$$T(g)\operatorname{Exp} x = e^{-(1/2)|r(g)|^2+(x,r(g^{-1}))}\operatorname{Exp}(U(g)x + r(g)), \qquad g \in G, \quad x \in H,$$

$$T(g_1)T(g_2) = e^{-i\operatorname{Im}(r(g_1^{-1}),r(g_2))}T(g_1g_2). \tag{6.3}$$

If the resulting 2-cocycle

$$(g_1, g_2) \mapsto e^{-i\operatorname{Im}(r(g_1^{-1}),r(g_2))} \tag{6.4}$$

is trivial, i.e., can be represented in the form $\lambda(g_1g_2)(\lambda(g_1)\lambda(g_2))^{-1}$, then the operators $g \mapsto \lambda(g)T(g)$ form a usual (not a projective) representation. In particular, consider an orthogonal representation $g \mapsto O(g)$ in a real Hilbert space V and a 1-cocycle $r\colon G \mapsto V$ of this representation. Then we obtain a representation $g \mapsto U(g)$ in the complexification $H = V_{\mathbb{C}}$ and the cocycle $r\colon G \to H$. In this case the cocycle (6.4) for the representation $T^{U,r}$ is equal to 1.

Now we return to an arbitrary operator $W(h, U)$ defined by formula (6.0). To make the notation consistent we will write $T^{U,h}$ instead of $W(h, U)$.

6.3. Realization in the space $L_2(\Omega, \mu)$. Let V be a real Hilbert space, $g \mapsto O(g)$, $g \in G$, an orthogonal representation of G. Let $r_k\colon G \to V$, $k = 1, 2$, be two cocycles. We get a unitary representation $g \mapsto U(g)$, $g \in G$, in the Hilbert space $H = V_{\mathbb{C}} = V + iV$ and a 1-cocycle $g \mapsto r(g) = r_1(g) + ir_2(g)$. Consider the corresponding representation $T = T^{U,r}$ in the space $\overline{S}(H)$. The isomorphism $\Phi_c\colon \overline{S}(H) \to L_2(\Omega, \mu)$ from 3.3 yields a representation in the space $L_2(\Omega, \mu)$; we denote it by T_c. This representation is given by the formula

$$\begin{aligned} T_c(g)f)(\omega) = \exp\{-\tfrac{1}{2}|r(g)|^2 - (c^2/2)(r(g), \overline{r(g)}) + c(r(g), \omega)\}\\ \times f\big((U(g))^*\omega + 2\operatorname{Re}(cr(g^{-1}))\big). \end{aligned}$$

§7. Irreducibility and equivalence of representations $T^{U,r}$

In this section we follow Vershik, Gelfand, and Graev [**6, 7**]. Theorem 7.1 (irreducibility of Fock representations of CCR in the Weil form) is well known; see, e.g., Guichardet's book [**54**].

7.1. Irreducibility of the representation $h \mapsto K(h)$, $h \in H$.

THEOREM 7.1. *The representation $h \mapsto K(h)$, $h \in H$, is irreducible.*

PROOF. We call a function $f\colon H \to \mathbb{C}$ holomorphic if $f(x) = \sum_{n\geq 0}(x^{*n}, f_n)$, where $f_n \in \overline{S}^n(H)$. A function f is called antiholomorphic if the function $x \mapsto \overline{f(x)}$ is holomorphic. To prove the theorem consider a bounded operator A in the space $\overline{S}(H)$ that commutes with all operators $K(h)$. Introduce the notation $f(x,y) = (A \operatorname{Exp} x, \operatorname{Exp} y) \exp(-(x,y))$. One can easily see that $f(x,y) = f(x+h, y+h)$ for each $h \in H$. Hence, $f(x,y) = f(0, y-x) = f(x-y, 0)$. On the other hand, the function $x \mapsto f(x,0) = (A \operatorname{Exp} x, \operatorname{Exp} 0)$ is holomorphic, whereas the function $y \mapsto f(0,y)$ is antiholomorphic. Therefore, $f(x,0) = f(0,y) = \lambda = \text{const}$. Hence, $f(x,y) = \lambda$, i.e., $(A \operatorname{Exp} x, \operatorname{Exp} y) = \lambda \exp(x,y)) = \lambda(\operatorname{Exp} x, \operatorname{Exp} y)$ for all $x, y \in H$. Therefore, $A = \lambda I$, and the representation $h \mapsto K(h)$ is irreducible.

7.2. An irreducibility condition for the representation $T^{U,r}$. Let G, H, U, and r be as in 3.1. Here we formulate simple sufficient conditions for irreducibility of the representation $T^{U,r}$.

THEOREM 7.2. *Let there exist a subgroup $G_0 \subset G$ such that*
(1) $r(g) = 0$ *for all* $g \in G_0$;
(2) *the restriction of U to G_0 does not have nontrivial finite-dimensional invariant subspaces;*
(3) *the set* $\{\operatorname{Exp} r(g)\}$ *is total in* $\overline{S}(H)$.
Then the representation $T^{U,r}$ is irreducible.

Condition (3) in this theorem is not always easy to verify. In the next theorem this condition is replaced by a simpler one.

THEOREM 7.3. *Let there exist a subgroup $G_0 \subset G$ such that*
(1) $r(g) = 0$ *for all* $g \in G_0$;
(2) *symmetric powers U_0^n, $n = 0, 1, \ldots$, of the representation $U_0 = U \mid G_0$ are pairwise disjoint;*
(3) *for each $n = 1, 2, \ldots$ the set of vectors $(r(g))^{*n}$ is total in the space $\overline{S}^n(H)$.*
Then the representation $T^{U,r}$ is irreducible.

PROOF OF THEOREM 7.2. Condition (1) implies that each of the spaces $\overline{S}^n(H)$ is invariant under the representation $g \mapsto T(g)$, $g \in G_0$. Moreover, the decomposition $\overline{S}(H) = \bigoplus \overline{S}^n(H)$ gives the decomposition $T \mid G_0 = \bigoplus U^n \mid G_0$.

LEMMA 7.1. *For $n \geq 1$ the space $\overline{S}^n(H)$ does not contain nonzero vectors invariant under $T(g)$, $g \in G_0$.*

PROOF OF THE LEMMA. We prove Lemma 7.1 by induction on n. It holds for $n = 1$ (this follows from condition (2) in Theorem 7.2). Let us assume that it holds for the subspace $\overline{S}^{n-1}(H)$. We have a natural surjection $p\colon H \otimes \overline{S}^{n-1}(H) \to \overline{S}^n(H)$, which is the composition of tensor multiplication with subsequent symmetrization. This surjection intertwines the tensor product $U_0 \otimes U_0^{n-1}$ with the representation U_0^n. Let us assume that the space $\overline{S}^n(H)$ contains a nonzero vector that is invariant under U_0^n. Then the space $H \otimes \overline{S}^{n-1}(H)$ contains a nonzero vector that is invariant under the representation $U_0 \otimes U_0^{n-1}$. Then there exists a Hilbert-Schmidt operator $A\colon \overline{S}^{n-1}(H) \to H$ satisfying the condition $AU_0^{n-1}(g) = U_0(g)A$, $g \in G_0$. Denote $S = AA^*$. Then $U_0(g)S = SU_0(g)$ and S is a nonzero compact nonnegative definite operator. Therefore, it has a positive eigenvalue such that the corresponding eigenspace is finite-dimensional. This eigenspace is invariant under all operators $U_0(g)$, $g \in G$, contrary to the conditions of Theorem 7.2. The lemma is proved.

Now we can prove Theorem 7.2. Let B be a bounded operator that commutes with all operators $T(g)$, $g \in G$. Lemma 7.1 implies that $\operatorname{Exp} 0$ is the unique (up to a factor) vector that is invariant under all $T(g)$, $g \in G_0$. Therefore, $B \operatorname{Exp} 0 = \lambda \operatorname{Exp} 0$, $\lambda \in \mathbb{C}$. Hence, $B \operatorname{Exp} r(g) = BT(g)e^{(1/2)|r(g)|^2} \operatorname{Exp} 0 = \lambda \operatorname{Exp} r(g)$. Using condition (3) of Theorem 7.2, we get $B = \lambda I$. The proof is completed.

PROOF OF THEOREM 7.3. Let B be as in the proof of Theorem 7.2. Repeating the previous arguments, we get $B \operatorname{Exp} r(g) = \lambda \operatorname{Exp} r(g)$. Condition (2) of Theorem 7.3 implies that each space $\overline{S}^n(H)$ is invariant under B. Therefore, $B(r(g))^{*n} = \lambda(r(g))^{*n}$. Using condition (3) we get $B = \lambda I$, and Theorem 7.3 is proved.

7.3. Spherical functions. Condition for equivalence of representations. Let the assumptions of Theorem 7.2 be satisfied. The unique (up to a factor) vector that is invariant under all operators $T(g)$, $g \in G_0$, is $\operatorname{Exp} 0$. The corresponding spherical function is defined by the formula

$$(T(g) \operatorname{Exp} 0, \operatorname{Exp} 0) = e^{-(1/2)|r(g)|^2}. \tag{7.1}$$

Let us take two representations T^{U_k, r_k} acting in spaces $\overline{S}(H_k)$, $k = 1, 2$, and assume that both these representations satisfy the conditions of Theorem 7.2. We give the condition for their equivalence, considering, for simplicity, the case of a connected group G.

THEOREM 7.4. *Representations $T_k = T^{U_k, r_k}$ are equivalent (as projective representations) if and only if there exists a unitary isomorphism $V\colon H_1 \to H_2$ satisfying the conditions $VU_1(g) = U_2(g)V$, $Vr_1(g) = r_2(g)$, $g \in G$.*

PROOF. The equivalence of projective representations T_1 and T_2 means that $UT_1(g) = f(g)T_2(g)U$ for some unitary operator U and function $f\colon G \to \mathbb{C}$, $|f(g)| = 1$. It is clear that $U \operatorname{Exp} 0 = \lambda \operatorname{Exp} 0$, $|\lambda| = 1$. Since the spherical functions (7.1) are positive, we have $f(g) = 1$. The equality of spherical functions implies that $|r_1(g)| = |r_2(g)|$. Together with (6.1), this implies that

$$\operatorname{Re}(r_1(g_1^{-1}), r_1(g_2)) = \operatorname{Re}(r_2(g_1^{-1}), r_2(g_2)).$$

Next, from (6.3) it follows that

$$\exp\{i \operatorname{Im}(r_1(g_1^{-1}), r_1(g_2))\} = \exp\{i \operatorname{Im}(r_2(g_1^{-1}), r_2(g_2))\}.$$

Since the group G is connected, we have

$$\operatorname{Im}(r_1(g_1^{-1}), r_1(g_2)) = \operatorname{Im}(r_2(g_1^{-1}), r_2(g_2)).$$

Hence, we get $(r_1(g_1^{-1}), r_1(g_2)) = (r_2(g_1^{-1}), r_2(g_2))$. Since the set $\{r_k(g),\, g \in G\}$ is total in H_k, $k = 1, 2$, there exists a unitary isomorphism $V\colon H_1 \to H_2$ such that $Vr_1(g) = r_2(g)$, $g \in G$. Using (6.1), we obtain $VU_1(g) = U_2(g)V$, and Theorem 7.4 is proved.

§8. Another irreducibility criterion for the representations $T^{U,r}$

To formulate this criterion we need certain preparations.

8.1. Representations of abelian groups. The spectral type. Let A be a topological abelian group; multiplication in A will be denoted ab, and the identity element is e. Let $\widehat{A}$ be the group of all continuous homomorphisms (characters) to $\mathbb{T}^1 = \{z \in \mathbb{C} \mid |z| = 1\}$. The value of a character $\chi \in \widehat{A}$ at an element $a \in A$ will be denoted by $\langle a, \chi \rangle$. We endow the group $\widehat{A}$ with the weak topology, which is generated by the functions $\chi \mapsto \langle a, \chi \rangle$, $a \in A$. We assume that A contains a countable dense subset and that the σ-algebra of Borel subsets in $\widehat{A}$ is standard (see 11.3).

Let τ be a nonnegative Borel measure on $\widehat{A}$, H^0 a Hilbert space, and $H = L_2(\widehat{A}, \tau, H^0)$ the Hilbert space of H^0-valued functions on $\widehat{A}$. We get a unitary representation $a \mapsto U(a)$, $a \in A$, acting in H by the formula

$$(U(a)f)(\chi) = \langle a, \chi \rangle f(\chi). \tag{8.1}$$

Since equivalent measures generate unitary equivalent representations, we can and will assume that $\tau(\widehat{A}) = 1$, i.e., τ is a probability measure. Note that $(\widehat{A}, \tau)$ is a Lebesgue space.

In this section we consider only those representations of the group A that can be decomposed into a finite or countable direct sum of representations of the form (8.1). In the case when the group G is not locally compact (this is the case we are interested in), this condition significantly narrows the class of admissible representations.

Let $a \mapsto U(a)$ be a representation of the above class, $U = \bigoplus U_k$ its decomposition into the direct sum of representations of the form (8.1). To each U_k there corresponds a probability measure τ_k, defined uniquely up to equivalence. Take $c_k > 0$ such that $\sum c_k = 1$, and consider the measure $\tau = \sum c_k \tau_k$. The equivalence class of the measure τ is called the spectral type of the representation U. Any measure from this equivalence class will also be called the spectral type of U. Let us present several examples of the computation of the spectral type.

EXAMPLE 1. Let the spectral types of representations U_k, $1 \le k \le n$, be given by measures τ_k. Then the spectral type of the representation $U_1 \otimes \cdots \otimes U_n$ is given by the measure $\tau_1 * \cdots * \tau_n$.

EXAMPLE 2. Let $a \mapsto U(a)$, $a \in A$, be a representation in a Hilbert space H, with the spectral type τ. Then the spectral type of the representation $T^{U,0}$ (see 6.2) is given by the measure $\operatorname{Exp} \tau = \sum_{k \ge 0} \tau^{*k}/k!$, where $\tau^{*k} = \tau * \cdots * \tau$ (k times).

EXAMPLE 3. Let H be as in Example 2 and let V be the range of a positive Hilbert-Schmidt operator $K \colon H \to H$. Assume that we are given a mapping $r \colon A \to V$ satisfying the following conditions: (1) $r(a_1 a_2) = r(a_1) + r(a_2)$ for all $a_1, a_2 \in A$; (2) $(r(a_1), r(a_2))$ is real for all $a_1, a_2 \in A$; (3) the mapping r is continuous with respect to the topology on A and the norm $v \mapsto |K^{-1}v|$, $v \in V$. According to 1.3, we have a Gaussian measure γ in the space Ω of linear functionals on V_0, $V = V_0 + iV_0$, $r(a) \in V_0$. Next, according to 6.3, we can find an isomorphism $\overline{S}(H) \simeq L_2(\Omega, \gamma)$ such that the operators $T^{I,h}$ in $\overline{S}(H)$ become the operators $f(\omega) \mapsto \exp\{i\langle h, \omega \rangle\} f(\omega)$, $h \in V_0$. In particular, the operators $U(a) = T^{I,r(a)}$ become the operators of multiplication by $\exp\{i\langle r(a), \omega \rangle\}$, $a \in A$. We obtain the mapping $r_* \colon \Omega \to \widehat{A}$ transforming a functional $\omega \in \Omega$ into the character $a \mapsto \exp\{i\langle r(a), \omega \rangle\}$. It transforms the measure γ on Ω into the measure $r_*(\gamma)$ on $\widehat{A}$. This measure is the spectral type of the representation $a \mapsto U(a)$, $a \in A$.

EXAMPLE 4. This is a special case of the previous example. Let the group A be a real linear topological space and the mapping r be linear. Then the spectral type of the representation U is defined by the Gaussian measure on the dual space A' corresponding to the inner product $(r(a_1), r(a_2))$ on A.

8.2. Singularly related measures. Let B be a topological group (later we will take $B = \widehat{A}$), ν_1, ν_2 two probability measures on B. Consider the spaces with measure $(B \times B, \nu_1 \times \nu_2)$ and $(B, \nu_1 * \nu_2)$; here $\nu_1 \times \nu_2$ is the direct product of the measures and $\nu_1 * \nu_2$ is the convolution. We have the mapping of spaces with measures

$$s\colon (B \times B, \nu_1 \times \nu_2) \to (B, \nu_1 * \nu_2) \tag{8.2}$$

given by $s(b_1, b_2) = b_1 b_2$. It preserves the measures, i.e., $(\nu_1 \times \nu_2)(s^{-1}(E)) = \nu_1 * \nu_2(E)$ for any measurable set $E \subset B$.

DEFINITION 8.1. Measures ν_1, ν_2 are called *singularly related* if the mapping (8.2) is an isomorphism of spaces with measures (of course, after deleting some sets of measure zero).

In other words, there exists a measurable set $\Omega_0 \subset B \times B$ of full $\nu_1 \times \nu_2$ measure such that if $(a_1, a_2), (b_1, b_2) \in \Omega_0$ and $a_1 a_2 = b_1 b_2$, then $a_1 = a_2$, $b_1 = b_2$.

Another equivalent formulation of this definition goes as follows. Any Borel function $f\colon B \times B \to \mathbb{C}$ coincides almost everywhere (with respect to the measure $\nu_1 \times \nu_2$) with a function of the form $\varphi(b_1 b_2)$. The equivalence of these conditions follows from the properties of Lebesgue spaces. The next lemma explains, to some extent, the use of the term in the above definition.

LEMMA 8.1. *Let measurable sets $B_i \subset B$, $i = 1, 2$, be given such that*
(1) $\nu_i(B_i) = 1$;
(2) *if $a \in B_1$, $b \in B_1$, $a \neq b$, then $aB_2 \cap bB_2 = \varnothing$.*
Then the measures ν_1 and ν_2 are singularly related.

Let us note that the statement of Lemma 8.1 remains true if we replace the condition (2) with the following condition:

B_1 is a subgroup in B and $aB_2 \cap B_2 = \varnothing$ for any $a \in B_1$, $a \neq e$. Hence, the translation by a is a singular transformation with respect to the measure ν_2.

PROOF OF LEMMA 8.1. Let us take $\Omega_0 = B_1 \times B_2$. If $(a_1, a_2), (b_1, b_2) \in \Omega_0$, $a_1 a_2 = b_1 b_2 = c$, then $c \in a_1 B_2$, $c \in b_1 B_2$. By the conditions of the lemma, $a_1 = b_1$, hence $a_2 = b_2$. Therefore, for $(a_1, a_2), (b_1, b_2) \in \Omega_0$ the equality $a_1 a_2 = b_1 b_2$ can happen only when $a_1 = a_2$, $b_1 = b_2$. The lemma is proved.

Now we get an important property of unitary representations of an abelian group A, which is related to the above notion. Let $a \mapsto U_k(a)$, $a \in A$, $k = 1, 2$, be unitary representations in Hilbert spaces H_k belonging to the class introduced in 8.1. Let τ_1, τ_2 be the spectral types of these representations. Consider the tensor product $a \mapsto U_1(a) \otimes U_2(a)$, $a \in A$, acting in $H_1 \otimes H_2$. Denote by M the von Neumann algebra generated by the operators $U_1(a) \otimes U_2(a)$, $a \in A$.

LEMMA 8.2. *Let τ_1, τ_2 be singularly related. Then the operators $U_1(a) \otimes I_2$, $I_1 \otimes U_2(a)$ belong to the algebra M for each $a \in A$; here I_1, I_2 are the identity operators in the spaces H_1, H_2 respectively.*

PROOF. Let us realize each representation U_k, $k = 1, 2$, by operators of multiplication with functions $\chi \mapsto \langle a, \chi \rangle$, $\chi \in \widehat{A}$, in the Hilbert spaces of vector-valued functions corresponding to the measures τ_k, $k = 1, 2$. Then the representation $U_1 \otimes U_2$ is realized by operators of multiplications with the functions $(\chi_1, \chi_2) \mapsto \langle a, \chi_1\chi_2 \rangle$ in the space of vector-valued functions corresponding to the measure $\tau_1 \times \tau_2$. Therefore, the algebra M consists of operators of multiplication with functions of the form $(\chi_1, \chi_2) \mapsto f(\chi_1\chi_2)$, where f is a measurable function on $\widehat{A}$. In this realization the operators $U_1(a) \otimes I_2$, $U_1 \otimes U_2(a)$ become the operators of multiplication with the functions $(\chi_1, \chi_2) \mapsto \langle a, \chi_1 \rangle$, $(\chi_1, \chi_2) \mapsto \langle a, \chi_2 \rangle$ respectively. Since the measures τ_1, τ_2 are singularly related, these two functions coincide almost everywhere on $\widehat{A} \times \widehat{A}$ (with respect to the measure $\tau_1 \times \tau_2$) with functions of the form $(\chi_1, \chi_2) \mapsto f_1(\chi_1\chi_2)$, $(\chi_1, \chi_2) \mapsto f_2(\chi_1\chi_2)$, where f_1, f_2 are Borel functions. This proves the lemma.

8.3. An irreducibility condition. Returning back to our main theme, let us consider a topological group G, an orthogonal representation $g \mapsto O(g)$, $g \in G$, in a real Hilbert space V, and a cocycle $r\colon G \to V$. Passing to the complexification $H = V_{\mathbb{C}}$, we get a unitary representation $g \mapsto U(g)$ and a cocycle $r\colon G \to H$. We consider the representation $T^{U,r}$ of the group G in the space $\overline{S}(H)$. Our goal is to find sufficient conditions for the irreducibility of $T^{U,r}$. For that, we will need some abelian subgroups of G that are called special subgroups. They are defined as follows.

Let A be an abelian subgroup in G. Let us assume that the space V can be decomposed into the direct sum

$$V = V^0 \oplus V^1 \tag{8.3}$$

in such a way that the spaces V^0 and V^1 are invariant under the representation $a \mapsto O(a)$, $a \in A$, and on the space V^0 the representation is trivial. Next, $r(a) = r^0(a) + r^1(a)$, $a \in A$, where $r^k(a) \in V^k$, $k = 0, 1$. It is clear that $r^0(a_1a_2) = r^0(a_1) + r^0(a_2)$ and the mapping $r^1\colon A \to V^1$ is a 1-cocycle of the representation O. Let us impose the following conditions.

(α) The vectors $r^0(a)$, $a \in A$, belong to the range of a positive Hilbert-Schmidt operator $K\colon V^0 \to V^0$; moreover, the mapping $r^0\colon A \to V^0$ is continuous with respect to the norm $v \mapsto |K^{-1}v|$, $v \in V^0$.

(β) The cocycle $r^1\colon A \to V^1$ is trivial (i.e., $r^1(a) = O(a)v^1 - v^1$ for some $v^1 \in V^1$).

Under these conditions, let us introduce probabilistic measures ν_0, ν_1 on the character group $\widehat{A}$ as follows. The measure ν_0 is a measure determining the spectral type of the representation T^{I,r^0} in the space $\overline{S}(H^0)$, where $H^0 = V^0_{\mathbb{C}}$ (this spectral type is described in Example 3 from 8.2). The measure ν_1 is a measure that determines the spectral type of the representation $T^{U^1,0}$ in the space $\overline{S}(H^1)$, where $H^1 = V^1_{\mathbb{C}}$ and U^1 is the complexification of the representation $O \mid V^1$ (this spectral type is described in Example 2 from 8.2). Let us impose the following condition.

(γ) The measures ν_0 and ν_1 are singularly related.

An abelian subgroup $A \subset G$ is called special if it satisfies conditions (α), (β), (γ), and, in addition, is infinitely divisible. (The last condition means that if $a \in A$, $n \in \mathbb{Z} \setminus \{0\}$, then $a = b^n$ for some $b \in A$.)

After all these preparations, let us return to the problem about the irreducibility of the representation $T = T^{U,r}$.

Let us be given a family of special subgroups $\{A_\alpha,\ \alpha \in \mathfrak{A}\}$ in G. To each subgroup A_α we associate the above decomposition $V = V_\alpha^0 \oplus V_\alpha^1$ and the homomorphism $r_\alpha^0\colon A_\alpha \to V$ with values in V_α^0.

THEOREM 8.1. *Let the following conditions be satisfied.*

(1) *The family of vectors*

$$\{O(g)r_\alpha^0(a) \mid g \in G, a \in A_\alpha, \alpha \in \mathfrak{A}\}$$

is total in V.

(2) *The representation $g \mapsto O(g)$, $g \in G$, does not contain nontrivial finite-dimensional invariant subspaces.*

Then the representation $T^{U,r}$ is irreducible.

PROOF. Denote by M the von Neumann algebra generated by the operators $T(g)$, $g \in G$. We must prove that M is the algebra of all bounded operators.

Let us choose one of the special subgroups A_α. For brevity below we will write A and r^0 instead of A_α and r_α^0.

LEMMA 8.3. $T^{I,r^0(a)} \in M$ *for all* $a \in A$.

PROOF OF LEMMA 8.3. Let us consider the following two representations of the group A in the space $\overline{S}(H)$: the representation T_1 obtained by restricting $T = T^{U,r}$ to A and the representation $T_0 = T^{U,r^0}$. Denote by M_1, M_0 the von Neumann algebras generated by these representations. Let us recall that $r(a) = r^0(a) + O(a)v^1 - v^1$, $a \in A$, for some $v^1 \in V^1$. This implies that $T_1 = T^{I,v^1} T_0 T^{I,-v^1}$ and $M_1 = T^{I,v^1} M_0 T^{I,-v^1}$. Therefore, to prove the lemma it suffices to verify that

$$T^{I,r^0(a)} \in M_0, \qquad a \in A \tag{8.4}$$

(here we use the fact that the operators T^{I,v^1} and $T^{I,r^0(a)}$ commute).

Let us prove (8.4). To the subgroup A we associate the decomposition $H = H^0 \oplus H^1$, $H^k = V_{\mathbb{C}}^k$, where the subspaces V^k, $k = 1, 2$, are defined by formula (8.3). This decomposition yields the isomorphism of Hilbert spaces $\overline{S}(H) \simeq \overline{S}(H^0) \otimes \overline{S}(H^1)$. Under this isomorphism, the operators $T^{I,r^0(a)}$, $T^{U(a),0}$, and $T_0(a) = T^{U(a),r^0(a)}$ become the operators $T^{I,r^0(a)} \otimes I_1$, $I_0 \otimes T^{U(a),0}$, and $T^{I,r^0(a)} \otimes T^{U^1(a),0}$; here I_0 and I_1 are the identity operators in $\overline{S}(H^0)$ and $\overline{S}(H^1)$. By Lemma 8.2, the operator $T^{I,r^0(a)} \otimes I_1$ belongs to the von Neumann algebra generated by the operators $T^{I,r^0(a)} \otimes T^{U^1(a),0}$ (this is where we use that the maximal spectral types ν_0, ν_1 of representations $a \mapsto T^{I,r^0(a)}$, $a \mapsto T^{U^1(a),0}$, $a \in A$, are singularly related). Returning from the space $\overline{S}(H^0) \otimes \overline{S}(H^1)$ to the space $\overline{S}(H)$, we obtain (8.4). Lemma 8.3 is proved.

LEMMA 8.4. $T^{I,v} \in M$ *for all* $v \in V$, *and* $T^{U(g),0} \in M$ *for all* $g \in G$.

PROOF OF LEMMA 8.4. Denote by L the set of all vectors $v \in V$ such that $T^{I,v} \in M$. By Lemma 8.3, $r^0(a) \in L$ for all $a \in A$. Since the group A is infinitely divisible, $\lambda r^0(a) \in L$ for all rational λ; hence $\lambda r^0(a) \in L$ for all $\lambda \in \mathbb{R}$. Therefore, L is a closed linear real subspace containing all vectors of the form $r^0(a)$, $a \in A$. The relation $T(g)T^{I,v} = T^{I,O(g)v}T(g)$ implies that L is invariant under the representation O of the group G. By condition (1) of Theorem 8.1, $L = V$. The first statement of Lemma 8.4 is proved, and it implies the second statement.

Now we prove Theorem 8.1. Let D be an operator in $\overline{S}(H)$ commuting with all operators of the representation $T = T^{U,r}$. Lemma 8.4 implies that D commutes with operators $T^{U(g),0}$, $g \in G$. Condition (2) of Theorem 8.1 and Lemma 7.1 imply that $\operatorname{Exp} 0$ is a unique (up to a factor) vector invariant under $T^{U(g),0}$, $g \in G$. But then $D \operatorname{Exp} 0 = \lambda \operatorname{Exp} 0$, $\lambda \in \mathbb{C}$. Since D commutes with all operators $T^{I,v}$, $v \in V$, we have

$$\begin{aligned} D \operatorname{Exp} v = D\left(e^{(1/2)|v|^2} T^{I,v} \operatorname{Exp} 0\right) &= e^{(1/2)|v|^2} T^{I,v} D \operatorname{Exp} 0 \\ = e^{(1/2)|v|^2} T^{I,v} (\lambda \operatorname{Exp} 0) &= \lambda \operatorname{Exp} v, \qquad v \in V. \end{aligned}$$

Hence, $D \operatorname{Exp} v = \lambda \operatorname{Exp} v$ for $v \in V$. But the set $\{\operatorname{Exp} v,\ v \in V\}$ is total in $\overline{S}(H)$ (Lemma 3.1). Therefore $D = \lambda I$, implying the irreducibility of the representation T. Theorem 8.1 is proved.

CHAPTER 2

Representations of the Group of Measurable Currents

In this chapter we consider representations of the group of measurable currents in the Fock space (see Araki [**42**], Vershik, Gelfand, Graev [**6**]). We present a new simple approach to the irreducibility problem that gives its complete solution.

Another topic of this chapter is the study of representations in continuous tensor products of Hilbert spaces (see Berezin [**3**]).

§9. Factorizable operators in the Fock space $\overline{S}(H)$, $H = L_2(X, \mu, H^0)$; the convexity theorem

9.1. Operators $T(A, b, c)$. Let H be a complex Hilbert space, $\overline{S}(H)$ the corresponding Fock space (see §3), $\mathcal{L}$ the linear span of all vectors of the form $\operatorname{Exp} h$, $h \in H$. By $L(H)$ we denote the algebra of all bounded operators in H. For any triple $(A, b, c) \in L(H) \times H \times H$ define the linear operator $T = (A, b, c)\colon \mathcal{L} \to \mathcal{L}$ by the formula

$$T(\operatorname{Exp} h) = e^{(h,c)} \operatorname{Exp}(Ah + b), \qquad h \in H. \tag{9.1}$$

This formula can be also written in the form

$$T(\operatorname{Exp} h) = \operatorname{Exp}(Ah + b + (h, c)\mathbf{1}),$$

where $\mathbf{1}$ is the unit of the algebra $\widehat{S}(H)$; see 3.1.

To describe the properties of these operators in a more convenient form, we introduce the following operation in the set $L(H) \times H \times H$:

$$(A_1, b_1, c_1)(A_2, b_2, c_2) = (A_1A_2, A_1b_2 + b_1, A_2^*c_1 + c_2)$$

and define $(A, b, c)^* = (A^*, c, b)$. The set $L(H) \times H \times H$ becomes a semigroup with involution, and the map $(A, b, c) \mapsto T(A, b, c)$ yields a projective $*$-representation of this semigroup in the space $\overline{S}(H)$. More precisely,

$$T(A_1, b_1, c_1)T(A_2, b_2, c_2) = e^{(b_2,c_1)}T(A_1 + A_2, b_1 + b_2, c_1 + c_2), \tag{9.2}$$

$$(T(A, b, c))^* = T((A, b, c)^*). \tag{9.3}$$

(Note that the last equality means that the operators $(T(A, b, c)^*)$ and $T(A, b, c)$ in the space $\mathcal{L}$ are formally adjoint to each other.)

Formula (9.3) shows that the operator $T(A, b, c)$ admits a closure. From now on we will assume, allowing a certain freedom of speech, that $T(A, b, c)$ is just this closure. Then the asterisk on the right-hand side of (9.3) means the adjoint operator in the sense of the theory of operators in Hilbert spaces.

The next theorem gives the boundedness conditions for the operator $T(A, b, c)$. In this theorem we assume that $\|A\| \le 1$ (one can easily see that for $\|A\| > 1$ the operator $T(A, b, c)$ is unbounded).

THEOREM 9.1. *The operator $T(A, b, c)$ is bounded if and only if*

$$A^*b + c \in (I - A^*A)^{1/2}H,$$

so that $A^*b + c = (I - A^*A)^{1/2}d$, where $d \in \overline{(I - A^*A)H}$. We have

$$\|T(A, b, c)\| = e^{(1/2)(|b|^2+|d|^2)} \tag{9.4}$$

Let us note that if $\|A\| < 1$, then formula (9.4) can be rewritten in the following more symmetric way:

$$\|T(A, b, c) = e^{(1/2)(|b_1|^2+|c_1|^2)+\mathrm{Re}(Ac_1,b_1))},$$
$$b_1 - (I - A^*A)^{-1/2}b, \quad c_1 = (I - AA^*)^{-1/2}c.$$

PROOF. (1) First, let $A = A^*$, $\|A\| < 1$. Let us prove that

$$\|T(A, b, c)\| = e^{|(I-A)^{-1/2}b|^2}. \tag{9.5}$$

Denote $x = (A)^{-1}b$, $U = e^{-(1/2)|x|^2}T(I, x, -x)$. Formulas (9.2) and (9.3) imply that U is a unitary operator and

$$T(A, b, c) = e^{|(I-A)^{-1/2}b|^2}UT(A, 0, 0)U^{-1}.$$

Clearly, $\|T(A, 0, 0)\| = 1$, and (9.5) follows.

(2) Let $A = A^*$, $\|A\| \le 1$. By (1), the boundedness of the operator $T(A, b, c)$ is equivalent to the fact that $b \in (I - A)^{-1/2}H$, i.e., that $b = (I - A)^{1/2}b_1$ for some $b_1 \in \overline{(I - A)H}$. Furthermore,

$$\|T(A, b, c)\| = e^{|b_1|^2}.$$

(3) Now let us consider an arbitrary operator $T(A, b, c)$ with $\|A\| \le 1$. The boundedness of $T(A, b, c)$ is equivalent to the boundedness of the operator

$$(T(A, b, c))^*T(A, b, c) = e^{|b^2|}T(A^*A, A^*b + c, A^*b + c).$$

Applying (2) to the last operator, we complete the proof.

9.2. Factorizable operators in $\overline{S}(H)$, $H = L_2(X, \mu, H^0)$. Let (X, μ) be a space with measure, H^0 a Hilbert space, $H = L_2(X, \mu, H^0)$ the Hilbert space of H^0-valued functions. Consider triples (A, b, c), where $b, c \in H$ and the operator $A\colon H \to H$ is given by the formula

$$(Af)(x) = A(x)f(x), \qquad f \in H, \quad x \in X;$$

here $x \mapsto A(x)$ is a weakly measurable map of X to the algebra $L(H^0)$ with $\|A(x)\| \le 1$ for $x \in X \bmod 0$. Sometimes we will write such triples in the form $\{A(x), b(x), c(x)\}$. Any such triple generates an operator $T(A, b, c)$ in the space $\overline{S}(X)$ according to formula (9.1). Theorem 9.1 implies the following formula:

$$\|T(A, b, c)\| = \exp\left\{\int_X \log \|T(A(x), b(x), c(x))\| \, d\mu\right\}. \tag{9.6}$$

In particular, this formula gives a condition for the boundedness of the operator $T(A, b, c)$.

Let us note an important property of the above operator. To formulate it, we introduce the notion of a factorizable operator in $\overline{S}(H)$, $H = L_2(X, \mu, H^0)$.

An arbitrary partition

$$X = X_1 \cup \cdots \cup X_p, \qquad p < \infty, \quad X_j \text{ are measurable sets,} \tag{9.7}$$

yields a natural identification of Hilbert spaces

$$\overline{S}(H) = \bigotimes_{k=1}^{p} \overline{S}(H_k), \qquad H_k = L_2(X_k, \mu, H^0)$$

A bounded linear operator $\mathcal{D}\colon \overline{S}(H) \to \overline{S}(H)$ is said to be *factorizable* if for any partition (9.7) it can be represented in the form

$$\mathcal{D} = \bigotimes_{k=1}^{p} \mathcal{D}_k,$$

where $\mathcal{D}_k$ is an operator in H_k, $1 \le k \le p$.

Let us return to the operator $T(A, b, c)$, where $(A, b, c) = \{(A(x), b(x), c(x))\} \in L(H) \times H \times H$, and assume that this operator is bounded. For a partition of the form (9.7) we have

$$T(A, b, c) = \bigotimes_{k=1}^{p} T(A_k, b_k, c_k),$$

where (A_k, b_k, c_k) is the restriction of the triple (A, b, c) to the subset X_k. Hence, $T(A, b, c)$ is a factorizable operator.

9.3. Convexity theorem (statement). Let us continue with the analysis of factorizable operators $T(A, b, c)$ acting in the space $\overline{S}(H)$, $H = L_2(X, \mu, H^0)$. The convexity theorem, which is the main result of this section, is formulated below.

Let us start with some preparations.

(1) From now on we assume that the measure μ is normalized and nonatomic, i.e.,

$$\mu(X) = 1, \qquad \mu(\{x\}) = 0 \text{ for any } x \in X. \tag{9.8}$$

(2) For any Hilbert space K (below we take $K = H^0$ or $K = H$) we endow the space $L(K) \times K \times K$ with the weak topology; the base of neighborhoods of zero in this topology consists of sets of triples $(A, b, c) \in L(K) \times K \times K$ determined by the inequalities

$$|(Ah_1, h_2)| < \varepsilon, \quad |(b, h_1)| < \varepsilon, \quad |(c, h_2| < \varepsilon,$$

where $h_1, h_2 \in K$, $\varepsilon > 0$. Furthermore, define the norm of an element (A, b, c) to be $\|A\| + |b| + |c|$. The mapping $x \mapsto (A(x), b(x), c(x))$ from $X \bmod 0$ to $L(K) \times K \times K$ is said to be measurable if the functions $x \mapsto (A(x)h_1, h_2)$, $x \mapsto (b(x), h_1)$, $x \mapsto (c(x), h_2)$, $x \in X \bmod 0$, are measurable for all $h_1, h_2 \in K$. This mapping will be denoted by (A, b, c).

(3) For any subset M of a linear topological space, denote by $\overline{\mathrm{co}} M$ the closure of the convex hull of M.

Now that we are prepared, let us turn to the statement of the main result. Let there be given a countable set S of measurable maps

$$x \mapsto ((A_k(x), b_k(x), c_k(x)), \qquad x \in X, \quad k = 1, 2, \dots, \tag{9.9}$$

from $X \bmod 0$ to the space $L(H^0) \times H^0 \times H^0$. Using S, we construct two families of measurable maps

$$x \mapsto ((A(x), b(x), c(x)) \tag{9.10}$$

from $X \bmod 0$ to $L(H^0) \times H^0 \times H^0$. First, denote by $F_0 = F_0(S)$ the set of all measurable maps of the form (9.10) satisfying the following condition: there exists a partition $X = X_1 \cup \cdots \cup X_n$ (here $n < \infty$, $X_i \cap X_k = \varnothing$ for $i \neq k$, and all the X_i are measurable) such that on each set X_i the map (9.10) coincides with one of the maps (9.9). Second, denote by $F_1 = F_1(S)$ the set of all measurable maps of the form (9.10) such that

$$((A(x), b(x), c(x)) \in \overline{\mathrm{co}}\{((A_k(x), b_k(x), c_k(x)), \ k = 1, 2, \dots\}$$

for each $x \in X \bmod 0$. For any subset $E \subset L(H) \times H \times H$ denote

$$T(E) = \{T(A, b, c) \mid (A, b, c) \in E\}.$$

In particular, we can consider the sets of operators $T(F_0)$ and $T(F_1)$.

THEOREM 9.2. *Let there exist* M_0 *such that*

$$\|T(A_k(x), b_k(x), c_k(x))\| \leq M_0 \quad \textit{for all } x \in X \bmod 0 \textit{ and all } k = 1, 2, \dots. \tag{9.11}$$

Then

$$\overline{F_0} = F_1, \tag{9.12}$$

$$\overline{T(F_0)} = T(F_1) \tag{9.13}$$

(*here the bar denotes the weak closure*).

The equality (9.13) will be especially important in the sequel; in particular, it asserts that the weak closure of the set of operator $T(A, b, c)$, $(A, b, c) \in F_0$ is unexpectedly large: it contains all the operators $T(A, b, c)$, $(A, b, c) \in F_1$.

We have formulated the convexity theorem in the form that is convenient in applications to representations of groups of measurable currents. Let us give a slightly different formulation of this theorem, in which the set F_0 is replaced by another set K which has a more "invariant" definition that does not appeal to a distinguished set of maps. To any point $x \in X \bmod 0$ let us associate a subset $K(x) \subset L(H^0) \times H^0 \times H^0$. We will denote this family of subsets by $\{K(x),\ x \in X\}$, and sometimes simply by K. Denote by $\overline{\mathrm{co}}K$ the family $\{\overline{\mathrm{co}}K(x),\ x \in X\}$; let us recall that $\overline{\mathrm{co}}K(x)$ is the weak closure of the convex hull of the set $K(x)$. A section of the family K is an arbitrary map $x \mapsto (A(x), b(x), c(x))$, $x \in X \bmod 0$, such that $(A(x), b(x), c(x)) \in K(x)$ for each $x \in X \bmod 0$. For simplicity, we denote such a section by (A, b, c). A family $K = \{K(x)\}$ is said to be measurable if there exists a countable set of measurable sections (A_k, b_k, c_k), $k = 1, 2, \dots$, such that for each $x \in X \bmod 0$ the set $K(x)$ is the weak closure of the set $\{(A_k(x), b_k(x), c_k(x)),\ k = 1, 2, \dots\}$. One can easily see that if a family K is measurable, then the family $\overline{\mathrm{co}}K$ is also measurable. For a measurable family $K = \{K(x)\}$ denote by $F(K)$ the set of all measurable sections of K.

THEOREM 9.3. *Let $K = \{K(x)\}$ be a measurable family. Let there exist M_0 such that for any $x \in X \bmod 0$ and any $(A^0, b^0, c^0) \in K(x)$ the inequality $\|T(A^0, b^0, c^0)\| \leq M_0$ holds. Then*

$$\overline{F(K)} = F(\overline{\text{co}}K),$$
$$\overline{T(F(K))} = T(F(\overline{\text{co}}K))$$

(the bar denotes the weak closure).

One can see that this theorem is equivalent to Theorem 9.2. Theorem 9.2 will be proved in 9.4 below.

9.4. Proof of Theorem 9.2. First of all, let us remark that the norms of triples $(A, b, c) \in F_0$ and of operators $T(A, b, c)$ for $(A, b, c) \in F_0$ are uniformly bounded, i.e.,

$$\|A\| + |b| + |c| \leq R_0, \qquad (A, b, c) \in F_0, \tag{9.14}$$
$$\|T(A, b, c\| \leq R_0, \qquad (A, b, c) \in F_0, \tag{9.15}$$

for some constant R_0. These inequalities follow from the condition (9.11) and formulas (9.4), (9.6). Let us note also the following obvious equality:

$$(T(A, b, c) \operatorname{Exp} x, \operatorname{Exp} y) = \exp\{(x, c) + (Ax, y) + (b, y)\}. \tag{9.16}$$

Our further arguments are divided into several steps.

Step 1. Let us prove that (9.13) follows from (9.12). Indeed, take an operator of the form $T(A, b, c)$, where $(A, b, c) \in F_1$. It follows from (9.12) that $(A, b, c) \in \overline{F_0}$. Since the norms of triples from F_0 are uniformly bounded, (A, b, c) is the weak limit of a sequence of triples $(A'_k, b'_k, c'_k) \in F_0$. It follows from (9.15) and (9.16) that $T(A'_k, b'_k, c'_k) \to T(A, b, c)$ (in the weak topology). Therefore, $T(A, b, c) \in \overline{T(F_0)}$. Hence, $T(F_1) \subset \overline{T(F_0)}$. Just as easily one can prove that $T(F_0) \subset \overline{T(F_1)}$.

Hence, (9.13) follows from (9.12) and the proof is reduced to the proof of the equality (9.12).

Step 2. The next lemma is the core of the proof.

LEMMA 9.2. *Let there be given a finite set of measurable complex-valued functions $f_n^m(x)$, $x \in X$, $1 \leq m \leq M$, $1 \leq n \leq N$, each belonging to the space $L_1(X, \mu)$. Let there also be given measurable functions $\lambda_n(x)$, $x \in X$, $1 \leq n \leq N$, such that*

$$\lambda_n(x) \geq 0, \quad \sum_{n=1}^{N} \lambda_n(x) = 1 \quad \textit{for any } x \in X \bmod 0. \tag{9.17}$$

Then for any $\varepsilon > 0$ there exists a partition $X = E_1 \cup \cdots \cup E_N$, $E_i \cap E_k = \varnothing$ for $i \neq k$, with all the E_i measurable, such that

$$\left| \sum_{n=1}^{N} \int_X \lambda_n(x) f_n^m(x)\, d\mu - \sum_{n=1}^{N} \int_{E_n} f_n^m(x)\, d\mu \right| < \varepsilon \textit{ for each } m \in \{1, \ldots, M\}. \tag{9.18}$$

PROOF OF THE LEMMA. (a) First let there exist a partition $X = X_1 \cup \cdots \cup X_P$ with X_i measurable such that each of the functions f_n^m is constant on each set X_i. Let us take measurable sets $Y_{kj} \subset X_j$, $1 \leq k \leq N$, $1 \leq j \leq P$, such that

$$\mu(Y_{kj}) = \int_{X_j} \lambda_k(x)\, d\mu, \tag{9.19}$$

$$Y_{kj} \cap Y_{sj} = \varnothing \text{ for } k \neq s, \quad \bigcup_{k=1}^{N} Y_{kj} = X_j. \tag{9.20}$$

Such sets Y_{kj} exist since, according to (9.17), the right-hand sides of the equalities in (9.19) are nonnegative and their sum over $k \in \{1, \ldots, N\}$ equals $\mu(X_j)$. Denote

$$E_k = \bigcup_{j=1}^{P} Y_{kj}, \qquad 1 \leq k \leq N.$$

Since the function f_n^m takes a constant value $f_m^n(X_j)$ on each set X_j, we have

$$\begin{aligned}\sum_{n=1}^{N} \int_X \lambda_n(x) f_n^m(x)\, d\mu &= \sum_{n=1}^{N}\sum_{j=1}^{P} \int_{X_j} \lambda_n(x) f_n^m(x)\, d\mu \\ &= \sum_{n=1}^{N}\sum_{j=1}^{P} f_n^m(X_j)\mu(Y_{nj}) = \sum_{n=1}^{N} \int_{E_n} f_n^m(x)\, d\mu.\end{aligned}$$

Comparing the first and the last term in this chain we see that (9.18) holds (with an arbitrary $\varepsilon > 0$).

(b) To prove the lemma in the general case it suffices to approximate the functions f_n^m (arbitrarily closely) in the L_1-metric by functions that are constant on elements of a certain finite partition of the set X.

Step 3. Let us take a triple $(A_0, b_0, c_0) \in F_1$ of the form

$$(A_0(x), b_0(x), c_0(x)) = \sum_{k=1}^{N} \lambda_k(x)(A_k(x), b_k(x), c_k(x)), \quad x \in X \mod 0, \tag{9.21}$$

where $\lambda_k(x)$, $1 \leq k \leq N$, are measurable functions satisfying the conditions (9.17). We prove that

$$(A_0, b_0, c_0) \in \overline{F_0}. \tag{9.22}$$

Take an arbitrary finite family $E \subset H$ of vector-valued functions. For any $\psi_1, \psi_2 \in E$ we have

(9.23)

$$\begin{aligned}&(\psi_1, c_0) + (A_0\psi_1, \psi_2) + (b_0, \psi_2) \\ &\quad = \sum_{k=1}^{N} \int_X \lambda_k(x)((\psi_1(x), c_k(x)) + (A_k(x)\psi_1(x), \psi_2(x)) + (b_k(x), \psi_2(x))\, d\mu.\end{aligned}$$

For each pair of elements $\psi_1, \psi_2 \in E$ we construct the function

$$\begin{gathered}x \mapsto (\psi_1(x), c_k(x)) + (A_k(x)\psi_1(x), \psi_2(x)) + (b_k(x), \psi_2(x)), \\ x \in X \bmod 0, \quad 1 \leq k \leq N, \quad \psi_1, \psi_2 \in E.\end{gathered}$$

Applying Lemma 9.2 to this family of functions we obtain, for each $\delta > 0$, a partition $X = E_1 \cup \cdots \cup E_N$ such that the left-hand side of (9.23) takes the form

$$(9.24)\quad (\psi_1, c_0) + (A_0\psi_1, \psi_2) + (b_0, \psi_2) = r(\psi_1, \psi_2) + \sum_{k=1}^{N} \int_{E_k} ((\psi_1(x), c_k(x)) + (A_k(x)\psi_1(x), \psi_2(x)) + (b_k(x), \psi_2(x)))\, d\mu,$$

where $|r(\psi_1, \psi_2)| \leq \delta$ for all $\psi_1, \psi_2 \in E$. Let $x \mapsto (A'(x), b'(x), c'(x))$ be the function defined on each set E_k by the formula

$$(A'(x), b'(x), c'(x)) = (A_k(x), b_k(x), c_k(x)), \qquad x \in E_k, \quad 1 \leq k \leq N.$$

This function belongs to the set F_0. Formula (9.24) takes the form

$$(\psi_1, c_0) + (A_0\psi_1, \psi_2) + (b_0, \psi_2) = (\psi_1, c') + (A'\psi_1, \psi_2) + (b', \psi_2) + r(\psi_1, \psi_2), \qquad \psi_1, \psi_2 \in E.$$

We see that in any weak neighborhood of the triple (A_0, b_0, c_0) of the form (9.21) there exists a triple (A', b', c') belonging to the set F_0. This proves (9.22).

Step 4. We take an arbitrary triple $(A_0, b_0, c_0) \in F_1$ and prove that $(A_0, b_0, c_0) \in \overline{F_0}$. To do that we need one auxiliary result.

Let L be a finite-dimensional Hilbert space, and let $f_k\colon X \to L$, $k = 1, 2, \ldots$, be measurable functions satisfying the condition

$$(9.25)\qquad |f_k(x)| \leq M_k < \infty \quad \text{for each } x \in X \bmod 0 \text{ and each } k = 1, 2, \ldots.$$

Next, let us take a measurable function $f \in L_2(X, \mu, L)$,

$$(9.26)\qquad f(x) \in \overline{\text{co}}\{f_1(x), f_2(x), \ldots\} \quad \text{for each } x \in X \bmod 0.$$

LEMMA 9.3. *For any $\varepsilon > 0$ there exists a vector valued function $\varphi\colon X \to L$ of the form*

$$(9.27)\qquad \varphi(x) = \sum_{k=1}^{N} \lambda_k(x) f_k(x), \qquad N < \infty,$$

where $\lambda_k(x)$, $1 \leq k \leq N$, are measurable functions satisfying conditions (9.17) *and*

$$\int_X |f(x) - \varphi(x)|^2\, d\mu < \varepsilon.$$

PROOF OF THE LEMMA. (a) First, let

$$f(x) \in \overline{\text{co}}\{f_1(x), \ldots, f_N(x)\}, \qquad x \in X \bmod 0,$$

for some $N < \infty$. Let us prove that in this case the function $f(x)$ can be represented in the form (9.27). If for any $x \in X \bmod 0$ the points $f_k(x)$, $1 \leq k \leq N$, are affine independent, this is clear: the functions $\lambda_k(x)$ are determined uniquely, hence are measurable. If this is not the case, then there exists a partition $X = X_1 \cup \cdots \cup X_m$ with all X_j measurable satisfying the following condition: at each set X_j there exists a family $f_{s_1}, \ldots f_{s_k}$ (depending on j) such that for each point $x \in X_j \bmod 0$ the vectors $f_{s_1}(x), \ldots f_{s_k}(x)$ are affine independent and generate the same affine plane as all vectors $f_k(x)$, $1 \leq k \leq N$. This proves our claim.

(b) Now let f be an arbitrary map satisfying the condition (9.26). For an integer $N > 0$ denote by $K_N(x)$ the convex hull of the vectors $\{f_k(x),\ 1 \le k \le N\}$, Denote by $\varphi_N(x)$ the point in $K_N(x)$ closest to $f(x)$; $\varphi_N(x)$ is well defined and $x \mapsto \varphi_N(x)$ is a measurable map. For each point $x \in X \bmod 0$ we have $\varphi_N(x) \to f(x)$ as $N \to \infty$. Furthermore, as follows from (9.25), the functions $x \mapsto |\varphi_N(x)|$ are uniformly bounded. By the Lebesgue convergence theorem, we have

$$\int_X |f(x) - \varphi_N(x)|^2 \, d\mu < \varepsilon$$

whenever N is sufficiently large. Applying part (a) to the function φ_N, we see that it can be represented in the form (9.27). The proof of the lemma is completed.

Now we complete the proof of (9.22) for an arbitrary triple $(A_0, b_0, c_0) \in F_1$. It suffices to prove that (A_0, b_0, c_0) belongs to the closure of the set of triples of the form (9.21) (since, as we have already proved, each triple of the form (9.21) belongs to $\overline{F_0}$).

Let us take an arbitrary finite set $E \subset H = L_2(X, \mu, H^0)$. Using Lemma 9.3 we can easily find a triple (A', b, c') of the form

$$(A'(x), b'(x), c'(x)) = \sum_{k-1}^{N} \lambda_k(x)(A_k(x), b_k(x), c_k(x)),$$

$$\lambda_k(x) \ge 0, \quad \sum_{k=1}^{N} \lambda_k(x) = 1,$$

such that all numbers

$$(\psi_1, c_0 - c') + ((A - A')\psi_1, \psi_2) + (b_0 - b', \psi_2), \qquad \psi_i \in E,$$

are arbitrarily small. This means that (A_0, b_0, c_0) belongs to the closure of the set of triples of the form (9.21). Hence, (9.22) is proved for all triples $(A_0, b_0, c_0) \in F_1$. Therefore, we proved that $F_1 \subset \overline{F_0}$.

Step 5. Let us prove that $\overline{F_0} \subset F_1$. Take a function $x \mapsto (A(x), b(x), c(x))$ from $\overline{F_0}$. It can be represented in the form

$$(A, b, c) = \lim_{n \to \infty} (A^{(n)}, b^{(n)}, c^{(n)}) \ \ \text{(weak limit)}, \qquad (A^{(n)}, b^{(n)}, c^{(n)}) \in F_0.$$

Therefore (using a known property of weak convergence) there exists a sequence of finite convex combinations

$$\sum_n \lambda_n^m (A^{(n)}, b^{(n)}, c^{(n)}), \qquad m = 1, 2, \dots, \tag{9.28}$$

that converges to (A, b, c) strongly. One can assume (taking an appropriate subsequence $m_1 \le m_2 \le \dots$) that the sequence (9.28) converges to (A, b, c) strongly at any point $x \in X \bmod 0$. This means that $(A, b, c) \in F_1$, which is the required result.

The equality (9.12) is proved, and the proof of Theorem 9.2 is completed.

9.6. Irreducibility condition for a family of factorizable operators in the space $\overline{S}(H)$, $H = L_2(X, \mu, H^0)$; statement of the result. Similarly to 9.2, let us consider a Hilbert space H^0, a measurable space (X, μ) (we assume that $\mu(X) = 1$ and $\mu(\{x\}) = 0$ for all $x \in X$), and the Hilbert spaces $H = L_2(X, \mu, H^0)$ and $\overline{S}(H)$. Let us take a countable family of measurable maps

$$x \mapsto (A_k(x), b_k(x), c_k(x)) \tag{9.29}$$

taking values in $L(H^0) \times H^0 \times H^0$. We assume that the following conditions are satisfied.

(1) The family (9.29) contains the map $x \mapsto (I, 0, 0)$, where I is the identity operator in H^0.

(2) The family (9.29) is selfadjoint, i.e., together with any map

$$x \mapsto (A_k(x), b_k(x), c_k(x))$$

it contains the map

$$x \mapsto ((A_k(x), b_k(x), c_k(x)))^* = ((A_k(x))^*, c_k(x), b_k(x)).$$

(3) For any $x \in X \bmod 0$ the zero vector is the only vector in H^0 invariant under all operators $A_k(x)$, $k = 1, 2, \dots$.

(4) For each k we have

$$\operatorname*{ess\,sup}_{x \in X} \|T(A_k(x), b_k(x), c_k(x))\| = M_k < \infty.$$

(This condition is less restrictive than (9.11).)

Using (9.29), we define the set F_0 similarly to 9.3. Let us recall that F_0 consists of all maps $x \mapsto (A(x), b(x), c(x))$ satisfying the following condition: there exists a partition $X = X_1 \cup \dots \cup X_m$ with measurable X_k's such that on X_k the map $(A(x), b(x), c(x))$ coincides with one of the maps (9.29). In this subsection we analyze the irreducibility conditions for the family of operators

$$T(F_0) = \{T(A, b, c) \mid (A, b, c) \in F_0\}. \tag{9.30}$$

We start with some preparations. For any $x \in X$ denote by F^x the set of values of (9.29) at x, so that F^x is a countable set of triples in $L(H^0) \times H^0 \times H^0$. To any triple $(A^0, b^0, c^0) \in F^x$ we associate an affine transformation of the space H^0 defined by the formula

$$h \mapsto A^0 h + b^0, \qquad h \in H^0. \tag{9.31}$$

By an affine plane in H^0 we mean any subset of the form $h + K$, where $h \in H^0$ and K is a linear subspace in H^0. We say that the action of the family F^x in H^0 is affine irreducible if H^0 does not contain closed affine subspaces that are invariant under all transformations (9.31). In the next theorem, which contains the main result of this section, we assume all conditions (1)–(4) listed at the beginning of this subsection to be satisfied.

THEOREM 9.3. *For each $x \in X \bmod 0$ let the action of the set F^x on H^0 be affine irreducible. Then the family of operators* (9.30) *is irreducible in the space $\overline{S}(H)$, $H = L_2(X, \mu, H^0)$.*

The proof of this theorem is based on the convexity theorem and a result of V. S. Shul'man (see [**38**]) obtained in the analysis of operator algebras in spaces with indefinite metric. This result is not formulated explicitly in [**38**]. Therefore here we repeat the proof, following [**38**].

9.7. Naimark algebras; Shul′man's theorem. Let H be a complex Hilbert space, $L(H)$ the algebra of all bounded linear operators in H. We make the set $L(H) \times H \times H$ an algebra with involution over $\mathbb{C}$ as follows:

$$\begin{aligned} \lambda_1(A_1, b_1, c_1) + \lambda_2(A_2, b_2, c_2) &= (\lambda_1 A_1 + \lambda_2 A_2, \lambda_1 b_1 + \lambda_2 b_2, \overline{\lambda_1} c_1 + \overline{\lambda_2} c_2), \\ (A_1, b_1, c_1)(A_2, b_2, c_2) &= (A_1 A_2, A_1 b_2, A_2^* c_1), \\ (A, b, c)^* &= (A^*, c, b) \end{aligned} \tag{9.32}$$

A Naimark algebra is any involutive subalgebra L of this algebra. These algebras first appeared in Naimark's work on operator theory in spaces with indefinite metric [**27, 28**].

Introduce the topology in $L(H) \times H \times H$ (hence in any Naimark algebra $L \subset L(H) \times H \times H$) as follows. Define the topologies s, s^*, u in $L(H)$, respectively, by the systems of seminorms $A \mapsto |Ax|$, $A \mapsto |A^*x|$ for $x \in X$ and the norm $A \mapsto \|A\|$. Endow the space H with the strong topology (corresponding to the norm $x \mapsto |x|$). If τ is one of three topologies s, s^*, u on $L(H)$, then the spaces $L(H) \times H$ and $L(H) \times H \times H$ are endowed with the product topology; we denote this product topology again by τ. Therefore, the above spaces (hence also the Naimark algebras) are endowed with one of three topologies s, s^*, u.

To any Naimark algebra L we associate the involutive algebra of operators $L_0 \subset L(H)$ of the form

$$L_0 = \{A \mid (A, b, c) \in L\}. \tag{9.33}$$

A Naimark algebra L is called nondegenerate of

$$\ker L_0 \equiv \{h \in H \mid Ah = 0 \text{ for any } A \in L_0\} = \{0\}. \tag{9.34}$$

THEOREM 9.4 (Shul′man [**38**]). *Let L be a nondegenerate Naimark algebra that is closed in $L(H) \times H \times H$ with respect to the u-topology, and for any $h \in H$ let the set of vectors*

$$\{Ah + b \mid (A, b, c) \in L\} \tag{9.35}$$

be dense in H. Then $L = L_0 \times H \times H$.

PROOF. Below by L we denote a Naimark algebra satisfying the conditions of the theorem. Define the linear subspace $L_1 \subset L(H) \times H$ by the formula

$$L_1 = \{(A, b) \mid (A, b, c) \in L\}. \tag{9.36}$$

Formula (9.32) implies that the algebra L_0 acts on the space L_1 by the formula

$$A_0(A, b) = (A_0 A, A_0 b), \qquad A_0 \in L_0, \quad (A, b) \in L_1.$$

First of all, we analyze the properties of the space L_1 as an L_0-module. In the next lemma L_0' denotes the centralizer of the algebra L_0 in the algebra $L(H)$.

LEMMA 9.4. *Let $x_0 \in H$, $T_0 = T_0^* \in L_0'$ and*

$$Ax_0 = T_0 b \quad \textit{for all } (A, b) \in L_1. \tag{9.37}$$

Then $T_0 = 0$, $x_0 = 0$.

PROOF. Let $\Delta \in \mathbb{R}$ be a closed interval not containing 0 and $P = P_\Delta$ the corresponding spectral projection of the operator T_0. Clearly, there exists a vector $h_0 \in P(H)$ such that $PT_0h_0+Px_0 = 0$. Using (9.37) and the fact that the operators PT_0 and A commute (for any $(A,b) \in L_1$), we see that

$$PT_0(Ah_0 + b) = PT_0Ah_0 + PAx_0 = A(PT_0h_0 + Px_0) = 0, \qquad (A,b) \in L_1.$$

Therefore $PT_0(Ah_0 + b) = 0$ for all $(A,b) \in L_1$. Since the set of vectors (9.35) is dense in H, we have $PT_0 = 0$ for any spectral projection $P = P_\Delta$ with $0 \notin \Delta$. Hence, $T_0 = 0$. Formulas (9.37) and (9.34) imply that $x_0 = 0$. Lemma 9.4 is proved.

Denote the closures of the space L_i in the topologies s, s^*, u by L_i^s, $L_i^{s^*}$, L_i^u respectively.

LEMMA 9.5.

$$L_1^u = L_0^u \times H. \tag{9.38}$$

The proof is divided into several steps.

(1) Let us prove that

$$L_1^s = L_0^s \times H. \tag{9.39}$$

We fix vectors $x_k \in H$, $1 \leq k \leq n$, and consider the linear subspace $R \subset H^{n+1} = H \oplus \cdots \oplus H$ ($n+1$ times), defined as the closure of the set of all vectors $(Ax_1, \ldots Ax_n, b)$ with $(A,b) \in L_1$. Let us prove that

$$(0, \ldots, 0, h) \in R \tag{9.40}$$

for each $h \in H$. Let $P = (P_{jk})$, $1 \leq j,k \leq n+1$, be the projection from H^{n+1} to R. Since the subspace R is invariant with respect to the natural representation of the algebra L_0 in H^{n+1}, we have $P_{jk} \in L_0'$. Clearly,

$$\sum_{k=1}^{n} P_{n+1,k}Ax_k + P_{n+1,n+1}b = b$$

for each $(A,b) \in L_1$. Therefore,

$$A\Big(\sum_{k=1}^{n} P_{n+1,k}x_k\Big) = (I - P_{n+1,n+1})b$$

for each $(A,b) \in L_1$, and $P_{n+1,n+1} = I$ by Lemma 9.4. Hence, $P(0,\ldots,0,h) = (0,\ldots,0,h)$ for each $h \in H$, and (9.40) is proved. From (9.40) it follows that any s-neighborhood of the element $(0,h)$ in $L(H) \times H$ contains an element $(A,b) \in L_1$. Therefore, $(0,h) \in L_1^s$ for all $h \in H$, and (9.39) follows.

(2) Let us prove that

$$L_1^{s^*} = L_0^{s^*} \times H. \tag{9.41}$$

Again, we fix vectors x_k, $1 \leq k \leq n$, and consider the subspace $S \subset H^{n+1}$, defined as the closure of the set of all vectors $(A^*x_1, \ldots A^*x_n, b)$ with $(A,b) \in L_1$. Let us prove that

$$(0, \ldots, 0, h) \in S \tag{9.42}$$

for each $h \in H$. Take an arbitrary vector $(y_1, \ldots, y_{n+1}) \in H^{n+1}$ orthogonal to S. Then

$$\sum_{k=1}^{n}(x_k, Ay_k) + (b, y_{n+1}) = 0 \quad \text{for all } (A, b) \in L_1. \tag{9.43}$$

It follows from (9.39) that $(A, b) \in L_1$ can be chosen in such a way that all numbers $|Ay_k|$, $|b - y_{n+1}|$, $1 \leq k \leq n$, are arbitrarily small. But then (9.43) implies that $y_{n+1} = 0$. Hence, any vector from the orthogonal complement to S is of the form $(y_1, \ldots, y_n, 0)$. This implies (9.42). Formula (9.41) follows from (9.42) (using the same arguments that lead from (9.40) to (9.39)).

(3) Let us prove that

$$L_1^u = L_0^u \times H. \tag{9.44}$$

Fix $h \in H$ and prove that $(0, h) \in L_1^u$; clearly, this implies (9.44).

It follows from (9.41) that there exists a sequence $(A_n, b_n) \in L_1$, $n \geq 1$, such that

$$|A_n^* h| \to 0, \quad |b_n - h| \to 0.$$

Take a number sequence γ_n, $n \geq 1$, such that

$$\gamma_n |A_n^* h| \to 0, \quad \gamma_n \to \infty.$$

Next, take a sequence of real polynomials $P_n(t)$, $n \geq 1$, such that

$$0 \leq P_n(t) \leq 1 \quad \text{for } t \in [0, \|A_n\|^2], \tag{9.46}$$

$$\max\{tP_n(t) \mid 0 \leq t \leq \|A_n\|^2\} \to 0 \quad \text{as } n \to \infty, \tag{9.47}$$

$$|1 - P_n(t)| \leq \gamma_n t \quad \text{for } t \in [0, \|A_n\|^2]. \tag{9.48}$$

The existence of such a sequence is clear: it suffices to take, for each n, a continuous function equal to 1 on $[0, \gamma_n^{-1}]$, to zero on $[2\gamma_n^{-1}, \infty)$, and linear on $[\gamma_n^{-1}, 2\gamma_n^{-1}]$, and to approximate it on the interval $[0, \|A_n\|^2]$ by a polynomial $P_n(t)$ (with sufficient accuracy). Now let us take elements $B_n = P_n(A_n A_n^*)$ in the algebra L_0. Then $(B_n A_n, B_n b_n) \in L_1$ and it remains to verify that this sequence u-converges to the element $(0, h) \in L(H) \times H$. Using properties (9.46) and (9.47) of the polynomials $P_n(t)$, we get

$$\begin{aligned}\|B_n A_n\|^2 &= \|B_n A_n A_n^* B_n\| = \|A_n A_n^* B_n^2\| \\ &= \|A_n A_n^* P_n^2(A_n A_n^*)\| \leq \max_{0 \leq t \leq \|A_n\|^2} tP_n^2(t) \to 0.\end{aligned}$$

Therefore, $\|B_n A_n\| \to 0$. Next, using properties (9.45), (9.46), and (9.48), we get

$$\begin{aligned}&|B_n h - h|^2 = ((I - P_n(A_n A_n^*))^2 h, h) \leq \gamma_n (A_n A_n^* h, h) = \gamma_n |A_n^* h|^2 \to 0, \\ &|B_n b_n - h|^2 \leq |B_n(b_n - h)| + |B_n h - h| \leq |b_n - h| + |B_n h - h| \to 0.\end{aligned}$$

Therefore, $|B_n b_n - h| \to 0$. We have proved that the sequence $(B_n A_n, B_n b_n)$ of elements from L_1 converges to $(0, h)$ in u-topology. This proves (9.44), and the proof of Lemma 9.5 is completed.

Now we prove Theorem 9.4. Take an arbitrary vector $h \in H$. By (9.38), there exists a sequence $(A_n, b_n, c_n) \in L$ such that

$$\|A_n\| \to 0, \quad |b_n - h| \to 0. \tag{9.49}$$

Take an arbitrary element $(A, b, c) \in L$. Then

$$(A, b, c)(A_n, b_n, c_n) = (AA_n, Ab_n, A_n^* c) \in L. \tag{9.50}$$

By (9.49), $\|AA_n\| \to 0$, $|Ab_n - Ah| \to 0$, $|A_n^* c| \to 0$. Therefore, the sequence (9.50) converges in the u-topology to the element $(0, Ah, 0)$. Since the algebra L is closed, we have $(0, Ah, 0) \in L$ for all $A \in L_0$, $h \in H$. The nondegeneracy condition (9.34) implies that the algebra L contains all elements of the form $(0, h, 0) \in L(H) \times H \times H$. Applying the involution in L, we see that L also contains all elements of the form $(0, 0, h)$, $h \in H$, and Theorem 9.4 follows.

9.8. Proof of Theorem 9.3. Let us recall that in 9.1 we made the set $L(H) \times H \times H$ a semigroup with involution, and the map $(A, b, c) \mapsto T(A, b, c)$ became a projective $*$-representation of this semigroup.

Let us consider the family (9.29) and the corresponding family $F_0 \subset L(H) \times H \times H$. Let Γ_0 be the subsemigroup in $L(H) \times H \times H$ generated by F_0, and let $\overline{\mathrm{Aff}\Gamma_0}$ be the smallest affine plane in $L(H) \times H \times H$ containing Γ_0 and closed in the u-topology (recall the definition of the u-topology in 9.7). Hence, $\overline{\mathrm{Aff}\Gamma_0}$ is the u-closure of the set of all elements of the form

$$\sum_{k=1}^{n} \lambda_k (A^{(k)}, b^{(k)}, c^{(k)}), \qquad (A^{(k)}, b^{(k)}, c^{(k)}) \in \Gamma_0, \quad \lambda_k \in \mathbb{C}, \quad \sum_{k=1}^{n} \lambda_k = 1. \tag{9.51}$$

Let us note that $\overline{\mathrm{Aff}\Gamma_0}$ is also a subsemigroup in $L(H) \times H \times H$; moreover, for any element of the form (9.51) and any $(A, b, c) \in \Gamma_0$ we have

$$\Big(\sum_{k=1}^{n} \lambda_k (A^{(k)}, b^{(k)}, c^{(k)})\Big)(A, b, c) = \sum_{k=1}^{n} \lambda_k (A^{(k)}, b^{(k)}, c^{(k)})(A, b, c).$$

Consider the set

$$L = \{(A - I, b, c) \mid (A, b, c) \in \overline{\mathrm{Aff}\Gamma_0}\}.$$

It is clear that L is a linear subspace in $L(H) \times H \times H$. Moreover, formulas defining the multiplication in the semigroup $L(H) \times H \times H$ easily show that L is a Naimark algebra.

LEMMA 9.6. *The Naimark algebra L is nondegenerate. For any vector $f \in H = L_2(X, \mu, H^0)$ the set of vectors of the form $Af + b$, $(A, b, c) \in L$, is dense in H.*

PROOF. The first statement follows from condition (3) at the beginning of 9.6. The second statement follows from the affine irreducibility of the action of the family F^x on H^0 for each $x \in X \bmod 0$.

Let D be a bounded operator in $\overline{S}(H)$ that commutes with all operators (9.30).

LEMMA 9.7. *For each $f_1, f_2 \in H$, $(A, b, c) \in \overline{\mathrm{Aff}\Gamma_0}$ we have*

$$(DT(A, b, c) \operatorname{Exp} f_1, \operatorname{Exp} f_2) = (D \operatorname{Exp} f_1, T(A^*, c, b) \operatorname{Exp} f_2). \tag{9.52}$$

PROOF. First let

$$(A,b,c)=\sum_{k=1}^{n}\lambda_k(A^{(k)},b^{(k)},c^{(k)}), \tag{9.53}$$

where $(A^{(k)},b^{(k)},c^{(k)})\in\Gamma_0$, $\lambda_k\geq 0$, $\sum_{k=1}^{n}\lambda_k=1$. By the convexity theorem, the operator $T(A,b,c)$ belongs to the von Neumann algebra generated by the operators (9.30). Therefore, it commutes with the operator D, and (9.52) follows.

Now let (A,b,c) be of the form (9.53), where $(A^{(k)},b^{(k)},c^{(k)})\in\Gamma_0$, $\lambda_k\in\mathbb{C}$, $\sum_{k=1}^{n}\lambda_k=1$. Then (9.52) holds because it holds for real λ_k and both part of (9.52) are holomorphic in λ_k. Since elements of the above form are dense in $\overline{\mathrm{Aff}\Gamma_0}$ (in the u-topology), the obvious passage to the limit shows that (9.53) holds for all $(A,b,c)\in\overline{\mathrm{Aff}\Gamma_0}$.

Now it is easy to complete the proof of Theorem 9.3. According to Shul′man's theorem, all elements of the form $(0,h_1,h_2)\in L(H)\times H\times H$ belong to the algebra L. Therefore, each element of the form $(I,h,-h)$, $h\in H$, belongs to the semigroup $\overline{\mathrm{Aff}\Gamma_0}$. Applying Lemma 9.7 to this element, we see that the operator D commutes with all operators of the form $T(I,h,-h)$, $h\in H$. But by Theorem 7.1, these operators form an irreducible family. Therefore, $D=\lambda I$, $\lambda\in\mathbb{C}$, and the proof of the irreducibility of the family (9.30) is completed.

9.9. Remark about the totalness of the family $\{\mathrm{Exp}\, f,\ f\in\mathcal{F}_0\}$. Let us consider once more the space $H=L_2(X,\mu,H^0)$ and a set of weakly measurable maps $f_k\colon X\to H^0$, $k=1,2,\dots$. We assume that for each k we have $|f_k(x)|\leq M_k$ for $x\in X \bmod 0$. Denote by $\mathcal{F}_0$ the set of all maps $f\colon X\to H^0$ that coincide with one of the f_k on each element of some finite partition $X=X_1\cup\dots\cup X_k$, X_i measurable; this is the simplest modification of the definition from 9.3.

THEOREM 9.5. *For any point $x\in X \bmod 0$ let the closure of the affine span of the set $\{f_k(x)\}$ coincide with H^0. Then the set of vectors*

$$\mathrm{Exp}\, f,\qquad f\in\mathcal{F}_0, \tag{9.54}$$

is total in the space $\overline{S}(H)$, $H=L_2(X,\mu,H^0)$.

PROOF (following the arguments from 9.2). Let K be the closure of the linear span of the set (9.54). Take a vector-valued function of the form

$$f(x)=\sum_{k=1}^{n}\lambda_k(x)f_k(x), \tag{9.55}$$

where the functions $\lambda_k(x)$ satisfy the conditions (9.17) and take only a finite number of values. By the convexity theorem, the vector $\mathrm{Exp}\, f$ belongs to the weak closure of the set (9.54), hence $\mathrm{Exp}\, f\in K$. The holomorphicity arguments (see 9.8) show that $\mathrm{Exp}\, f\in K$ for any vector-valued function (9.55), where $\lambda_k(x)$ are complex-valued functions taking finite number of values and satisfying the condition $\sum_{k=1}^{n}\lambda_k(x)=1$. The conditions of the theorem imply that the set of such functions is dense in H, and the theorem follows.

§10. Representations of the group G^X (the Araki construction)

Let G be a topological group (everywhere below we assume that G is separable and metrizable), (X, μ) a space with measure such that μ is normalized and X does not contains atoms (see 9.8). Denote by G^X the group of measurable maps $v\colon X \to G$ (more precisely, classes of almost everywhere coinciding maps) such that v takes a finite number of values on a set of measure 1. This group is the group of measurable currents. The group operation in G^X is pointwise multiplication. We present the construction of factorizable representations of the group G^X from the works of Araki [**42**] and Vershik, Gelfand, Graev [**6**].

10.1. Representations $T^{U,r}$. For a complex Hilbert space H^0, denote $H = L_2(X, \mu, H^0)$. Let us assume that to a point $x \in X \bmod 0$ we associate a unitary representation $g \mapsto U_x(g)$, $g \in G$, in the space H^0 and a 1-cocycle $r_x\colon G \to H^0$ of this representation, so that

$$r_x(g_1 g_2) = r_x(g_1) + U_x(g_1) r_x(g_2), \qquad x \in X \bmod 0, \quad g_1, g_2 \in G.$$

Let us assume that for each $h_1, h_2, h \in H$ the functions $(x, g) \mapsto (U_x(g)h_1, h_2)$, $(x, g) \mapsto (r_x(g), h)$ are continuous in g for a fixed $x \in X \bmod 0$ and are measurable in x for a fixed $g \in G$. Finally, let there exist a continuous positive function $w\colon G \to \mathbb{R}$ such that

$$|r_x(g)| \le w(g), \qquad g \in G, \quad x \in X \bmod 0.$$

Define the representation U of the group G^X in the space $H = L_2(X, \mu, H^0)$ and the cocycle $r\colon G^X \to H$ of this representation by the formulas

$$(U(v)f)(x) = U_x(v(x))f(x), \qquad r(v)(x) = r_x(v(x)).$$

Using these data, construct, similarly to 6.2, the representation $T = T^{U,r}$ of the group G^X in the Fock space $\overline{S}(H)$. In general, the constructed representation is projective; more precisely,

$$T(v_1 v_2) = e^{-i \operatorname{Im}(r(v_1^{-1}), r(v_2))} T(v_1) T(v_2), \qquad v_1, v_2 \in G^X.$$

If the 2-cocycle

$$(v_1, v_2) \mapsto e^{-i \operatorname{Im}(r(v_1^{-1}), r(v_2))}$$

is trivial, we get an ordinary (nonprojective) representation.

10.2. Irreducibility condition for representations $T^{U,r}$. Introduce a simple preliminary notion. Let $g \mapsto U(g)$, $g \in G$, be an arbitrary unitary representation in the space H^0, and $r\colon G \to H^0$ a cocycle of U. Let H^1 be a closed subspace in H^0 invariant under the representation U, and P^1 the orthogonal projection from H^0 to H^1. Then the function $g \mapsto P^1 r(g)$ is a cocycle of the representation U (and of its restriction to the subspace H^1). A cocycle $r\colon G \to H^0$ is called absolutely nontrivial if the cocycle $g \mapsto P^1 r(g)$ is nontrivial (i.e., does not take the form $g \mapsto U(g)h_0 - h_0$) for any nonzero invariant subspace $H^1 \subset H^0$.

Let us return to our representation $T^{U,r}$.

THEOREM 10.1. *Assume that for each point $x \in X \bmod 0$ the following two conditions are satisfied:*

1. *the representation $g \mapsto U_x(g)$ does not contain the trivial subrepresentation;*
2. *the cocycle $r_x\colon G \to H^0$ is absolutely nontrivial.*

Then the representation $T^{U,r}$ is irreducible.

COROLLARY 10.1 (Delorme). *Assume that for each point $x \in X \bmod 0$ the representation U_x is a direct finite or countable sum of pairwise nonequivalent irreducible nontrivial representations, $U_x = \bigoplus_k U_{x,k}$. In the corresponding decomposition $r_x = \bigoplus_k r_{x,k}$ let all cocycles $r_{x,k}$ be nontrivial. Then the representation $T^{U,r}$ is irreducible.*

PROOF OF THEOREM 10.1. Let us take a countable dense subset $\{g_k\} \subset G$. Apply Theorem 9.3, taking for (9.29) the family

$$x \mapsto (U_x(g_k), r_x(g_k), -r_x(g_k^{-1})), \qquad k = 1, 2, \ldots, \qquad x \in X \bmod 0.$$

The only condition we must verify is the affine irreducibility of the family of affine transformation $h \mapsto U_x(g_k) + r_x(g_k)$ for each point $x \in X \bmod 0$ (here we can take an arbitrary $g \in G$ instead of g_k because the set $\{g_k\}$ is dense in G). It suffices to prove that for any $h \in H$ the system of vectors $U_x(g)h - h + r_x(g)$, $g \in G$, is total. Assuming that this system is not total, we can find a projection P in H^0, $P \neq 0, I$, such that $P(U_x(g)h - h + r_x(g)) = 0$ for all $g \in G$. Then $Pr_x(g) = U_x(g)h_1 - h_1$, $h_1 = Ph$. This contradicts the assumption that the cocycle r_x is absolutely nontrivial. Theorem 10.1 is proved.

10.3. Condition for equivalence of representations. Let us take two families of representations $g \mapsto U_x^k(g)$, $g \in G$, in Hilbert spaces H_k^0 and the corresponding cocycles $g \mapsto r_x^k(g)$, $k = 1, 2$. For each of these families let the conditions from 10.1 and also the assumptions of Theorem 10.1 be satisfied, so that both representations $T_k = T^{U^k, r^k}$, $k = 1, 2$, are irreducible.

THEOREM 10.2. *For the representations T_1, T_2 to be equivalent the following condition is necessary and sufficient: there exist a weakly measurable family of isomorphisms of Hilbert spaces $M_x\colon H_1^0 \to H_2^0$ and a weakly measurable map $\beta\colon X \to H_0^2$ such that for each $g \in G$ and each $x \in X \bmod 0$ we have*

$$M_x U_x^1(g) = U_x^2(g) M_x, \qquad M_x r_x^1(g) - r_x^2(g) = U_x^2(g)\beta(x) - \beta(x).$$

PROOF. (1) An element $\varphi \in \overline{S}(H)$, $\varphi \neq 0$, is said to be factorizable if for any partition $X = X_1 \cup \cdots \cup X_m$ with measurable disjoint X_i it can be represented in the form $\varphi = \varphi_1 * \cdots * \varphi_m$, where $\varphi_k \in \widehat{S}(H_k)$, $H_k = L_2(X_k, \mu, H^0)$.

LEMMA 10.1. *A vector $\varphi \in \overline{S}(H)$ is factorizable if and only if $\varphi = \lambda \operatorname{Exp} f$, $\lambda \in \mathbb{C}$, $f \in H$.*

This lemma is a part of the factorization theory constructed in the papers of Araki and Woods [**43**] and Araki [**42**].

PROOF OF LEMMA 10.1. The vector φ is of the form

$$\varphi = \varphi_0 \oplus \varphi_1 \oplus \cdots, \qquad \varphi_k \in \overline{S}^k(H). \tag{10.1}$$

Let us prove that $\varphi_0 \neq 0$. Assume the contrary. Then $\varphi = \varphi_p \oplus \cdots$, $\varphi_p \neq 0$, $p > 0$. The conditions of the lemma imply that φ_p is a factorizable element. But then $\varphi_p \in \overline{S}^p(H')$, where $H' = L_2(X', \mu, H^0)$ and $\mu(X')$ is arbitrarily small. Hence, $\varphi_p = 0$, a contradiction. Hence, in the representation (10.1) we have $\varphi_0 \neq 0$, and we can assume that $\varphi_0 = 1$.

Define the element $\psi = \log\varphi \in \widehat{S}(H)$ (the algebra $\widehat{S}(H)$ was defined in 3.1) by the formula

$$\psi = \sum_{k=1}^{\infty} \frac{(-1)^{k-1}}{k} (\varphi_1 \oplus \varphi_2 \oplus \dots)^k, \quad \text{where } \varphi = 1 \oplus \varphi_1 \oplus \dots.$$

Then $\varphi = \operatorname{Exp}\psi$, $\psi = \psi_1 \oplus \psi_2 \oplus \cdots$. Since φ is a factorizable element, for any partition $X = X_1 \cup \cdots \cup X_m$ the vector ψ_k can be represented in the form $\psi_{k1} + \cdots + \psi_{km}$, $\psi_{ki} \in \overline{S}^k(H_i)$. Let us prove that $\psi_k = 0$ for $k > 1$. Indeed, take an arbitrary vector $f \in H$ and represent it in the form $f = f_1 + \cdots + f_m$, $f_i \in H_i$. Then

$$|(\psi_k, f^{*k})| = \left|\left(\psi_k, \sum_i f_i^{*k}\right)\right| \le |\psi_k| \cdot \left|\sum_i f_i^{*k}\right| = |\psi_k| \sqrt{\sum_i |f_i^{*k}|^2}$$
$$\le (\sqrt{k!})|\psi_k| \sqrt{\sum_i |f_i|^{2k}} \le (\sqrt{k!})|\psi_k| \max_i |f_i|^{k-1}|f|.$$

For a sufficiently fine partition $X = \bigcup X_i$ the last term can be made arbitrarily small. Therefore, $(\psi_k, f^{*k}) = 0$ for each $f \in H$, so that $\psi_k = 0$ for $k > 1$. Hence, $\varphi = \operatorname{Exp}\psi_1$, $\psi_1 \in H$. The lemma is proved.

LEMMA 10.2. *Let a unitary operator $U\colon \overline{S}(H) \to \overline{S}(H)$ satisfy the condition $U(\operatorname{Exp} h) = \alpha \operatorname{Exp} h'$ for each $h \in H$ (here the number α and the vector h' depend on h). Then $U = \lambda e^{-|b|^2/2} T(A, b, -A^*b)$ for a unitary operator $A\colon H \to H$ and a vector $b \in H$, $|\lambda| = 1$.*

PROOF. It is clear that $U(\operatorname{Exp} h) = e^{\beta(h)} \operatorname{Exp}\varphi(h)$, where the maps $\beta\colon H \to \mathbb{C}$ and $\varphi\colon H \to H$ are holomorphic and

$$(\varphi(h_1), \varphi(h_2)) = (h_1, h_2) - \beta(h_1) - \overline{\beta(h_2)}.$$

Denote $\varphi(h) - \varphi(0) = \psi(h)$. Then $(\psi(h_1), \psi(h_2)) = (h_1, h_2)$. This means that ψ is a unitary operator in H, and the lemma follows.

Now we prove the theorem. Let $M\colon \overline{S}(H_1) \to \overline{S}(H_2)$ be an isomorphism of Hilbert spaces intertwining representations $T_k = T^{U^k, r^k}$, $k = 1, 2$. Then

$$\begin{aligned}(10.2)\qquad &(T(U^2(v), r^2(v), r^2(v^{-1}))M \operatorname{Exp} f_1, M \operatorname{Exp} f_2)\\ &\quad = e^{(1/2)(|r^2(v)|-|r^1(v)|)^2}(T(U^1(v), r^1(v), r^1(v^{-1})) \operatorname{Exp} f_1, \operatorname{Exp} f_2)\end{aligned}$$

Denote $M \operatorname{Exp} f_i = \varphi_i$, $i = 1, 2$.

LEMMA 10.3. *$\varphi_i = \lambda_i \operatorname{Exp}\psi_i$, where $\lambda_i \in \mathbb{C}$, $\psi_i \in H$.*

PROOF. It suffices to prove that the elements φ_i are factorizable. Take a partition $X = X_1 \cup \cdots \cup X_m$. Then for any $v_i \in G^{X_i}$ we have

$$\left(\prod_{i=1}^m T(U^2(v_i), r^2(v_i), r^2(v_i^{-1}))\varphi_1, \varphi_2\right) = \prod_{i=1}^m (T(U^2(v_i), r^2(v_i), r^2(v_i^{-1}))\varphi_1, \varphi_2).$$

This equality follows from (10.2) since its right-hand side obviously satisfies a similar condition. Each of the groups of operators $\{T(U^2(v), r^2(v), r^2(v^{-1})),\ v \in G^{X_i}\}$

acts in $\overline{S}(H_i)$, $H_i = L_2(X_i, \mu, H^0)$ irreducibly (by Theorem 10.1). Hence we get the formula

$$\Big(\Big(\prod_{i=1}^{m} A_i\Big)\varphi_i, \varphi_2\Big) = \prod_{i=1}^{m}(A_i\varphi_i, \varphi_2),$$

where the operators $A_i\colon \overline{S}(H) \to \overline{S}(H)$ are of the form $A_i = I \otimes \cdots \otimes A_i^0 \otimes \cdots \otimes I$, and the A_i^0 are arbitrary. This implies that φ_i can be represented in the form $\varphi_i = \varphi_{i1} \otimes \cdots \otimes \varphi_{im}$, $\varphi_{ik} \in \overline{S}(H)$. We have proved that φ_i is a factorizable vector, and the lemma follows.

Using Lemma 10.2, we see that $M = e^{-|b|^2/2}T(A, b, -A^*b)$, where A is a unitary operator in H, $b \in H$. Therefore,

$$\begin{aligned} &e^{(1/2)|r^1(v)|^2}T(A,b,-A^*b)T(U^1(v), r^1(v), r^1(v^{-1})) \\ &\qquad = e^{(1/2)|r^2(v)|^2}T(U^2(v), r^2(v), r^2(v^{-1}))T(A,b,-A^*b). \end{aligned} \tag{10.3}$$

Hence, $AU^1(v) = U^2(v)A$, $v \in G^X$. Let us recall that for each $x \in X \bmod 0$ the representation U_x^k, $k = 1, 2$, does not have nonzero invariant vectors. Hence, A is determined by a measurable family of unitary operators M_x according to the formula $(Af)(x) = M_x f(x)$. Finally, the statement about the cocycles in the formulation of the theorem follows from (10.3). Theorem 10.2 is proved.

10.4. On the group G^X in the case $\mu(X) = \infty$. In our previous discussion we have assumed that $\mu(X) = 1$ and the space X does not contain atoms (i.e., points of positive measure). Now let $\mu(X) = \infty$ and again assume, for simplicity, that X does not contain atoms. In this case we define G^X as the group of all measurable maps $v\colon G \to X$ satisfying the following conditions:

1. v takes only a finite number of values (outside a set of measure zero);
2. $\{x \mid f(x) \neq e\} < \infty$, where e is the identity of the group G.

All results of this section can be extended to this case in an obvious manner.

10.5 Remark. For some groups G irreducibility and pairwise nonequivalence of factorizable representations can be obtained using the results from 7.2; examples of such groups are presented in the next section.

§11. Examples

In the paper [**42**] by Araki the structure of cocycles of representations of a solvable group was analyzed, but the irreducibility of representations of the current group was not mentioned there. In the paper by Gelfand, Graev, Vershik [**6**] the case $G = SL(2, \mathbb{C})$ was discussed in detail, and, in particular, the irreducibility of the representations $T^{U,r}$ was proved. In [**7**] this result was generalized to other semisimple groups. A detailed analysis of cocycles is contained in the book by Guichardet [**10**].

In this section we consider two simple examples: $G = SU(1,1)$ and $G = U(\infty)$. In discussing the first example we essentially follow [**6, 7**]. The second example is due to Olshanskiĭ [**32**].

11.1. The group G^X, $G = SU(1,1)$. The group $SU(1,1)$ is formed by the matrices

$$g = \begin{pmatrix} a & b \\ \overline{b} & \overline{a} \end{pmatrix}. \tag{11.1}$$

The formula

$$zg = \frac{az+\overline{b}}{bz+\overline{a}}$$

defines the action of the group G on the circle $T^1 = \{z, |z| = 1\}$, on the disk $D_+ = \{z, |z| < 1\}$, and on the complement to the disk $D_- = \{z, |z| > 1\}$.

Consider the representation of the group G in the space $C^\infty(T^1)$ given by the formula

$$(\pi(g)f)(z) = (\overline{a}+bz)^{-2}f(zg), \qquad z \in T^1.$$

Consider the following invariant subspaces in $C^\infty(T^1)$. Let $\{f_k, k \in \mathbb{Z}\}$ be the basis in $C^\infty(T^1)$ of the form $f_k(z) = z^k$. The subspaces L_+ and L_- spanned by the sets $\{f_k, k \geq 0\}$ and $\{f_k, k \leq -2\}$ respectively are invariant under the representation π; they can be identified with the spaces of functions that are holomorphic for $|z| \leq 1$ and $|z| \geq 1$ respectively (and $f(z) = O(z^{-2})$ for $f \in L_-$). Hence, $C^\infty(T^1) = L_+ \oplus L_- \oplus \mathbb{C}f_{-1}$.

The inner product

$$(f_1, f_2) = \frac{1}{\pi}\int_{D_+} f_1(z)\overline{f_2(x)}\,d\nu,$$

where $d\nu$ is the Lebesgue measure, is invariant under the representation in L_+. Completing L_+ with respect to this inner product, we obtain the Hilbert space H_+ and the unitary representation π_+ of the group G in H_+. Similarly one constructs the Hilbert space H_- and the unitary representation π_-. Finally, denote $H_0 = H_+ \oplus H_-$, $\pi_0 = \pi_+ \oplus \pi_-$. The vectors $|k+1|^{-1/2}f_k$, $k \neq -1$, form an orthonormal basis in H_0.

Now let us consider the cocycles. It is clear that the vector-valued function $g \mapsto r_0(g) = \pi_0(g)f_{-1} - f_{-1}$ is an 1-cocycle of the representation π_0. To the decomposition $H_0 = H_+ \oplus H_-$ there corresponds the decomposition $r_0(g) = r_+(g) + r_-(g)$, where $r_+(g) \in H_+$, $r_-(g) \in H_-$. The vector-valued functions $g \mapsto r_+(g)$, $g \mapsto r_-(g)$ are 1-cocycles of the representations π_+, π_-. An easy computation shows that

$$r_+(g)(z) = -\frac{b}{\overline{a}+bz}, \qquad z \in D_+,$$

$$r_-(g)(z) = -\frac{\overline{b}}{z(az+b)}, \qquad z \in D_-.$$

Below we will use the following equality, which is proved by an easy computation (based on the decomposition of functions with respect to the basis $\{f_k\}$):

$$\varphi(g_1, g_2) \equiv (r_+(g_1^{-1}), r_+(g_2)) = \log\frac{(g_1)_{11}(g_2)_{11}}{(g_1g_2)_{11}}, \qquad g_1, g_2 \in G. \tag{11.2}$$

One can easily see that the above function is a 2-cocycle on G (the Guichardet-Wigner cocycle). It allows us to construct the following central extension of the group G by $\mathbb{Z}$. Consider the pairs (g, s), where g is a matrix of the form (11.1) and $e^s = a$. Define the multiplication of such pairs by the formula $(g_1, s_1)(g_2, s_2) = (g_1g_2, s_1 + s_2 + \varphi(g_1, g_2))$. One can easily see that this multiplication defines a group, denoted by $\widetilde{G}$; this group is the desired central extension. The group $\widetilde{G}$ turns out to be the universal covering group for G.

Now let us consider representations of the group G^X. Let (X, μ) be a space with a normalized nonatomic measure. Apply the Araki construction from 10.1, taking for the Hilbert space H^0, the representations U_x, and the cocycles r_x used in this construction the space H_+, the representation π_+, and the cocycle r_+ respectively. (Note that here U_x and r_x do not depend on x.) We obtain the representation $T_+ = T^{\pi_+, r_+}$. In a similar way we construct the representations $T_- = T^{\pi_-, r_-}$ and $T_0 = T^{\pi_0, r_0}$. The next theorem summarizes the properties of these representations.

THEOREM 11.1. *T_+ and T_- are projective representations; more precisely,*

$$T_\pm(v_1 v_2) = \exp\left\{i \int_X \operatorname{Im}(\varphi(v_1(x), v_2(x)))\, d\mu\right\} T_\pm(v_1) T_\pm(v_2), \qquad v_1, v_2 \in G^X$$

(the function φ is defined by (11.2)), and T_0 is an ordinary (nonprojective) representation. All three representations are irreducible. The following formulas hold:

$$(T_\pm(v) \operatorname{Exp} 0, \operatorname{Exp} 0) = \exp\left\{-\int_X \log|v_{11}(x)|\, d\mu\right\},$$

$$(T_0(v) \operatorname{Exp} 0, \operatorname{Exp} 0) = \exp\left\{-2\int_X \log|v_{11}(x)|\, d\mu\right\}.$$

The proof follows from the results of §10. The representation T_+ gives the following (ordinary) representation of the group $\widetilde{G}^X$:

$$T(v, s) = \exp\left\{i \operatorname{Im} \int_X s(x)\, d\mu\right\} T_+(v), \qquad (v, s) \in \widetilde{G}^X. \tag{11.3}$$

The corresponding spherical function is given by the formula

$$(T(v, s) \operatorname{Exp} 0, \operatorname{Exp} 0) = \exp\left\{-\int s(x)\, d\mu\right\}. \tag{11.4}$$

11.2. Representations of the group G^X, $G = U(\infty)$. The group of all unitary matrices $g = (g_{ij})$, $i, j = 1, 2, \dots$, such that $g_{ij} = \delta_{ij}$ for sufficiently large i, j, is denoted by $U(\infty)$. Let J_2^0 be the Hilbert space of matrices $s = (s_{ij})$ such that $s_{ij} = s_{ji}$ and $\operatorname{Tr} ss^* < \infty$; the inner product in J_2^0 is given by the formula $(s_1, s_2) = \operatorname{Tr} s_1 s_2^*$. Define the representation U of the group $U(\infty)$ in the space J_2^0 by the formula $U(g)s = gsg'$, where $g' = \overline{g^*}$. This representation is irreducible.

Take an arbitrary matrix $s^0 = (s_{ij}^0)$ satisfying the condition $(s')^0 = s^0$. Then the formula $r(g) = gs^0 g' - s^0$ defines a cocycle of the representation U. If the matrix s^0 does not belong to the set J_2^0, then this cocycle is nontrivial. Finally, consider the group G^X, $G = U(\infty)$, and the representation $T^{U,r}$ of this group. Theorem 10.1 implies that this representation is irreducible. Let us note that in the paper by Olshanskiĭ **[32]**, from which this example is taken, the matrix $s^0 = (\delta_{ij})$ is used.

§12. Continuous tensor products (CTP) of Hilbert spaces and representation of the group G^X

12.1. Positive definite kernels and tensor powers of Hilbert spaces. The main object of this section is a triple (H, S, φ), where H is a Hilbert space, S is a set, and $\varphi\colon S \to H$ is a map such that $\varphi(S)$ is total in H. The function $K(s_1, s_2) = (\varphi(s_1), \varphi(s_2))$, $s_1, s_2 \in S$, is a positive definite kernel, i.e., $\sum K(s_i, s_j)\xi_i \overline{\xi_j} \geq 0$ for all finite sets $\{s_i\} \subset S$, $\{\xi_i\} \subset \mathbb{C}$. Any positive definite kernel can be obtained using this construction. This allows us to identify the triple (H, S, φ) with the pair (S, K).

Consider a triple (H^1, S, φ^1), where S is a arcwise connected simply connected topological space, and the map φ^1 is continuous. Let the function $K^1(s_1, s_2) = (\varphi^1(s_1), \varphi^1(s_2))$ not vanish anywhere on $S \times S$. Then we can define single-valued functions $\log K^1(s_1, s_2)$ and $K^\alpha(s_1, s_2) = \exp(\alpha \log K^1(s_1, s_2))$, where log is the continuous branch of the base e logarithm satisfying $\log(1) = 0$. Let us assume that the function $K^\alpha(s_1, s_2)$ is positive definite for all $\alpha > 0$; in this case the kernel K^1 and the triple (H^1, S, φ^1) are called *infinitely divisible.* As we have explained earlier, to the kernel $K^\alpha(s_1, s_2)$ we can associate some triple. We denote it by $(H^{\otimes\alpha}, S, \varphi^{\otimes\alpha})$ and it call the αth tensor power of the triple (H^1, S, φ^1). The Hilbert space $H^{\otimes\alpha}$ is called the αth tensor power of the space H^1; the inner product in $H^{\otimes\alpha}$ is denoted by $(\cdot, \cdot)_\alpha$. We have

$$(\varphi^{\otimes\alpha}(s_1), \varphi^{\otimes\alpha}(s_2))_\alpha = (\varphi^1(s_1), \varphi^1(s_2))^\alpha = K^\alpha(s_1, s_2).$$

If α is a positive integer, then $H^{\otimes\alpha}$ is the usual symmetric power of the space H^1.

12.2. Tensor powers of operators. Consider a triple (H^1, S, φ^1) and a unitary operator $U^1 \colon H^1 \to H^1$ that acts on vectors $\varphi(s)$ by the formula

$$U^1(\varphi(s)) = e^{a(s)}\varphi(g(s)),$$

where $s \mapsto a(s)$ is a continuous function and $g \colon S \to S$ is a homeomorphism. The function $a(s)$ is defined up to a summand of the form $2\pi in$, $n \in \mathbb{Z}$. Then there exists a unique unitary operator $U^{\otimes\alpha} \colon H^{\otimes\alpha} \to H^{\otimes\alpha}$ satisfying the condition

$$U^{\otimes\alpha}\varphi^{\otimes\alpha}(s) = e^{\alpha a(s)}\varphi^{\otimes\alpha}(g(s)). \tag{12.1}$$

The proof of the existence of $U^{\otimes\alpha}$ is reduced to a simple verification of the fact that the mapping defined by (12.1) on vectors of the form $\varphi^{\otimes\alpha}(s)$ preserves the inner product. The operator $U^{\otimes\alpha}$ is called the αth tensor power of the operator U^1. Let us note that the definition of the operator $U^{\otimes\alpha}$ contains an ambiguity: if the function $a(s)$ is replaced by $a(s) + 2\pi in$, the operator $U^{\otimes\alpha}$ is multiplied by $\exp(2\pi in\alpha)$.

12.3. Continuous tensor products (CTP). Let (X, μ) be a space with a normalized measure without atoms, S an arcwise connected simply connected topological space. Let us assume that to each point $x \in X \bmod 0$ there corresponds an infinitely divisible triple (H_x, S, φ_x) such that the function $\log(\varphi_x(s_1), \varphi_x(s_2))$ is measurable and bounded on $X \bmod 0$ for all $s_1, s_2 \in S$. Our goal is to define the continuous tensor product (below we write CTP) of the family $\{(H_x, S, \varphi_x)\}$.

Denote by Γ the set of all mappings $f \colon X \to S$ taking a finite number of values on a set of full measure; as usual, we identify mappings that coincide almost everywhere. Introduce the function

$$K(f_1, f_2) = \exp\left\{\int_X \log K_x(f_1(x), f_2(x))\, d\mu\right\}, \qquad f_1, f_2 \in \Gamma, \tag{12.2}$$

where

$$K_x(f_1, f_2) = (\varphi_x(s_1), \varphi_x(s_2)).$$

One can easily prove that the function (12.2) is positive definite. As we explained in 12.1, to this function there corresponds the triple (H_*, S, φ_*) such that $K(f_1, f_2) =$

$(\varphi_*(f_1), \varphi_*(f_2))$, $f_1, f_2 \in \Gamma$. By definition, this triple is the CTP of the family $\{(H_x, S, \varphi_x)\}$, and the Hilbert space H_* is the CTP of the family of spaces $\{H_x\}$. Below we will use the notation $\varphi_*(f) = f_*$, $f \in \Gamma$.

Now let us define the CTP of a family of operators. Let us assume that unitary operators $U_x : H_x \to H_x$ are defined by the formula

$$U_x \varphi_x(s) = e^{a(x,s)} \varphi_x(\alpha_x(s)).$$

Here $x \mapsto \alpha_x$ is a mapping from X into the group of homeomorphisms of the space S satisfying the following condition: there exists a partition $X = X_1 \cup \cdots \cup X_r$ such that all the X_i are measurable and the mapping $x \mapsto \alpha_x$ is constant on each X_i. The function $a(x, s)$ is assumed to be measurable in x for each $s \in S$ and continuous in s for each $x \in X \bmod 0$. For an element $f \in \Gamma$ define $\alpha(f) \in \Gamma$ by the formula $\alpha(f)(x) = \alpha_x(f(x))$. Under these conditions, there exists a unique unitary operator $U_* : H_* \to H_*$ such that

$$U_* f_* = \exp\left\{\int_X a(x, f(x))\, d\mu\right\} (\alpha(f))_*. \tag{12.3}$$

(Recall that $f_* = \varphi_*(f)$, $(\alpha(f))_* = \varphi_*(\alpha(f))$. The proof of the existence of the operator U_* reduces to a simple verification that the mapping (12.3) preserves the inner product.

12.4. Constriction of the CTP of group representations. Let us generalize the constructions described for unitary operators to families of group representations. We start with the αth tensor power of a given representation.

Let us be given an infinitely divisible triple (H^1, S, φ^1) and a unitary representation U_1 of a group G in the space H^1. Let us assume that U^1 is defined by the formula

$$U^1(g)\varphi(s) = e^{a(g,s)} \varphi(g(s)),$$

where $s \mapsto g(s)$ denotes the action of G on S (by homeomorphisms) The function $a(g, s)$ is continuous and satisfies the condition

$$a(g_1 g_2, s) = a(g_2, s) a(g_1, g_2(s)).$$

Following the definition (12.1), for any $\alpha > 0$ we obtain a representation $U^{\otimes\alpha}$ of the group G acting in the space $H^{\otimes\alpha}$; it is defined by the formula

$$H^{\otimes\alpha}(g)\varphi^{\otimes\alpha}(s) = e^{\alpha a(g,s)} \varphi^{\otimes\alpha}(g(s)).$$

Hence, we have constructed the αth tensor product of our representation.

Now we consider the CTP of unitary representations. Let (H_x, S, φ_x) be infinitely divisible triples satisfying the condition formulated at the beginning of 12.2. They define the CTP denoted by (H_*, S, φ_*). Let $g \mapsto U_x(g)$, $g \in G$, be unitary representations defined in H_x by the formula

$$U_x(g)\varphi_x(s) = e^{a(x,g,s)} \varphi_x(g(s)).$$

Here $s \mapsto g(s)$ is the action of the group G on S (by homeomorphisms) and the function $a(x, g, s)$ satisfies the conditions from 12.3 for each $g \in G$. Define the representation U_* of the current group G^X by the formula (using the definition (12.3))

$$U_*(v) f_* = \exp\left\{\int_X a(x, v(x), f(x))\, d\mu\right\} (v(f))_*,$$

where the element $v(f) \in \Gamma$ is defined by $v(f)(x) = v(x)(f(x))$.

In conclusion, we consider the CTP of a constant family $\{U_x = U\}$; we concentrate on irreducibility conditions for the representation U_*.

THEOREM 12.1. *If $U^{\otimes\alpha}$ is irreducible for each α, then U_* is also irreducible.*

PROOF. Let $\tau\colon X = X_1 \cup \dots \cup X_n$ be a partition, where $X_i \cap X_k = \varnothing$ for $i \neq k$, and all the X_i are measurable. Denote by G_τ^X the subgroup of functions $v \in G^X$ that are constant on each set X_i, so that $G_\tau^X = G \times \dots \times G = G^n$. Denote by $\Gamma_\tau \subset \Gamma$ the subset of functions that are constant on each X_i. Finally, denote by H_τ the smallest subspace in H_* that contain all vectors $\varphi_*(f)$, $f \in \Gamma_\tau$, and is invariant under G_τ^X. One can easily see that we get a natural isomorphism of Hilbert spaces $H_\tau \simeq \bigotimes_k H^{\otimes\alpha_k}$, where $\alpha_k = \mu(X_k)$, which maps the representation $U^{\otimes n}$ of the group $G_\tau^X \simeq G^n$ to the representation $\bigotimes_k H^{\otimes\alpha_k}$; we recall that the last representation acts by the formula $\bigotimes_k x_k \mapsto \bigotimes_k U^{\otimes\alpha_k}(g_k)x_k$, $(g_1, \dots, g_n) \in G^n$. Since each representation $U^{\otimes\alpha_k}$ is irreducible, the representation G_τ^X in H_τ is irreducible. Refining the partition τ, we see that the representation U_* is irreducible.

12.5. Example (Berezin's construction). Denote by $G = \widetilde{SU(1,1)}$ the universal covering group of $SU(1,1)$; it is described in 11.1. Berezin noted that this group admits irreducible infinitely divisible representations; see [**3**]. This allows us to define a representation of the group G^X as the CTP of irreducible representations. Moreover, Berezin carried out a similar construction for the group of holomorphic automorphisms of the n-dimensional complex ball (this group is isomorphic to $SU(n,1)$). Here we restrict ourself to the case $n = 1$.

Let us start with representations of the group G. For any $\alpha > 0$ define the Hilbert space H^α consisting of functions that are holomorphic in the disk $D = \{z, |z| < 1\}$, with the inner product given by the formula

$$(f_1, f_2)_\alpha = \frac{\alpha - 1}{\pi} \int_D f_1(z)\overline{f_2(z)}(1 - |z|^2)^{\alpha-2}\, d\nu \tag{12.4}$$

(so that H^α is specified by the condition $(f, f)_\alpha < \infty$). A simple computation shows that the functions f_k, $k = 0, 1, 2, \dots$, of the form $f_k(z) = z^k$ are pairwise orthogonal and

$$(f_k, f_k)_\alpha = \frac{\Gamma(k+1)\Gamma(\alpha-1)(\alpha-1)}{\Gamma(k+\alpha)}, \qquad k = 0, 1, 2 \dots.$$

This shows that $(f_k, f_k)_\alpha > 0$ for $\alpha > 0$, $k = 0, 1, 2, \dots$. Therefore, the inner product (12.4) can be defined for each $\alpha > 0$, remaining positive definite.

Now we consider representations of the group $G = \widetilde{SU(1.1)}$. Let us recall that elements of G are pairs (g, s), where

$$g = \begin{pmatrix} a & b \\ \bar{b} & \bar{a} \end{pmatrix} \in SU(1,1), \quad s \in \mathbb{C}, \quad e^s = a.$$

The group operation in G is given by the formula

$$(g_1, s_1)(g_2, s_2) = (g_1 g_2, s_1 + s_2 + \varphi(g_1, g_2)),$$

where $\varphi(g_1, g_2)$ is the Guichardet-Wigner cocycle (11.2). Define the representation T^α of the group G in H^α by the formula

$$(T^\alpha(g,s)f)(z) = \exp\left\{-\alpha\bar{s} + \log(1 + b\bar{a}^{-1}z)\right\} f(zg).$$

Let us note that for elements $(g, s) \in G$ with $|a-1| < 1/4$, the factor on the right-hand side takes a more standard form $(\bar{a} + bz)^{-\alpha} = \exp\{-\alpha\bar{s} - \alpha\log(1 + b\bar{a}^{-1}z)\}$, where a continuous branch of the logarithm is specified by the condition $\log 1 = 0$.

To any $\zeta \in D$ we associate the vector $\varphi_\zeta \in H^\alpha$ of the form

$$\varphi_\zeta^\alpha(z) = (1 - \bar{\zeta}z)^{-\alpha}, \qquad z \in D. \tag{12.5}$$

One can easily prove that $(\varphi_\zeta^\alpha, f)_\alpha = \overline{f(z)}$ for $f \in H^\alpha$. In particular,

$$(\varphi_{\zeta_1}^\alpha, \varphi_{\zeta_2}^\alpha) = (1 - \bar{\zeta}_1\zeta_2)^{-\alpha}. \tag{12.6}$$

Furthermore,

$$T^\alpha(g,s)\varphi_\zeta^\alpha = \exp\left\{\alpha(\bar{s} + \log(1 - \bar{b}\bar{a}\bar{\zeta}))\right\} \varphi_{\zeta g^{-1}}^\alpha \tag{12.7}$$

(as was explained earlier, for $|a-1| < 1/4$ the factor on the right-hand side simplifies). Formulas (12.6) and (12.7) can be interpreted as follows. Consider the triples $(H^\alpha, D, \varphi^\alpha)$, where the mapping $\varphi^\alpha \colon D \to H^\alpha$ is of the form (12.5). Then the triple $(H^\alpha, D, \varphi^\alpha)$ and the representation T^α are the αth tensor product of the triple (H^1, D, φ^1) and the representation T^1 respectively.

Finally, using the infinitely divisible representation T^1 we construct, following the method from 12.4, a representation of the group G^X as the CTP of the constant family $\{U_x = T^1\}$. Let us present the corresponding formulas. Following 12.3, denote by Γ the set of all measurable mappings $f \colon X \to D$ taking a finite number of values for $x \in X \bmod 0$. Introduce the function

$$K(f_1, f_2) = \exp\left\{\int_X \log(1 - f_1(x)\overline{f_2(x)})\, d\mu\right\}.$$

It defines the inner product on a linear space L with the basis f_*, $f \in \Gamma$, by the formula $(f_{1*}, f_{2*}) = K(f_1, f_2)$. Define a representation of the group G^X in the space L by the formula

$$T(v,s)f_* = \exp\left\{\int_X \left(\overline{s}(x) + \log(1 - \overline{b(x)}\,\overline{a^{-1}(x)}\,\overline{f(x)})\right) d\mu\right\}(vf)_*,$$

where vf is defined by the formula $(vf)(x) = f(x)(v(x))^{-1}$. Completing L, we obtain a Hilbert space H_* and a unitary representation T_* of the group G^X. The irreducibility of this representation follows from Theorem 12.1.

Denote by $f_0 \in \Gamma$ the function of the form $f_0(x) = 0$, $x \in X$. For the corresponding vector $f_{0*} \in H_*$, one can easily compute the spherical function of the representation T_*. It coincides with the function (11.4). Therefore, our representation T_* is equivalent to the representation constructed using the Araki scheme.

§13. The group $\Gamma^X \cdot M(X,\mu)$

In this section we define a group that will be used in §23; representations of this group will be considered in §33.

Let (X,μ) be a space with a measure (without atoms), $M(X,\mu)$ the group of all measure preserving transformations $x \mapsto x\sigma$ that coincide with the identity transformation outside a set of finite measure. As usual, we identify transformations that coincide almost everywhere. Endow $M(X,\mu)$ with the weak topology, setting $\sigma_n \to \sigma$ (weakly) if all σ_n coincide with the identity transformation outside a fixed set of finite measure and $\mu(A\sigma \triangle A\sigma_n) \to 0$ for each measurable A with $\mu(A) < \infty$.

Let Γ be a group and Γ^X the group of measurable currents. The group $M(X,\mu)$ acts of Γ^X by translations, $(\sigma\varphi)(x) = \varphi(x\sigma)$, $\varphi \in \Gamma^X$, $\sigma \in M(X,\mu)$. This action defines the semidirect product $\Gamma^X \cdot M(X,\mu)$; an element of this group is written in the form (φ,σ), $\varphi \in \Gamma^X$, $\sigma \in M(X,\mu)$. To simplify the exposition (and keeping in mind further applications) we restrict ourselves to the case of a discrete countable group Γ. Introduce the convergence in Γ^X, assuming that $\varphi_n \to \varphi$ if all φ_n take values in a fixed finite subset of Γ and $\mu\{x \mid \varphi_n(x) \neq \varphi(x)\} \to 0$ as $n \to \infty$.

Consider the set of double cosets $M(X,\mu)\backslash\Gamma^X \cdot M(X,\mu)/M(X,\mu)$. First assume that $\mu(X) < \infty$; we can in fact assume that $\mu(X) = 1$. Denote by $P(\Gamma)$ the set of all functions $p\colon \Gamma \to \mathbb{R}$ satisfying the conditions

$$p(\gamma) \geq 0, \qquad \sum_\gamma p(\gamma) = 1.$$

The set $P(\Gamma)$ can be identified with the semigroup of probability measures on Γ (with the convolution as the semigroup operation). To any function $f \in \Gamma^X$ we associate its distribution function $p^f \in P(\Gamma)$ defined by the formula

$$p^f(\gamma) = \mu\{x \mid f(x) = \gamma\}, \qquad \gamma \in \Gamma. \tag{13.1}$$

Define the mapping $q\colon \Gamma^X \cdot M(X,\mu) \to P(\Gamma)$ by the formula

$$q(\varphi,\sigma) = p^\varphi, \qquad \varphi \cdot \sigma \in \Gamma^X \cdot M(X,\mu).$$

One can easily see that this mapping is two-side invariant under $M(X,\mu)$ and determines a bijection of the sets $M(X,\mu) \setminus \Gamma^X \cdot M(X,\mu)/M(X,\mu)$ and $P(\Gamma)$.

Now let $\mu(X) = \infty$. Denote by $P_1(\Gamma)$ the set of all nonnegative functions $p\colon \Gamma \setminus \{e_\Gamma\} \to \mathbb{R}$ with finite support. To a function $f \in \Gamma^X$ we associate the function $p^f \in P_1(\Gamma)$ defined by (13.1) for $\gamma \neq e_\Gamma$. We have an obvious bijection

$$M(X,\mu) \setminus \Gamma^X \cdot M(X,\mu)/M(X,\mu) \to P_1(\Gamma).$$

CHAPTER 3

Representations of the Group of Smooth Currents

The main topic of this chapter is the study of certain unitary representations of the group of smooth currents (i.e., of the group of smooth mappings of a manifold X to a Lie group), called energy representations. These representations were first studied by Albeverio and Høegh-Krohn [**40**], Vershik, Gelfand, Graev [**51**], Parthasarathy and Schmidt [**62**], and the author [**17**]. As was shown by the author in [**17**], the irreducibility of these representations can be reduced to certain nontrivial problems related to Gaussian measures in function spaces and depends essentially on the dimension of the manifold X. The cases $\dim X \geq 5$, $\dim X \geq 4$, and $\dim X \geq 2$ were studied in [**17**], [**51**], and [**40**] respectively. The case $\dim X = 2$ still is not completely analyzed. In §15 we present a rather simple proof of the corresponding properties of the Gaussian measure; this proof uses known estimates for norms of random trigonometric polynomials (see [**23**, Chapter VI]. For $\dim X = 2$ our estimates are not best possible. We have chosen them because we wanted to restrict ourselves to elementary techniques; sharper estimates can be extracted from the papers of Kahane [**55**] and Kusuoka [**57**].

In §§17 and 18 we consider modifications of energy representations that yield unitary representations of extentions of the current group; here we reproduce the results of the author [**21**]. The chapter ends with the analysis of abelian extensions of current groups related to the generalized determinant bundles (see §2), the known Mickelsson–Rajeev extensions. Here we follow Mickelsson and Rajeev [**60**].

§14. Energy representations of the group $C_0^\infty(X,G)$

In this section we use the following notation: G is a compact semisimple Lie group with the Lie algebra $\mathfrak{g}$, X is a C^∞ closed manifold of dimension $n < \infty$, $C_0^\infty(X,G)$ is the group of all compactly supported C^∞ mappings $u\colon X \to G$; recall that the support of $u\colon X \to G$ is the closure of the set $\{x \mid u(x) \neq e\}$, where e is the unit element of G. For a compact manifold X we will write $C^\infty(X,G)$ instead of $C_0^\infty(X,G)$. The tangent space to X at a point x is denoted by T_xX. We will also need a Riemannian metric on X; it makes each space T_xX a real Hilbert space. The measure on X generated by the Riemannian metric is denoted by dx.

In this section we will construct unitary representations of the group $C_0^\infty(X,G)$ using the general construction from §6. To do this we construct an orthogonal representation of $C_0^\infty(X,G)$ and the corresponding 1-cocycle. Using these data, we obtain a representation acting in the Fock space.

14.1. The space $E^1(X,\mathfrak{g})$. By definition, the space $E^1(X,\mathfrak{g})$ consists of all compactly supported C^∞ 1-forms on X with values in the Lie algebra $\mathfrak{g}$. We introduce the inner product in $E^1(X,\mathfrak{g})$ as follows. Fix an invariant positive definite inner product in $\mathfrak{g}$, a Riemannian metric r on X, and a positive C^∞-function $\rho(x)$, $x \in X$. Let $\varphi \in E^1(X,\mathfrak{g})$. For any point $x \in X$ we have a linear map $\varphi(x)\colon T_xX \to \mathfrak{g}$ and the adjoint map $(\varphi(x))^*\colon \mathfrak{g} \to T_xX$. Define the inner product in $E^1(X,\mathfrak{g})$ by the formula

$$(\varphi_1,\varphi_2) = (\varphi_1,\varphi_2)_{r,\rho} = \int_X \operatorname{Tr}(\varphi_1(x)(\varphi_2(x))^*)\rho(x)\,dx; \tag{14.1}$$

here Tr is the trace of an operator in $\mathfrak{g}$. This makes $E^1(X,\mathfrak{g})$ a pre-Hilbert space.

The group $C_0^\infty(X,G)$ acts in $E^1(X,\mathfrak{g})$ by the formula

$$(u\varphi)(x) = \operatorname{Ad}_{u(x)}\varphi(x), \tag{14.2}$$

where Ad is the adjoint representation of the group G in $\mathfrak{g}$. If G is a matrix group (this can be achieved by passing to an exact matrix representation of G), then this action takes the form $(u\varphi)(x) = u(x)\varphi(x)(u(x))^{-1}$. The invariance of the inner product in $\mathfrak{g}$ implies that formula (14.2) defines an orthogonal representation of the group $C_0^\infty(X,G)$ in the space $E^1(X,\mathfrak{g})$.

Finally, consider the complexification $E^1_{\mathbb{C}}(X,\mathfrak{g})$, which can be identified with the space of compactly supported 1-forms taking values in the complexification $\mathfrak{g}_{\mathbb{C}}$ of the Lie algebra $\mathfrak{g}$. The inner product (14.1) can be extended to an Hermitian positive definite inner product in $E^1_{\mathbb{C}}(X,\mathfrak{g})$. The action (14.2) also extends to $E^1_{\mathbb{C}}(X,\mathfrak{g})$, and the extended action preserves the above inner product in $E^1_{\mathbb{C}}(X,\mathfrak{g})$.

14.2. The Maurer–Cartan cocycle. For any $g \in G$ define the right translation R_g on G by the formula $R_g(g_1) = g_1 g$. The right translation generates the corresponding map of the tangent spaces $dR_g\colon T_{g_1}G \to T_{g_1 g}G$, $g, g_1 \in G$. In particular, identifying the tangent space T_eG with the Lie algebra $\mathfrak{g}$, we get the map $dR_{g^{-1}}\colon T_gG \to \mathfrak{g}$, $g \in G$. Using this identification, for any element $u \in C_0^\infty(X,G)$ we define the element $\dot u \in E^1(X,\mathfrak{g})$ by the formula

$$\dot u(x) = dR_{(u(x))^{-1}}(du)(x).$$

If G is a matrix group, this definition takes the form $\dot u = du \cdot u^{-1}$. Since, clearly, $(uv)^{\cdot} = \dot u + u\dot v$, the mapping $u \mapsto \dot u$ is a 1-cocycle of the action (14.2). It is called the Maurer–Cartan cocycle.

14.3. Energy representations. Denote by H the completion of the space $E^1_{\mathbb{C}}(X,\mathfrak{g})$ with respect to the Hilbert norm introduced in 14.1. The action (14.2) is extended to a unitary representation of the group $C_0^\infty(X,G)$ in the space H; we continue to denote this action by $u\varphi$, $\varphi \in H$, $u \in C_0^\infty(X,G)$. We can assume that the cocycle $u \mapsto \dot u$ takes values in H. Applying the general construction from §6, we obtain a representation of the group $C_0^\infty(X,G)$. It acts in the Fock space $\overline{S}(H)$ by the formula

$$T(u)\operatorname{Exp} h = \exp\left\{-\frac{1}{2}|\dot u|^2 + (h,(u^{-1})^{\cdot})\right\}\operatorname{Exp}(uh + \dot u),$$
$$u \in C_0^\infty(X,G), \quad h \in H.$$

Fix an inner product in $\mathfrak{g}$, so that the representation T depends on a Riemannian metric r and a function ρ. Sometimes we will denote this representation by $T_{r,\rho}$.

We will study conditions for irreducibility and pairwise nonequivalence of representations $T_{r,\rho}$. This study is based on certain nontrivial properties of Gaussian measures in function spaces; these properties are studied in §15. We return to the representations themselves in §16.

§15. Properties of Gaussian measures related to Dirichlet forms

In this section we denote by $C_0^\infty(X,\mathbb{R})$ the space of real compactly supported C^∞-functions $f\colon X \to \mathbb{R}$ and by $\Omega = (C_0^\infty(X,\mathbb{R}))'$ the dual space. The value of a functional $\omega \in \Omega$ at a function $f \in C_0^\infty(X,\mathbb{R})$ is denoted by $\langle f, \omega\rangle$. For any point $x \in X$ define the functional δ_x by the formula $\langle f, \delta_x\rangle = f(x)$.

15.1. Statement of the theorem. Let B be an open bounded region in $\mathbb{R}^n$; assume that some neighborhood of the closure $\overline{B}$ admits a diffeomorphic embedding into $\mathbb{R}^n$ that maps B into an open ball or a cube. Fix a real matrix function $x \mapsto a(x) = (a_{ij}(x))$, $x \in B$, $1 \le i,j \le n$, such that:

(1) all functions $x \mapsto a_{jk}(x)$ admit a C^∞-extension to the entire space $\mathbb{R}^n$;

(2) $a_{jk}(x) = a_{kj}(x)$ for $x \in B$, $1 \le j,k \le n$;

(3) there exists $c > 0$ such that

$$(a(x)\xi,\xi) = \sum_{j,k} a_{jk}(x)\xi_j\xi_k \ge c|\xi|^2 = c\sum_{k=1}^n \xi_k^2, \qquad \xi = (\xi_1,\dots,\xi_n) \in \mathbb{R}^n \tag{15.1}$$

(we assume $\mathbb{R}^n$ to be endowed with the standard inner product). Introduce the following bilinear form in $C_0^\infty(B,\mathbb{R})$:

$$(f_1,f_2)_a = \int_B \sum_{j,k} a_{jk}(x)\frac{\partial f_1}{\partial x_j}\frac{\partial f_2}{\partial x_k}\,dx. \tag{15.2}$$

It is called the Dirichlet form. This form is positive definite (by (15.1)) and, following general rules, it defines a Gaussian measure μ in the dual space $\Omega = (C_0^\infty(B,\mathbb{R}))'$ by the formula

$$\int_\Omega e^{\langle F,\omega\rangle}\,d\mu(\omega) = e^{(1/2)|f|_a^2}, \qquad f \in C_0^\infty(B,\mathbb{R}). \tag{15.3}$$

Let us formulate the main results of this section.

THEOREM 15.1. (1) *Let $n \ge 3$. Then there exists a measurable set $E \subset \Omega$ such that*

$$\mu(E) = 1, \tag{15.4}$$

$$E \cap (E + t\delta_x) = \varnothing \tag{15.5}$$

for all $x \in B$ and all real $t \neq 0$.

(2) *Let $n = 2$. Then there exists a measurable set $E \subset \Omega$ satisfying the condition* (15.4) *and the condition* (15.5) *for all points $x \in B$ and all real t such that*

$$|t| > 88\sqrt{M}, \tag{15.6}$$

where

$$M = \sup\{(a(x)\xi,\xi) \mid x \in B,\ \xi \in \mathbb{R}^n,\ |\xi| = 1\}.$$

The singularity of the translation $\omega \mapsto \omega + t\delta_x$, $\omega \in \Omega$ for $t \neq 0$ (i.e., the fact that it transforms a set E of full measure into a set of zero measure) follows from general properties of the Gaussian measure. Theorem 15.1 gives a substantial strengthening of this known result: it claims that the set E can be chosen to be the same for all $x \in B$ and all $t \neq 0$.

To formulate the second main theorem of this section, consider two matrix functions $x \mapsto a_k(x)$, $k = 1, 2$, satisfying the conditions listed at the beginning of this subsection. Denote by μ_k, $k = 1, 2$, the Gaussian measures on Ω generated by these functions.

THEOREM 15.2. *Assume that $a_1(x_0) \neq a_2(x_0)$ for some $x_0 \in B$. Then there exist measurable sets $F_i \subset \Omega$ such that $\mu_i(F_i) = 1$, $F_1 \cap F_2 = \varnothing$, $F_i + t\delta_x = F_i$ for all $x \in B$ and all $t \in \mathbb{R}$. (Here $n = \dim X$ is arbitrary, $n \geq 1$),*

These two theorems represent the main results of this section. The next two theorems contain some generalizations (these are the results that will be used in the later sections).

Once again, let X be a manifold with a Riemannian metric r and a positive C^∞-function $f \colon X \to \mathbb{R}$. Introduce the form

$$(f_1, f_2)_{r,\rho} = \int_X (df_1, df_2)\rho\, dx, \qquad f_1, f_2 \in C_0^\infty(X, \mathbb{R}). \tag{15.7}$$

Here the function $x \mapsto (df_1, df_2)(x)$ is determined using the inner product in the tangent space T_xX, corresponding to the given Riemannian metric. The form (15.7) generates a Gaussian measure μ on the space $\Omega = (C_0^\infty(X, \mathbb{R}))'$ by the formula (15.3) with $f \in C_0^\infty(X, \mathbb{R})$. The next theorem generalizes Theorem 15.1.

THEOREM 15.3. (1) *Let* $\dim X \geq 3$. *Then there exists a measurable set $E \subset \Omega$ such that*

$$\mu(E) = 1, \tag{15.8}$$

$$E \cap \left(E + \sum_k t_k \delta_{x_k}\right) = \varnothing \tag{15.9}$$

for any finite set of points $\{x_k\} \subset B$ and any finite family of real numbers $t_k \neq 0$.

(2) *Let* $\dim X = 2$. *Then there exists a measurable set $E \subset \Omega$ satisfying the condition* (15.8) *and the condition* (15.9) *under the additional condition*

$$\max_k |t_k| (\rho(x_k))^{-1/2} > 88. \tag{15.10}$$

Finally, let us generalize Theorem 15.2. For any point $x \in X$ introduce the space $R(x)$ of all quadratic forms on the cotangent space at x and the space $P(x)$ of all 1-densities in the tangent space. (Recall that a 1-density on a real finite-dimensional space is a function on the space of bases that is multiplied by the modulus of the determinant of the transition matrix in passing from one basis to another.) The Riemannian metric r and the function ρ generate the section of the bundle $x \to R(x) \otimes P(x)$ by the formula $x \mapsto \rho(x) r^0(x) \otimes p(x)$; here the quadratic form $r^0(x)$ is generated by the Riemannian metric r and the 1-density $p(x)$ takes the value 1 on all orthonormal bases of the space T_xX.

Now we pass to a generalization of Theorem 15.2. Let there be given two Riemannian metrics r_k and two positive functions ρ_k, $k = 1, 2$, on X. Denote by

μ_k the corresponding Gaussian measures on the space $\Omega = (C_0^\infty(X,\mathbb{R}))'$. Finally, consider the above fields $x \mapsto \rho_k(x) r_k^0(x) \otimes p_k(x)$.

THEOREM 15.4. *Assume that the fields $x \mapsto \rho_k(x) r_k^0(x) \otimes p_k(x)$ are distinct (i.e., their values at some point of X are distinct). Then there exist measurable sets $F_k \subset \Omega$ such that $\mu_k(F_k) = 1$, $F_1 \cap F_2 = \varnothing$, $F_k + t\delta_x = F_k$ for any $x \in B$ and any real t.*

15.2. Proof of Theorem 15.1. We break the proof into several steps.

Step 1. We reproduce (in a form that is convenient for further use) a well-known estimate for the norm of a random trigonometric polynomial; we follow Kahane [**23**, Chapter 6]. Denote $\mathbb{Z}_+^n = \{(k_1, \dots, k_n) \in \mathbb{Z}^n,\ k_i > 0\}$. Let $\omega_k = \alpha_k - i\beta_k$, $k \in \mathbb{Z}_+^n$, be independent complex random variables such that α_k, β_k are independent and have the Gaussian distribution with the density $(2\pi)^{-1/2}\exp(-x^2/2)$. Let $E \subset \mathbb{Z}_+^n$ be a finite subset and a_k, $k \in E$, complex numbers. Consider the random trigonometric polynomial

$$T(x) = \sum_{k \in E} \omega_k a_k \exp(i(k,x)), \qquad x \in \mathbb{R}^n.$$

Let $\mathcal{N} = \max\{k_1 + \cdots + k_n,\ k = (k_1, \dots, k_n) \in E\}$, $\|T\|_\infty = \max_x |T(x)|$. In the next lemma Pr denote the probability (the Gaussian measure) corresponding to the random variables $\{\omega_k\}$.

LEMMA 15.1. *For any $H > 0$ the following inequality holds*:

$$\Pr\{\|T\|_\infty > H\} \le D_n \mathcal{N}^{2n} \exp\left\{-\frac{H^2}{36|a|^2}\right\}, \tag{15.11}$$

where

$$D_n = \pi^n 2^{n+3}, \qquad |a| = \left(\sum_{k \in E} |a_k|^2\right)^{1/2}.$$

This lemma easily follows from the estimates presented in [**23**, Chapter VI, Sect. 2].

Step 2. For an integer $r > 1$ consider the set $E_r = \{k \mid k_i > 0, \sum_{i=1}^n k_i \le r\}$. Let the random variables ω_k be as above and

$$T_r(x) = \sum_{k \in E_r} \frac{\omega_k}{|k|} e^{i(k,x)}, \qquad x \in \mathbb{R}^n. \tag{15.12}$$

LEMMA 15.2. (1) *Let $n \ge 3$. Then*

$$\limsup_{r \to \infty} \|T_r\|_\infty r^{-n/2+3/4} \le 1$$

with probability 1.

(2) *Let $n = 2$. Then*

$$\limsup_{m \to \infty} m^{-1} \|T_{2^m}\|_\infty \le \sqrt{72\pi}\log 2$$

with probability 1.

PROOF. (1) Let $n \geq 3$. Apply Lemma 15.1 to the polynomial (15.12), taking $H = r^{n/2-3/4}$. Notice that if $r \to \infty$, then

$$|a|^2 = \sum_{k \in E_r} \frac{1}{|k|^2} \sim C_n r^{n-2}, \quad a = \{1/|k|,\ k \in E_r\},$$

where C_n is a positive number depending only on n. It follows from (15.11) that

$$\Pr\{\|T_r\|_\infty > r^{n/2-3/4}\} \leq D_n e^{-b_n \sqrt{r}}, \qquad b_n = \text{const} > 0.$$

Now the lemma follows (for $n \geq 3$) from the Borel–Cantelli lemma.

(2) Let $n = 2$. Take an arbitrary $\gamma > \sqrt{72\pi}$ and apply Lemma 15.1 with $H = \gamma \log r$. Note that if r is sufficiently large, then

$$|a|^2 = \sum_{k \in E_r} \frac{1}{|k|^2} \leq \iint_{K_r} \frac{dx\,dy}{x^2+y^2} \leq \frac{\pi}{2} \log r,$$

where $K_r = \{(x,y) \in \mathbb{R}^2 \mid x \geq 1,\ y \geq 1,\ x^2+y^2 \leq r^2\}$. Hence, by Lemma 15.1,

$$\Pr\{\|T_r\|_\infty > \gamma \log r\} \leq \text{const}\, r^{4-\frac{\gamma^2}{18\pi}}.$$

(Note that the exponent $4 - \frac{\gamma^2}{18\pi}$ on the right-hand side is negative.) Taking $r = 2^m$ in this inequality and applying the Borel–Cantelli lemma, we obtain the second part of the lemma.

Step 3. Now we return to the space $C_0^\infty(B)$ with the norm (15.2). Denote by W the completion of this space with respect to the norm $|f|_a = \sqrt{(f,f)_a}$. We describe some simple properties of the space W (in more detail, the analysis of similar spaces can be found in any book on partial differential equations).

Let a selfadjoint operator A in $L_2(B)$ be defined by the formula

$$Af = \sum_{j,k} \frac{\partial}{\partial x_j}\left(a_{jk}(x)\frac{\partial f}{\partial x_k}\right)$$

with the boundary condition $f\,|_{\partial B} = 0$. Then $(f_1, f_2)_a = (A^{1/2} f_1, A^{1/2} f_2)$, where $(\cdot,\cdot)$ is the standard inner product in $L_2(B)$. We have a unitary isomorphism of Hilbert spaces $A^{-1/2}\colon L_2(B) \to W$.

LEMMA 15.3. *There exist bounded linear operators $S_k\colon W \to W$, $k = 1, 2, \ldots$, such that*

(1) for any $f \in W$ we have $|S_k f - f|_a \to 0$ as $k \to \infty$;

(2) $S_k(W) \subset C_0^\infty(B)$ for any k;

(3) if a function $f \in W$ is continuous in B, then $(S_k f)(x) \to f(x)$ as $k \to \infty$ for all $x \in B$.

(Here we identify the space W with the nonclosed linear subspace $A^{-1/2}(L_2(B))$; this allows us to identify elements of W with certain functions from $L_2(B)$.)

PROOF. Take the functions $\theta_k \in C_0^\infty(B)$, $\gamma_k \in C_0^\infty(\mathbb{R}^n)$, $k = 1, 2, \ldots$, such that: (1) $\theta_k(x) = 1$ when $\operatorname{dist}(x, \partial B) > 1/k$, $\theta_k(x) = 0$ when $\operatorname{dist}(x, \partial B) < 1/2k$;

(2) $|\partial\theta_k/\partial x_j| \le c_0 k$, where c_0 does not depend on k; (3) $\gamma_k(x) \ge 0$ for $x \in \mathbb{R}^n$, $\gamma_k(x) = 0$ for $|x| > 1/3k$, and

$$\int_{\mathbb{R}^n} \gamma_k(x)\, dx = 1.$$

Define the operator S_k in W by the formula

$$S_k f = \gamma_k * (\theta_k f), \qquad f \in W, \quad k = 1, 2, \ldots.$$

It is clear that these operators satisfy conditions (2) and (3) of the lemma, and condition (1) is satisfied for $f \in C_0^\infty(B)$. To verify condition (1) for an arbitrary function $f \in W$, it suffices to prove that $\|S_k\| \le \text{const}$. Also, it is sufficient to consider the case $a_{jk} = \delta_{jk}$, when the form (15.2) reduces to the standard Dirichlet form. It is easy to see that the norm of the operator $f \mapsto \gamma_k * f$ equals 1 (it suffices to consider the operator in $L_2(\mathbb{R}^n)$ and to pass to the unitarily equivalent realization of this operator using the Fourier transform). It remains to verify that

$$|\theta_k f|_a \le c|f|_a, \qquad f \in C_0^\infty(B), \quad k = 1, 2, \ldots, \tag{15.13}$$

with c independent of k. We have

$$\begin{aligned} |\theta_k f|_a^2 &= \int_B \sum_{k,j} \left|\frac{\partial}{\partial x_j}(\theta_k f)\right|^2 dx \\ &\le 2\int_B \sum_{k,j} \left|\frac{\partial}{\partial x_j}\theta_k\right|^2 |f|^2\, dx + 2\int_B \sum_{k,j} \theta_k \left|\frac{\partial}{\partial x_j} f\right|^2 dx \\ &\le 2c_0 k^2 \int_{D_k} |f|^2 dx + 2|f|_a^2, \end{aligned} \tag{15.14}$$

where $D_k = \{x \mid x \in B,\ \operatorname{dist}(x, \partial B) < k^{-1}\}$. Assuming that k is large enough, take a point $x \in D_k$ and introduce the coordinates (s,t), where s is the point in ∂B that is closest to x, and $t = |x - s|$. Then

$$|f(x)|^2 = |f(s,t)|^2 = \left|\int_0^t \frac{\partial}{\partial\xi} f(s,\xi)\, d\xi\right|^2 \le k^{-1} \int_0^{k^{-1}} \left|\frac{\partial}{\partial\xi} f(s,\xi)\right|^2 d\xi,$$
$$x = (s,t) \in D_k.$$

Therefore,

$$\begin{aligned} \int_{D_k} |f(x)|^2 dx &\le \text{const} \int_{\partial B} \int_0^{k^{-1}} |f(s,t)|^2 ds\, dt \\ &\le \frac{\text{const}}{k} \int_{\partial B} \int_0^{k^{-1}} \int_0^{k^{-1}} \left|\frac{\partial f(s,\xi)}{\partial\xi}\right|^2 ds\, dt\, d\xi \\ &\le \frac{\text{const}}{k^2} \int_{\partial B} \int_0^{k^{-1}} \left|\frac{\partial f(s,\xi)}{\partial\xi}\right|^2 ds\, d\xi \le \frac{\text{const}}{k^2} |f|_a^2. \end{aligned}$$

Now (15.13) follows from (15.14). The lemma is proved.

Step 4. For a point $b = (b_1, \ldots, b_n) \in B$ and a number $h > 0$ consider the cube

$$C_{b,h} = \{x \in \mathbb{R}^n, \mid |x_i - b_i| < h/2,\ 1 \le i \le n\};$$

the number h is assumed to be so small that $\overline{C_{b,h}} \subset B$ (here $\overline{C_{b,h}}$ is the closure of the cube $C_{b,h}$). Consider the functions

$$\psi_k(x) = \begin{cases} \sqrt{2}h^{-n/2} \exp\{2\pi i(k,x)h^{-1}\}, & x \in C_{b,h}, \\ 0, & x \notin C_{b,h}, \end{cases}$$

where $k = (k_1, \dots, k_n) \in \mathbb{Z}_+^n$. The real and imaginary parts of these functions together form an orthonormal family in $L_2(B)$. Let

$$\varphi_k = A^{-1/2}\psi_k, \qquad k \in \mathbb{Z}_+^n.$$

We obtain a family of functions $\{\varphi_k\} \subset W$, $k \in \mathbb{R}_+^n$, and the family $\{\operatorname{Im}\varphi_k, \operatorname{Re}\varphi_k\}$ is orthonormal in W.

LEMMA 15.4. *The functions φ_k are continuous in B. Furthermore, for $x \in C_{b,h}$ we have*

$$\varphi_k(x) = \frac{1}{\pi\sqrt{2}} h^{-n/2+1}(a(x)k,k)^{-1/2} \exp\left\{\frac{2\pi i(k,x)}{h}\right\}(1+\varepsilon_k(x)), \tag{15.15}$$

where $\varepsilon_k(x) \to 0$ as $|k| \to \infty$, $k \in \mathbb{Z}_+^n$.

PROOF. We recall some general facts about fractional powers of the operator A; they can be found, for example, in Seeley's paper [**63**] (in a much more general form). For $s < 0$ the operator A^s is given by a kernel $K(x,y)$. For a fixed $x \in B$ the function $y \mapsto K(x,y)$ is continuous in y for $y \neq x$. As $y \to x$, this function has a singularity, which can be conveniently described in terms of the Fourier transform. Namely, we have an asymptotic expansion of the form

$$\int_{|y|\le r} K(x,y)e^{-i(\xi,y)}\,dy \sim ((a(x)\xi,\xi))^s e^{-i(\xi,x)}, \qquad |\xi| \to \infty;$$

here $0 < r < \operatorname{dist}(x, \partial B)$.

Return to the function φ_k from Lemma 15.4. Using the above properties of the kernel $K(x,y)$ (for $s = -1/2$), we see that the functions φ_k are continuous in B and if $x \in C_{b,h}$, then

$$\varphi_k(x) = \int_{C_{b,h}} K(x,y)\psi_k(y)\,dy \sim \frac{1}{\pi\sqrt{2}} h^{-n/2+1}(a(x)k,k)^{-1/2} \exp\left\{\frac{2\pi i(k,x)}{h}\right\} \tag{15.16}$$

as $|k| \to \infty$. Lemma 15.4 is proved.

Step 5. Now we return to the space Ω with the Gaussian measure μ. Take an element $f \in W$ and construct the subset $\Omega_f \subset \Omega$ and the functions $\omega \mapsto (\omega, f)$, $\omega \in \Omega_f$, as follows. Using the operator S_k from Lemma 15.3, we have $|S_k f - f|_a \to 0$ as $k \to \infty$, and $S_k f \in C_0^\infty(B)$. This implies that the sequence of functions $\omega \mapsto \langle S_k f, \omega\rangle$, $\omega \in \Omega$, converges in the metric of the space $L_2(\Omega, \mu)$. Therefore, for a subsequence of indices $\{m_k\}$ the sequence $\omega \mapsto \langle S_{m_k} f, \omega\rangle$ converges almost everywhere with respect to the measure μ on Ω. Denote

$$\Omega_f = \{\omega \mid \text{the limit } \lim_{k\to\infty} \langle S_{m_k} f, \omega\rangle \text{ exists}\}$$

(we recall that $\mu(\Omega_f) = 1$), and define the function $\omega \mapsto (f, \omega)$, $\omega \in \Omega_f$, by

$$(f, \omega) = \lim_{k\to\infty} \langle S_{m_k} f, \omega\rangle.$$

If we are given a countable family of functions $\{f_j\} \subset W$, then an obvious modification of the above arguments allows us to construct a sequence $m_k \to \infty$ such that the sequence of functions $\langle S_{m_k} f_j, \omega\rangle$ converges almost everywhere on Ω for each j. Denote by $\Omega_{\{f_j\}}$ the set of all ω for which this convergence takes place. Define the functions

$$\omega \mapsto (f_j, \omega) \stackrel{\text{def}}{=} \lim_{k\to\infty} \langle S_{m_k} f_j, \omega\rangle.$$

Notice the following important fact: if all functions f_j are continuous in B, then Ω_{f_j} contains all delta-functionals δ_x, $x \in B$, and $(f_j, \delta_x) = f_j(x)$.

Step 6. Fix a cube $C_{b,h} \subset B$ (see Step 4) and consider the corresponding family of functions $\{\varphi_k,\ k \in \mathbb{Z}^n_+\}$. Following Step 5, consider the set Ω_{φ_k}. On this set, which has the full μ-measure, we have the functions $\omega \mapsto (\varphi_k, \omega)$; we denote these functions (random variables) simply by ω_k. It is clear that the random variables $\{\operatorname{Re}\omega_k, \operatorname{Im}\omega_k\}$ are independent and each has the distribution $(2\pi)^{-1/2}\exp(-x^2/2)$. Construct the random polynomials $T_r(x)$ by formula (15.12). Define the subset $D_{b,h} \subset \Omega_{\{\varphi_k\}}$ as follows:

(15.17)
$$D_{b,h} = \{\omega \mid \omega \in \Omega_{\{\varphi_k\}},\ \limsup_{r\to\infty} \|T_r\|_\infty r^{-n/2+3/4} \le 1\} \quad \text{for } n \ge 3,$$

(15.18)
$$D_{b,h} = \{\omega \mid \omega \in \Omega_{\{\varphi_k\}},\ \limsup_{m\to\infty} \|T_{2^m}\|_\infty m^{-1} \le \sqrt{72\pi}\log 2\} \quad \text{for } n = 2.$$

LEMMA 15.5. (1) $\mu(D_{b,h}) = 1$ *for all* $n \ge 2$.
(2) *For* $n \ge 3$

$$D_{b,h} \cap (D_{b,h} + t\delta_x) = \varnothing \tag{15.19}$$

for $x \in C_{b,h}$, $t \neq 0$.
(3) *If* $n = 2$, *then the condition* (15.19) *is satisfied for*

$$|t| > 88\sqrt{M},$$

where M *is the number defined in Theorem* 15.1.

PROOF. The equality $\mu(D_{b,h}) = 1$ follows from Lemma 15.2. Let us prove the second statement of the lemma. Let $n \ge 3$. Assuming that (15.19) does not hold, take an element $\overline{\omega} \in D_{b,h} \cap (D_{b,h} + t_0\delta_{x_0})$, $x_0 \in C_{b,h}$, $t_0 \neq 0$. By the definition of the set $D_{b,h}$ (see formula (15.17)) we have

$$\left|\sum_{k\in E_r} \frac{e^{i(k,s)}}{|k|}(\varphi_k, \overline{\omega})\right| < 2r^{n/2-3/4},$$

$$\left|\sum_{k\in E_r} \frac{e^{i(k,s)}}{|k|}(\varphi_k, \overline{\omega} - t_0\delta_{x_0})\right| < 2r^{n/2-3/4}, \quad r > r_0$$

for some $x_0 \in C_{b,h}$, $t_0 \neq 0$, and all $s \in \mathbb{R}^n$ (here r_0 depends on $\overline{\omega}$). Therefore,

$$|t_0|\left|\sum_{k\in E_r} \frac{e^{i(k,s)}}{|k|}\varphi_k(x_0)\right| < 4r^{n/2-3/4}, \qquad r > r_0, \quad s \in \mathbb{R}^n.$$

Substitute the right-hand side of (15.15) for $\varphi_k(x_0)$ and set $s = -2\pi x_0 h^{-1}$. We get

$$\sum_{k\in E_r} \frac{1}{|k|(a(x_0)k,k)^{1/2}} \leq C_1 r^{n/2-3/4} = \text{const} < \infty. \tag{15.20}$$

It is clear that the left-hand side of this inequality is greater than

$$C_2 \sum_{k\in E_r} \frac{1}{|k|^2} \sim C_3 r^{n-2}, \qquad C_2 > 0, \quad C_3 > 0, \quad r > r_0.$$

Since $n \geq 3$, the inequality (15.20) is violated for sufficiently large r, and (15.19) is proved.

Now let $n = 2$. Let us assume that for some $x_0 \in C_{b,h}$, $t_0 \neq 0$ there exists $\overline{\omega} \in D_{b,h} \cap (D_{b,h} + t_0\delta_{x_0})$. Take a number $\gamma > \sqrt{72\pi}\log 2$. It follows from the definition of the set $D_{b,h}$ (see formula (15.18)) that

$$\left| \sum_{k\in E_{2m}} \frac{e^{i(k,s)}}{|k|}(\varphi_k, \overline{\omega}) \right| < \gamma m,$$

$$\left| \sum_{k\in E_{2m}} \frac{e^{i(k,s)}}{|k|}(\varphi_k, \overline{\omega} - t_0\delta_{x_0}) \right| < \gamma m$$

for sufficiently large m. Therefore

$$|t_0| \left| \sum_{k\in E_{2m}} \frac{e^{i(k,s)}}{|k|}\varphi_k(x_0) \right| < 2\gamma m, \qquad m > m_0, \quad s \in \mathbb{R}^n.$$

Using (15.15) and setting $s = -2\pi x_o h^{-1}$, we get

$$|t_0| \sum_{k\in E_{2m}} \frac{1}{|k|(a(x_0)k,k)^{1/2}} < 2\sqrt{2}\pi\gamma m, \qquad m > m_0. \tag{15.21}$$

The sum on the left-hand side is greater than

$$\frac{1}{\sqrt{M}} \sum_{k\in E_{2m}} \frac{1}{|k|^2} > \frac{1}{\sqrt{M}} \cdot \frac{\pi \log 2}{2} m - d_0, \qquad m > m_0, \quad d_0 = \text{const}.$$

Therefore, (15.21) implies that $|t_0| < \frac{4\sqrt{2}}{\log 2}\gamma$. Recalling the choice of γ, we get $|t_0| \leq 88\sqrt{M}$. Lemma 15.5 is proved.

Step 7. Take open cubes C_i, L_i, $i = 1, 2, \dots$, satisfying the following conditions:

(1) $\overline{C_i} \subset L_i$ for all i;

(2) $\operatorname{diam} L_i \to 0$ as $i \to \infty$;

(3) $C_m \cup C_{m+1} \cup \dots = B$ for all m.

Let $\Omega_i = (C_0^\infty(L_i))'$. Formula (15.3) defines the Gaussian measure μ_i on Ω_i (in exactly the same way as it defined the measure μ on Ω). The natural projection $p_i \colon \Omega \to \Omega_i$ maps the measure μ into μ_i. Apply the construction described above, taking for B the cube L_i and for $C_{b,h}$ the cube C_i. We obtain Borel sets $D_i \subset \Omega_i$ satisfying the following conditions:

(1) $\mu_i(D_i) = 1$;

(2) $D_i \cap (D_i + t\delta_x) = \varnothing$ for $x \in C_i$, $t \neq 0$ (if $n \geq 3$) and for $x \in C_i$, $|t| > 88\sqrt{M}$ if $n = 2$.

Define $E \subset \Omega$ by the formula

$$E = \bigcap_{i=1}^{\infty} p_i^{-1}(D_i).$$

Let us prove that it satisfies the conditions of Theorem 51.1. The equality (15.4) is obvious. Let us prove (15.5). Assume that $\overline{\omega} \in E \cap (E + \omega_0)$ for some functional $\omega_0 = \sum t_k \delta_{b_k}$. Fix one of the points b_{k^0}; if $n = 2$ we choose it in such a way that $|t_{k^0}| > 88\sqrt{M}$ (recall that in the case $n = 2$ we imposed the restriction (15.6)). Then there exists a cube L_{i^0} such that b_{k^0} is the only point from the family $\{b_k\}$ that lies inside L_{i^0}, and we have $b_{k^0} \in C_{i^0}$. Take the restrictions of distributions ω, ω_0 to the cube L_{i^0} (in other words, apply to these distributions the projection $p_{i^0} : \Omega \to \Omega_{i^0}$). We obtain

$$\overline{\omega}\,|_{\Omega_{i^0}} \in D_{i^0} \cap (D_{i^0} + t_{k^0}\delta_{b_{k^0}}).$$

This contradicts the condition $D_{i^0} \cap (D_{i^0} + t_{k^0}\delta_{b_{k^0}}) \neq \varnothing$. Theorem 15.1 is proved.

15.3. Proof of Theorem 15.2. Let the assumptions of Theorem 15.2 be satisfied.

LEMMA 15.6. *There exists a sequence of elements $u_k \in C_0^\infty(B)$, $k = 1, 2, \dots,$ such that*

$$\operatorname{supp} u_k \cap \operatorname{supp} u_j = \varnothing \quad \textit{for } k \neq j, \tag{15.22}$$

$$|u_k|_{a_1} \to A_1, \quad |u_k|_{a_2} \to A_2, \quad A_1 \neq A_2. \tag{15.23}$$

(Recall that the norm $|\cdot|_{a_k}$, $k = 1, 2$, is defined by equality (15.2).)

PROOF. Recall that $a_1(x_0) \neq a_2(x_0)$ for some point $x_0 \in B$. Therefore, there exists a function $\varphi \in C_0^\infty(\mathbb{R}^n)$ such that

$$(\varphi, \varphi)_{a_k(x_0)} = A_k, \quad k = 1, 2; \qquad A_1 \neq A_2,$$
$$\varphi(x) = 0 \quad \text{for } |x| \geq 1.$$

Take points $c_k \in B$ and numbers $\lambda_k \to \infty$, $k = 1, 2, \dots,$ such that $c_k \to x_0$ and the λ_k^{-1}-neighborhoods of points c_k are pairwise disjoint. Denote $u_k(x) = \lambda_k^{n-2}\varphi(\lambda_k(x - c_k))$. Then the condition (15.22) is satisfied and, moreover,

$$(u_k, u_k)_{a_j} \to A_j, \qquad j = 1, 2, \quad k \to \infty,$$

where the numbers A_1, A_2 are defined by (15.23). It is clear that the functions u_k satisfy the conditions of the lemma.

Now we prove Theorem 15.2. Denote

$$F_j = \left\{ \omega \in \Omega, \left| \frac{1}{m} \sum_{i=1}^{m} |(u_i, \omega)|^2 \to A_j \right. \right\}, \qquad j = 1, 2.$$

Formula (15.23) and the law of large numbers imply that $\mu_j(F_j) = 1$. Moreover, since $(u_i, \delta_x) = u_i(x) \to 0$ as $i \to \infty$, we have $F_j + t\delta_x = F_j$ for all $x \in B$ and t. This proves Theorem 15.2.

15.4. Proof of Theorems 15.3 and 15.4. First, let $n \geq 3$. Take the ball $Y = Y_{c,h} = \{x \in X, | \ \rho(c,h) \leq h\}$ in X; here ρ is the Riemannian distance between points. The number h is assumed to be so small that in some neighborhood of the ball Y one can introduce local coordinates $x_1, \dots, x_n$ such that in these coordinates the ball is given by the inequality $x_1^2 + \cdots + x_n^2 \leq 1$. In these coordinates the Riemannian metric r is given by the formula

$$ds^2 = \sum_{j,k} g_{jk}(x)\, dx_j\, dx_k,$$

and the form $(f_1, f_2)_{r,\rho}$ introduced in Theorem 15.3 becomes

$$(f_1, f_2)_{r,\rho} = \int_B \sum_{j,k} \left((g(x)^{-1}\right)_{jk} \sqrt{\det g}\, \frac{\partial f_1}{\partial x_j} \frac{\partial f_2}{\partial x_k} \rho(x)\, dx, \qquad f_i \in C_0^\infty(Y),$$

where $B = \{x \mid x \in \mathbb{R}^n,\ (x,x) < 1\}$. Thus we have identified the space $C_0^\infty(Y)$ endowed with the form $(f_1, f_2)_{r,\rho}$ with the space $C_0^\infty(B)$ endowed with the norm (15.2). Denote by μ_Y the Gaussian measure on $\Omega(Y) = (C_0^\infty(Y))'$ corresponding to the form $(\cdot,\cdot)$. Using Theorem 15.1, we find a set $E(Y) \subset \Omega(Y)$ such that $\mu_Y(E(Y)) = 1$ and $E(Y) \cap (E(Y) + t\delta_x) = \varnothing$ for all $x \in Y$ and all real $t \neq 0$. The natural projection $p_Y : \Omega \to \Omega_Y$ transforms the measure μ on Ω to the measure μ_Y. Now we take a countable set $\mathcal{L}$ of balls such that for any $\varepsilon > 0$ the balls of radius $< \varepsilon$ in $\mathcal{L}$ cover the manifold X. Introduce the set

$$E = \bigcap_{Y \in \mathcal{L}} p_Y^{-1}(Y)).$$

and prove that it satisfies the conditions of Theorem 15.3. First of all, it is obvious that $\mu(E) = 1$. Now let $\omega \in E$, $\omega + \sum t_k \delta_{x_k} \in E$ for some nonempty finite set $\{x_k\} \subset E$ and numbers $t_k \neq 0$. Take a ball $Y \in \mathcal{L}$ that contains the point x_1 and does not contain the points x_j for $j \neq 1$. By the definition of the set E, we have $p_Y(\omega) \in E(Y)$, $p_Y(\omega + \sum t_k \delta_{x_k}) = p_Y(\omega) + t_1 \delta_{x_1} \in E(Y)$. Therefore, $p_Y(\omega) \in E(Y) \cap E(Y) - t_1\delta_{x_1}$, and so the set $E(Y) \cap E(Y) - t_1 \delta_{x_1}$ is nonempty. This contradicts the definition of the set $E(Y)$. Theorem 15.3 is proved in the case $\dim X = n \geq 3$.

Now let $n = 2$. In this case we repeat the above arguments with the following modifications. Fix $\varepsilon > 0$ and assume that the balls $Y \in \mathcal{L}$ are so small and the local coordinates in these balls are chosen in such a way that $|g_{ik}(x) - \delta_{ik}| < \varepsilon$, $|\rho(x) - \overline{\rho}(Y)| < \varepsilon$ for $x \in Y$, $Y \in \mathcal{L}$. Here g_{ik} are the coefficients of the Riemannian metric in local coordinates, and $\overline{\rho}(Y)$ is a constant depending on the ball Y. Using Theorem 15.1, we obtain sets $E(Y, \varepsilon) \subset \Omega(Y)$ such that

$$\mu_Y(E(Y,\varepsilon)) = 1,$$
$$E(Y,\varepsilon) \cap (E(Y,\varepsilon) + t\delta_x) = \varnothing, \qquad x \in Y, \quad Y \in \mathcal{L},$$

provided that $|t| > 88\sqrt{\overline{\rho}(Y) + \varepsilon_1}$, where ε_1 is small together with ε. Next, we construct the set

$$E = \bigcap_{m=1}^{\infty} \left(\bigcap_{Y \in \mathcal{L}} p_Y^{-1}(E(Y, 1/m)) \right).$$

Repeating (with obvious modifications) the previous arguments concerning the case $n \geq 3$, we see the E is the desired set. This completes the proof of Theorem 15.3.

Finally, to prove Theorem 15.4, it suffices to select a subset in X in which the fields $x \mapsto \rho_k(x) r_k^0(x) \otimes p_k(x)$ are distinct, and apply Theorem 15.2.

15.5. Properties of Gaussian measures related to the L_2-form on the interval. The results of this subsection will be used only in Chapter 7.

Consider the real space $C_0^\infty(a,b)$ and the space $\Omega = (C_0^\infty(a,b))'$. Introduce the inner product

$$(u,v) = \frac{1}{b-a} \int_a^b u(x)v(x)\,dx, \qquad u,v \in C_0^\infty(a,b).$$

It generates a Gaussian measure μ on Ω. The next theorem somewhat resembles the results obtained above for Gaussian measures generated by Dirichlet forms.

THEOREM 15.5. *There exists a Borel set $E \subset \Omega$ satisfying the following conditions*: (1) $\mu(E) = 1$; (2) *the equality* $\omega_1 - c\omega_2 = \sum c_k \delta_{t_k}$ *with* $\omega_1, \omega_2 \in E$, $t_1 < \cdots < t_m$ *points from the interval* (a,b), *is possible only for* $c = \pm 1$, $c_1 = \cdots = c_m = 0$.

Moreover, for any finite-dimensional subspace $V \subset L_2(a,b)$ the set E can be chosen so that the following additional condition be satisfied: (3) $E + v = E$ *for any* $v \in V$ *(here we use the natural embedding $V \subset \Omega$).*

We will need some corollaries of this theorem. Take a positive function $p \in C^\infty([a,b])$ and define the following inner product in $C_0^\infty(a,b)$:

$$(u,v) = \int_a^b u(x)v(x)p(x)\,dx. \tag{15.24}$$

Next, take a finite-dimensional subspace $V \subset L_2(a,b)$ and define $D_V = \{u \in C_0^\infty(a,b), u \perp V\}$; the orthogonality is understood with respect to the form (15.24). Finally, consider the dual space $\Omega_1 = D_V'$ and the Gaussian measure θ generated by tho form (15.24).

COROLLARY 15.1. *There exists a Borel subset $E_1 \subset \Omega_1$ such that* (1) $\theta(E_1) = 1$; (2) *the equality* $\omega_1 - c\omega_2 = \sum c_k \delta_{t_k}$ *with* $\omega_1, \omega_2 \in E_1$, $a < t_1 < \cdots < t_m < b$, *is possible only for* $c = \pm 1$, $c_1 = \cdots = c_m = 0$.

Here by δ_{t_k} we denote the restriction of the functional δ_{t_k} to the subspace D_V.

COROLLARY 15.2. *The Gaussian measure θ and any probabilistic measure γ on Ω_1 that is supported on the set of functionals of the form $\sum c_k \delta_{t_k}$ (finite sum) are singularly related—see* 8.2.

Next, for a number $c > 0$ take the Gaussian measure θ_c generated by the inner product of the form (15.24) with the function $p(x)$ replaced by the form $cp(x)$.

COROLLARY 15.3. *Let γ be a measure as in Corollary* 15.2. *Then the measures* $\theta_{c_k} * \gamma$, $k = 1, 2$, *are mutually singular for* $c_1 \neq c_2$.

PROOF OF THEOREM 15.5. We use the following statement about random trigonometric series. If $\{\omega_k\}_1^\infty$ are independent Gaussian random variables with zero average and equal variances, then with probability 1 the series

$$\sum_{k=1}^\infty \frac{\omega_k \cos kt}{k}, \qquad t \in \mathbb{R}^n,$$

represent a continuous function. This is a simple special case of a theorem from Kahane's book [**23**, Chapter VIII, Sect. 2].

Now let us return to the proof of Theorem 15.5. We assume that $[a,b] = [0,\pi]$. To a random function $\omega \in \Omega$ we associate the sequence $\{\omega_k\}_0^\infty$ of its Fourier coefficients in the basis $\{\pi^{-1/2}\cos kt\}_0^\infty$. Next, choose a basis $\{f_j\}_1^m$ in V and denote by $\{f_{jk}\}$ the Fourier coefficients of the function f_j, $1 \le j \le m$. Introduce the set

$$E = \Big\{\omega \Big| \text{ the function } \sum_{i=1}^{\infty} \frac{\omega_k \cos kt}{k} \text{ is continuous for } t \in \mathbb{R}, \text{ and}$$

$$\text{the series } \sum_{k=1}^{\infty} \omega_k f_{jk},\ j = 1,2,\dots,m, \text{ and } \sum_{k=1}^{\infty} \frac{\omega_k^2 - 1}{k} \text{ converge.}\Big\}$$

It is clear that conditions (1) and (3) of Theorem 15.5 are satisfied. Let us verify condition (2). Assume that $\alpha, \beta \in E$, $\alpha - c\beta = \sum c_k \delta_{t_k}$, $0 < t_1 < \dots < t_n < \pi$. Passing to Fourier transforms of the distributions α, β, δ_{t_k}, $1 \le k \le m$, we get that the series

$$\sum_{k=1}^{\infty} (c_1 \cos kt_1 + \dots + c_m \cos kt_m) \frac{\cos kt}{k}$$

is the Fourier series of a continuous function $t \mapsto \psi(t)$, $t \in \mathbb{R}$. Consider the function

$$\varphi(t) = \sum_{k=1}^{\infty} \frac{\cos tk}{k}, \qquad t \notin 2\pi\mathbb{Z}. \tag{15.25}$$

LEMMA 15.6. *The function φ is continuous for $t \notin 2\pi\mathbb{Z}$, and*

$$\varphi(t) > h \log(1/t) \quad \text{for } t \in (0,\delta),\ h > 0,\ \delta > 0. \tag{15.26}$$

PROOF. To proof the lemma we rewrite the series (15.25) using the Abel transform. We get

$$\varphi(t) = -\frac{1}{2} + \sum_{k=1}^{\infty} \frac{1}{k^2+2} \cdot \frac{\sin(k+1/2)t}{2\sin(t/2)}.$$

This implies the first statement of the lemma.

Let us prove the estimate (15.26). For any $t \in (0,\pi)$, choose an integer $N = N_t$ such that $tN \to \pi/2$ as $t \to 0$. Then

$$\frac{\sin(k+1/2)t}{2\sin(t/2)} \ge h_0 \quad \text{for } 1 \le k \le N,$$
$$\left|\frac{\sin(k+1/2)t}{2\sin(t/2)}\right| \le N \quad \text{for } k > N.$$

Therefore,

$$\varphi(t) = -\frac{1}{2} + \sum_{k=1}^{\infty} + \sum_{k>N} > h_0 \log N - h_1, \qquad h_0 = \text{const} > 0, \quad h_1 = \text{const}.$$

This implies the inequality (15.26). The lemma is proved

Returning to the function ψ, we notice that

$$\psi(t) = \sum_{i=1}^{m} c_i \left(\varphi(t+t_i) + \varphi(t-t_i)\right).$$

Lemma 15.6 implies that if $c_i \neq 0$ for some i, then the function ψ is discontinuous at the point t_i. Hence, $c_i = 0$ for all i, and we have $\alpha = c\beta$, $\alpha, \beta \in E$. It follows from the definition of the set E that the series

$$\sum_{i=1}^{k} \frac{\alpha_k^2 - 1}{k}, \quad \sum_{i=1}^{k} \frac{c^2\alpha_k^2 - 1}{k}$$

converge. Obviously, this is possible only if $c = \pm 1$. Theorem 15.5 is proved.

Now we prove Corollary 15.1. Obvious changes of variables reduce it to the case $[a, b] = [0, \pi]$, $p(t) = 1$ for $t \in [0, \pi]$. Take a set E that satisfies the conditions of Theorem 15.5 and is invariant under translations by elements from V. The natural projection $\Omega \to \Omega_1$ maps E into a Borel set E_1 (here we use the invariance under translations by elements from V). One can easily see that Ω_1 satisfies the conditions of Corollary 15.1.

Corollary 15.1 and Lemma 8.1 imply Corollary 15.2.

Finally, to prove Corollary 15.3, we again take the set E_1 from Corollary 15.1. Then the measure $\theta_c * \gamma$ is supported on the set $cE_1 + T$, where T is the set of functionals of the form $\sum c_k \delta_{t_k}$, $t_k \in [0, \pi]$. By Corollary 15.2, $(c_1E_1 + T) \cap (c_2E_1 + T) = \varnothing$ for $c_1 \neq c_2$. Corollary 15.3 is proved.

§16. Irreducibility and pairwise nonequivalence of energy representations of the group $C_0^\infty(X, G)$

16.1. Statements of theorems. Return to the representations $T = T_{r,\rho}$ of the group $C_0^\infty(X, G)$ constructed in §14. Fix a Cartan subalgebra $\mathfrak{g}^0 \subset \mathfrak{g}$.

THEOREM 16.1. (1) *Let* $\dim X \geq 3$. *Then the representation* T *is irreducible (for any Riemannian metric* r *and any positive function* $\rho \in C^\infty(X)$*).*

(2) *Let* $\dim X = 2$, *and for any root* l *of the Lie algebra* $\mathfrak{g}$ *let*

$$|l| > 88\sqrt{\rho(x)}, \qquad x \in X.$$

Then the representation T *is irreducible.*

Before formulating the second theorem, let us recall that for a Riemannian metric r we have constructed the field $x \mapsto \rho(x) r^0(x) \otimes p(x)$ on X. In the next theorem we consider the representations $T_k = T_{r_k, \rho_k}$ constructed from Riemannian metrics r_k and functions ρ_k, $k = 1, 2$.

THEOREM 16.2. *Assume that the fields* $x \mapsto \rho_k(x) r_k^0(x) \otimes p_k(x)$, $k = 1, 2$, *do not coincide identically. Then the representations* T_k, $k = 1, 2$, *are disjoint.*

Let us notice that in this theorem $\dim X = n \geq 1$ and we do not impose any conditions on r_k and ρ_k; therefore, for $\dim X \leq 2$ the representations T_k might be reducible.

16.2. Proof of Theorem 16.1. The proof is based on Theorem 8.1 from Chapter I. Fix a root vector $l \in \mathfrak{g}^0$ and denote $l_0 = l/|l|$. Take a region $Y \subset X$ diffeomorphic to $\mathbb{R}^n$ such that $Y \neq X$. The space $C_0^\infty(Y, \mathbb{R})$ is embedded into the space $C_0^\infty(X, \mathbb{R})$ (functions from $C_0^\infty(Y, \mathbb{R})$ are extended by 0 to all of X). Consider the subgroup A_l^Y consisting of all functions of the form $x \mapsto \exp(f(x)l_0)$; here we use the exponential mapping $\exp\colon \mathfrak{g} \to G$. It is clear that the subgroups A_l^Y and $C_0^\infty(X, \mathbb{R})$ are isomorphic. Let us show that A_l^Y is a special subgroup (in the sense of the definition given in 8.3).

Denote by $\mathfrak{g}^1$ the orthogonal complement to $\mathfrak{g}^0$ in the Lie algebra $\mathfrak{g}$ (with respect to the Cartan inner product in $\mathfrak{g}$). Then $E^1(X, \mathfrak{g}) = E^1(X, \mathfrak{g}^0) \oplus E^1(X, \mathfrak{g}^1)$, where $E^1(X, \mathfrak{g}^i)$ is the subspace of 1-forms with values in $\mathfrak{g}^i$, $i = 0, 1$. Denote by L, L^0, L^1 the real Hilbert spaces obtained by completing $E^1(X, \mathfrak{g})$ and $E^1(X, \mathfrak{g}^i)$, $i = 1, 2$. Then $L = L^0 \oplus L^1$. We have the orthogonal representation $g \to \mathrm{Ad}_g$, $g \in C_0^\infty(X, G)$, in the space L. Consider the restriction of this representation to the subgroup A_l^Y. Identifying this subgroup with $C_0^\infty(Y, \mathbb{R})$ via the mapping $f \mapsto \exp(f \cdot l_0)$, we obtain the representation

$$f \mapsto \mathrm{Ad}_{\exp(f l_0)}, \qquad f \in C_0^\infty(Y, \mathbb{R}),$$

in the space L. Its restriction to L^0 is the trivial representation. The complexification of its restriction to L^1 is decomposed into unitary representations equivalent to representations of the form

$$\varphi \mapsto e^{i(\lambda, l_0)f} \varphi, \qquad \varphi \in L_2(X), \quad f \in C_0^\infty(Y, \mathbb{R}), \tag{16.1}$$

where λ is a nonzero root of the Lie algebra $\mathfrak{g}$. Finally, the restriction of the Maurer–Cartan cocycle to the subgroup A_l^Y is given by the formula $f \mapsto df \cdot l_0$ (under the identification of the group A_l^Y with $C_0^\infty(Y, \mathbb{R})$). Therefore, this cocycle takes values in L^0.

Now we consider the measures ν_0 and ν_1 entering the definition of a special subgroup (see 8.3). Identifying A_l^Y with $C_0^\infty(Y, \mathbb{R})$, we can identify each functional $\omega \in (C_0^\infty(Y, \mathbb{R}))'$ with a character of the group $C_0^\infty(Y, \mathbb{R})$ of the form

$$f \mapsto e^{i\langle f, \omega\rangle}, \qquad f \in C_0^\infty(Y, \mathbb{R}),$$

It is clear that ν_0 is a Gaussian measure on $(C_0^\infty(Y, \mathbb{R}))'$ generated by the Dirichlet form (15.7). Next, (16.1) implies that the measure ν_1 is supported on the set of functionals of the form

$$\sum m_k (\lambda_k, l_0) \delta_{x_k}, \tag{16.2}$$

where $m_k \in \mathbb{Z}$, λ_k are roots of the algebra $\mathfrak{g}$, $\lambda_k \neq 0$, $x_k \in X$; the sum (16.2) is finite. It remains to prove that the measures ν_0 and ν_1 are singularly related. According to Lemma 8.1, it suffices to construct a measurable set $E \subset (C_0^\infty(Y, \mathbb{R}))'$ such that $\nu_0(E) = 1$ and $(\omega + E) \cap E = \varnothing$ for any nonzero functional of the form (16.2). If $n = \dim X \geq 3$, then such a set E is given by Theorem 15.3. If $n = 2$, then to apply Theorem 15.3 we must verify that all nonzero numbers (λ_k, l_0) from (16.2) satisfy the conditions

$$|(\lambda_k, l_0)| \, (\rho(x))^{-1/2} > 88 \quad \text{for all } x \in X. \tag{16.3}$$

Let us notice that

$$(\lambda_k, l_0) = \frac{2(\lambda_k, l)}{(l.l)} |l|$$

and $\frac{2(\lambda_k,l)}{(l,l)} \in \mathbb{Z}$. Therefore, (16.3) holds if $|l|(\rho(x))^{-1/2} > 88$ everywhere on X.

Hence, we have proved that the measures ν_0 and ν_1 are singularly related. Therefore, A_l^Y is a special subgroup. Consider now all subgroups of the form A_l^Y, where Y runs over all the domains described above and l runs over all nonzero roots. One can easily verify that this set of special subgroups satisfies condition (1) of Theorem 8.1. Condition (2) of this theorem is also obviously satisfied. Application of Theorem 8.1 completes the proof of Theorem 16.1.

16.3. Proof of Theorem 16.2. Again let us choose a root l and a domain $Y \subset X$, and consider the subgroup A_l^Y. Let the fields $x \mapsto \rho_k(x) r_k^0(x) \otimes p_k(x)$, $k = 1, 2$, not coincide in the domain Y. Consider the restrictions of representations $T_k = T_{r_k,\rho_k}$ to the subgroup $A_l^Y \simeq C_0^\infty(Y, \mathbb{R})$. By Theorem 15.4, spectral types of these two representations are supported on disjoint sets from $(C_0^\infty(Y,\mathbb{R}))'$. Therefore, the restrictions of the representations T_1, T_2 to A_l^Y are disjoint. Theorem 16.2 is proved.

Following the idea of this book, we did not discuss the case $\dim X = 1$. Let us note only that in this case the representations are reducible; moreover, they are factor representations of type III (see [**41**]).

§17. Projective unitary representations of the group $C_0^\infty(X,G)$; representations of the universal covering group

17.1. Constructions of representations. Similarly to §14, we use the action of the group $C_0^\infty(X,G)$ in the Hilbert space H defined as the completion of the space $E_{\mathbb{C}}^1$ in the norm described in 14.1. Starting with the Maurer–Cartan cocycle $u \mapsto \dot{u}$, $u \in C_0^\infty(X,G)$, we construct new cocycles as follows. Take linear operators $a(x)$ acting in tangent spaces T_xX and depending smoothly on x. To any 1-form $\alpha \in E(X, \mathfrak{g})$ we associate the form $\alpha \circ a$. Taking one more field of operators $b(x)\colon T_xX \to T_xX$, we construct the mapping $r\colon C_0^\infty(X,G) \to H$ of the form

$$r(u) = \dot{u} \circ a + i\dot{u} \circ b. \tag{17.1}$$

One can see that this is a cocycle. From the unitary representation of the group $C_0^\infty(X,G)$ in H and the cocycle (17.1) we construct a representation T of the group $C_0^\infty(X,G)$ in the space $\overline{S}(H)$ following the general scheme described in §6. This representation is projective; more precisely,

$$T(u_1)T(u_2) = e^{-f(u_1,u_2)}T(u_1u_2),$$

where the function $f(u_1,u_2)$ is defined by the formula

$$\begin{aligned} f(u_1,u_2) &= \left((u_1^{-1\cdot} \circ s, \dot{u}_2\right), \\ s(x) &= a(x)(b(x))^* - b(x)(a(x))^*. \end{aligned} \tag{17.2}$$

Note that $(s(x))^* = -s(x)$. Rewrite (17.2) in a different form. From now on we assume that the manifold X is oriented. Choose a positive volume form σ^n generating the measure dx from §14. There exists a unique exterior $(n-2)$-form ω^{n-2} satisfying the condition

$$\rho \cdot (\varphi_1 \circ s, \varphi_2) \cdot \sigma^n = \varphi_1 \wedge \varphi_2 \wedge \omega^{n-2}$$

for all real 1-forms φ_1, φ_2; here $(\cdot,\cdot)$ is the inner product in $(T_xX)^*$ generated by the Riemannian metric. Therefore,

$$f(u_1, u_2) = \int_X ((u_1^{-1})^{\cdot} \wedge \dot{u}_2) \wedge \omega^{n-2}; \tag{17.3}$$

here $(u_1^{-1})^{\cdot} \wedge \dot{u}_2$ is the real 2-form obtained from the inner product in $\mathfrak{g}$ and the exterior product of real 1-forms.

Consider the Lie algebra of the group $C_0^\infty(X,G)$. It can be identified with the space $C_0^\infty(X,\mathfrak{g}) = \bar{\mathfrak{g}}$ consisting of smooth functions $a\colon X \to \mathfrak{g}$, with the pointwise bracket operation. Our representation of $C_0^\infty(X,G)$ yields a projective representation of the Lie algebra $\bar{\mathfrak{g}}$, defined on a dense linear submanifold in $\overline{S}(H)$. The corresponding 2-cocycle on $\bar{\mathfrak{g}}$ is given by the formula

$$f_0(a_1, a_2) = -2\int_X (da_1 \wedge da_2) \wedge \omega^{n-2}, \qquad a_1, a_2 \in \bar{\mathfrak{g}}. \tag{17.4}$$

17.2. The case of a closed form ω^{n-2}. Let the form ω^{n-2} be closed. Then the integral (17.4) vanishes. Let us look at the group 2-cocycle (17.3). From now on we assume that X is compact. Consider the universal covering $p\colon \widetilde{C}^\infty(X,G) \to C^\infty(X,G)$; its kernel is the fundamental group $\Gamma = \pi_1(C^\infty(X,G))$. Lifting the function (17.3) to $\widetilde{C}^\infty(X,G)$, we obtain the 2-cocycle

$$F(\widetilde{u}_1, \widetilde{u}_2) = f(u_1, u_2) \qquad \widetilde{u}_i \in \widetilde{C}^\infty(X,G), \quad u_i = p(\widetilde{u}_i). \tag{17.5}$$

Let us prove that this cocycle is trivial, i.e., it can be written in the form

$$F(\widetilde{u}_1, \widetilde{u}_2) = M(\widetilde{u}_1\widetilde{u}_2) - M(\widetilde{u}_1) - M(\widetilde{u}_2), \qquad \widetilde{u}_i \in \widetilde{C}^\infty(X,G). \tag{17.6}$$

Define a trilinear form on the Lie algebra by the formula $([x,y],z)$. This form is antisymmetric and G-invariant. Therefore, it defines a closed 3-form τ^3 on G, which is invariant under left and right translations.

Now we construct the function $M\colon \widetilde{C}^\infty(X,G) \to \mathbb{R}$ as follows. Take an element $\widetilde{u} \in \widetilde{C}^\infty(X,G)$. It is determined by the element $u = p(\widetilde{u})$ and the homotopy class of paths in $C^\infty(X,G)$ from e to u. Choose a smooth path $s \mapsto \varphi(s)$, $0 \le s \le 1$, $\varphi(0) = e$, $\varphi(1) = u$, from this class. Define the map $\Phi\colon X \times [0,1] \to G$ by the formula $\Phi(x,s) = \varphi(s)(x)$, and consider the 3-form $\Phi^*\tau^3$ on $X \times [0,1]$. This form is represented as $a^3 + b^2 \wedge ds$; a^3, b^2 are the forms on X (of degrees 3 and 2 respectively) depending on the parameter s. Denote

$$\beta^2 = \int_0^1 b^2(s)\, ds. \tag{17.7}$$

Define the function $M\colon \widetilde{C}^\infty(X,G) \to \mathbb{R}$ by the formula

$$M(\widetilde{u}) = \int_X \beta^2 \wedge \omega^{n-2}, \tag{17.8}$$

It is clear that this formula can be also written in the form

$$M(\widetilde{u}) = \int_{X\times[0,1]} \Phi^*\tau^3 \wedge \pi^*\omega^{n-2}, \tag{17.9}$$

where $\pi\colon X \times [0,1]$ is the projection to X. Let us prove that this integral does not depend on the choice of the path $\varphi(s)$, $0 \le s \le 1$, defining the element $\widetilde{u} \in$

$\widetilde{C}^\infty(X,G)$. Take a smooth family of paths $s \mapsto \varphi_t(s)$, $0 \le s,t \le 1$, joining the points e and u. To the path φ_t there corresponds the mapping $\Phi_t\colon X \times [0,1] \to G$, $\Phi_t(x,s) = \varphi_t(s)(x)$. It is clear that $\Phi_t(x,0) = e$, $\Phi_t(x,1) = u(x)$ for all $t \in [0,1]$. For any point $(x,s,t) \in X \times [0,1] \times [0,1]$, introduce the vector $\xi_t(x,s) = (\partial/\partial t)\Phi_t(x,s)$. Introduce the form $\Phi_t^*(i_{\xi_t}\tau^3)$ on $X \times [0,1]$, defining its value at a pair of tangent vectors $\eta_1, \eta_2 \in T_{x,s}(X \times [0,1])$ by the formula

$$\Phi_t^*(i_{\xi_t}\tau^3)(\eta_1,\eta_2) = \tau^3((d\Phi_t)(\eta_1),(d\Phi_t)(\eta_2),\xi_t(x,s)).$$

Then

$$\begin{aligned}\frac{d}{dt}\int_{X\times[0,1]} \Phi_t\tau^3 \wedge \pi^*\omega^{n-2} &= \int_{X\times[0,1]} d\Phi_t^*(i_{\xi_t}\tau^3)\pi^*\omega^{n-2}\\ &= \int_X (\Phi_t^*(i_{\xi_t}\tau^3))\,|_{s=1} \wedge\omega^{n-2} - \int_X (\Phi_t^*(i_{\xi_t}\tau^3))\,|_{s=0} \wedge\omega^{n-2} = 0,\end{aligned}$$

because $\xi_t(x,0)=0$, $\xi_t(x,1)=0$. This proves that (17.9) does not depend on the choice of the path $s \mapsto \varphi(s)$ representing the element $\widetilde{u} \in \widetilde{C}^\infty(X,G)$. (Notice that in this computations we used the arguments from the book "Geometric asymptotics" by Guillemin and Sternberg ([**67**, Chapter VI, §1]).

THEOREM 17.1. *The functions* (17.5) *and* (17.8) *are related by formula* (17.6).

PROOF. To make the computations easier we assume that G is the matrix group and the inner product in $\mathfrak{g}$ has the form $\operatorname{Tr} AB$, where $A, B \in \mathfrak{g}$. If α, β are the exterior forms on some manifolds with values in the matrix algebra $\operatorname{Mat}(\mathbb{R})$, then the usual matrix multiplication and the exterior multiplication of real exterior forms allow us to define the product $\alpha \wedge \beta$. For any $u \in C^\infty(X,G)$ the following formulas are proved by simple computations:

$$\begin{aligned} u^*(\tau^3) &= \tfrac{1}{3}\operatorname{Tr}(\dot u \wedge \dot u \wedge \dot u) = \tfrac{1}{3}\operatorname{Tr}\left((\dot u)^{\wedge 3}\right),\\ d\dot u &= -\dot u \wedge \dot u = -(\dot u)^{\wedge 2},\\ \operatorname{Ad}_{u^{-1}} \dot u &= -(u^{-1})^{\cdot}. \end{aligned} \tag{17.10}$$

Here we use the notation $\alpha^{\wedge m} = \alpha \wedge \cdots \wedge \alpha$ (m times). Later formulas (17.10) will be applied for mappings $X \times [0,1] \to G$.

Passing to the proof of the theorem, let us take elements $\widetilde{u}_k \in \widetilde{C}^\infty(X,G)$, the corresponding paths $s \mapsto \varphi_k(s)$, $0 \le s \le 1$, $\varphi_k(0) = e$, $\varphi_k(1) = u_k$, and the mappings $\Phi_k\colon X \times [0,1] \to G$ of the form $\Phi_k(x,s) = \varphi_k(s)(x)$, $k = 1.2$. Let us prove formula (17.6). In the computations below we use formulas (17.10) and the invariance of the function $A \mapsto \operatorname{Tr} A$ under the transformations $A \mapsto BAB^{-1}$. We have

$$\begin{aligned} M(\widetilde u_1) + M(\widetilde u_2) - M(\widetilde u_1 \widetilde u_2) &= \operatorname{Tr}\int_{X\times[0,1]} (\Phi_1^*\tau^3 + \Phi_2^*\tau^3 - (\Phi_1\Phi_2)^*\tau^3) \wedge \pi^*\omega^{n-2}\\ &= \frac{1}{3}\operatorname{Tr}\int_{X\times[0,1]} \left((\dot\Phi_1)^{\wedge 3} + (\dot\Phi_2)^{\wedge 3} - ((\Phi_1\Phi_2)^{\cdot})^{\wedge 3}\right) \wedge \pi^*\omega^{n-2}\\ &= \frac{1}{3}\operatorname{Tr}\int_{X\times[0,1]} \left((\dot\Phi_1)^{\wedge 3} + (\dot\Phi_2)^{\wedge 3} - (\dot\Phi_1 + \operatorname{Ad}_{\Phi_1}\dot\Phi_2)^{\wedge 3}\right) \wedge \pi^*\omega^{n-2}\\ &= \operatorname{Tr}\int_{X\times[0,1]} \left((\Phi_1^{-1})^{\cdot} \wedge \dot\Phi_2 \wedge \dot\Phi_2 - (\Phi_1^{-1})^{\cdot} \wedge (\Phi_1^{-1})^{\cdot} \wedge \dot\Phi_2\right) \wedge \pi^*\omega^{n-2} \end{aligned}$$

$$= \int_{X\times[0,1]} \mathrm{Tr}((\Phi_1^{-1})^{\cdot} \wedge \dot{\Phi}_2) \wedge \pi^*\omega^{n-2}$$

$$= \int_X \mathrm{Tr}((\Phi_1^{-1})^{\cdot} \wedge \dot{\Phi}_2)\Big|_{s=1} \wedge \omega^{n-2} = \int_X \mathrm{Tr}((u_1^{-1})^{\cdot} \wedge \dot{u}_2) \wedge \omega^{n-2}.$$

Theorem 17.1 is proved.

Theorem 17.1 yields a usual (nonprojective) representation $\widetilde{T}$ of the universal covering group $\widetilde{C}^\infty(X,G)$; it can be expressed in terms of the projective representation T of the group $C^\infty(X,G)$ by the formula

$$\widetilde{T}(\widetilde{u}) = e^{-iM(\widetilde{u})}T(u), \qquad \widetilde{u} \in \widetilde{C}^\infty(X,G), \quad u = p(\widetilde{u}). \tag{17.11}$$

17.3. The restriction of the representation $\widetilde{T}$ to the group $\Gamma = \pi_1(C^\infty(X,G))$. The fundamental group Γ is embedded into $\widetilde{C}^\infty(X,G)$ as a central subgroup; let us agree to identify Γ with its image under this projection. Formula (17.11) shows that the restriction of the representation $\widetilde{T}$ to Γ has the form

$$\widetilde{T}(\gamma) = e^{-iM(\gamma)}, \qquad \gamma \in \Gamma.$$

The function $M(\gamma)$ can be easily expressed in terms of topological quantities. Namely, define the mapping $\theta\colon \Gamma \to H^2(X,\mathbb{R})$ with $\theta(\gamma)$ to be the cohomology class of the form β^2 defined by formula (17.7) above. Formula (17.8) shows that

$$M(\gamma) = \langle \theta(\gamma), [\omega^{n-2}] \rangle,$$

where the element $[\omega^{n-2}] \in H^{n-2}(X,\mathbb{R})$ corresponds to the form ω^{n-2} and $\langle \cdot, \cdot \rangle$ is the natural pairing (recall that X is compact and oriented).

Below we will need the group $\widetilde{C}^\infty(X,G)/\ker\theta = \widehat{C}^\infty(X,G)$. We have a covering (in general, not universal) $\widehat{C}^\infty(X,G) \to C^\infty(X,G)$, whose kernel is the additive group $\theta(\Gamma) \subset H^2(X,\mathbb{R})$. It is clear that under the appropriate choice of the form ω^{n-2} (more precisely, of the element $[\omega^{n-2}]$) we obtain an exact representation $\widehat{T}$ of the group $\widehat{C}^\infty(X,G)$.

Note that we might have $\ker\theta = \{1\}$. Here is the simplest example: $\dim X = 2$, $G = SU_2$. In this case we obtain an exact unitary representation of the universal covering group $\widetilde{C}^\infty(X,G)$.

17.4. The subrepresentation T^0. In general, the representation T of the group $C^\infty(X,G)$ constructed in 17.1 is reducible. Let us describe a subrepresentation T^0 of T (which might coincide with T). Consider the space of all $\mathfrak{g}_{\mathbb{C}}$-valued 1-forms of the form $\alpha \circ a + i\alpha \circ b$, where $\alpha \in E^1_{\mathbb{C}}$, and a, b are the fields of linear operators on T_xX introduced in 17.1. Let H^0 be the closure of this space in H. It is clear that H^0 is invariant under the adjoint representation $u \mapsto \mathrm{Ad}_u$, $u \in C^\infty(X,G)$, and the values of the cocycle (17.1) lie in H^0. Therefore, the subspace $\overline{S}(H^0) \subset \overline{S}(H)$ is invariant under T. Denote by T^0 the restriction of the representation T to $\overline{S}(H^0)$.

The irreducibility and pairwise nonequivalence of representations T^0 (for different r, ρ, a, b) is analyzed similarly to the case $a = I$, $b = 0$ studied in §16. Not going into details, we note only that T^0 is irreducible for $\dim X \geq 3$, $a = I$. If $\dim X = 2$, $a = I$, then it is irreducible for a sufficiently small function $\rho(x)$ (for fixed r, b).

If the operator $a(x) + ib(x)$ acting in the complexification of the tangent space T_xX is invertible for almost all $x \in X$, then $T = T^0$.

17.5. Example: the current group on a Riemann surface. Let X be a compact Riemann surface. Assume that the Riemannian metric on X is compatible with the complex structure (i.e., for any holomorphic local parameter $z = x + iy$, the vector fields $\partial/\partial x$ and $\partial/\partial y$ are orthogonal at any point). Define the operator fields $a(x)$, $b(x)$ from 1.2 as follows: $a(x) = I$ for all $x \in X$, and $b(x)$ is the operator of multiplication by $-i$ in the tangent space $T_x X$. Then the subspace $\{\alpha \circ a + i\alpha \circ b\}$ from 17.4 consists of all $\mathfrak{g}_{\mathbb{C}}$-valued form of type (1,0), i.e., of forms $A(z)\,dz$ (in local coordinates). Denote the completion of this space with respect to the Hilbert norm by $H(1,0)$. We obtain a representation (projective) of the group $C^\infty(X, G)$ in the space $\overline{S}(H(1,0))$. The conditions for irreducibility and pairwise nonequivalence of these representations are given in 17.4.

Concluding this section, we mention that in the case $\dim X = 2$ the 2-cocycle (17.2) is given by the formula

$$f(u_1, u_2) = c \int_X \left((u_1^{-2})^{\cdot} \wedge \dot{u}_2\right).$$

This cocycle is equivalent (for some c) to an integer (discontinuous) cocycle describing the universal extension of the current group. Similar formulas are known for semisimple Lie groups satisfying certain conditions (see [**10**, Chapter III, §7].

§18. Certain group extensions associated to dynamical systems and to canonical commutation relations (CCR); extensions of the current groups

In this section we consider noncentral (abelian) extensions and their representations. We begin with a general construction and then return to the current group.

18.1. Extensions generated by a dynamical system and CCR. Let (Ω, μ, X) be a dynamical system; here (Ω, μ) is a space with measure and a (topological) group G acts on Ω leaving the measure μ quasi-invariant. We assume that this action satisfies the necessary continuity conditions.

Further, let V be a real linear space with a skew-symmetric form τ, H^0 a complex Hilbert space, and $U(H^0)$ the group of all unitary operators in H^0. Assume that we have a representation of CCR in the Weyl form, i.e., a mapping $\Phi\colon V \mapsto U(H^0)$ satisfying the condition

$$\Phi(v_1 + v_2) = e^{i\tau(v_1, v_2)}\Phi(v_1)\Phi(v_2).$$

Finally, let there be given a cocycle $c\colon G \times \Omega \to V$. This mapping is measurable with respect to ω, continuous in g, and satisfies the condition

$$c(g_1 g_2, \omega) = c(g_1, \omega) + c(g_2, \omega g_1).$$

Introduce the function

$$f(g_1, g_2, \omega) = \tfrac{1}{2}\tau\left(c(g_1, \omega), c(g_2, \omega g_1)\right)$$

and denote by A a linear space of functions on Ω containing all functions of the form $\omega \mapsto f(g_1, g_2, \omega)$ and invariant under the action of the group G by the formula $(g\varphi)(\omega) = \varphi(\omega g)$. Then the set $A \times G$ becomes a group with the multiplication given by

$$(a_1, g_1)(a_2, g_2) = (a_1 + g_1 a_2 + f(g_1, g_2, \cdot), g_1 g_2).$$

Denote this group by $A \overset{f}{\times} G$. Take the Hilbert space of vector-valued functions $L_2(\Omega, \mu, H^0)$ and define the operators

$$(T(a,g)\varphi) = \sqrt{\frac{d\mu(\omega g)}{d\mu(\omega)}}\,\Phi(c(g,\omega))e^{ia(\omega)}\varphi(\omega g).$$

An easy computation shows that these operators form a unitary representation of the group $A \overset{f}{\times} G$.

18.2. Extensions of current groups and their representations. Let us apply the above construction to the current group. Consider the space Ω consisting of all continuous linear functionals on E^1. The inner product (14.1) generated, in a standard way, a Gaussian measure μ on Ω. We have a natural embedding $E^1 \hookrightarrow \Omega$. The group $C^\infty(X, G)$ acts on Ω by the formula

$$\omega u = (\mathrm{Ad}_u)^* \omega + \dot{u}, \qquad \omega \in \Omega, \quad u \in C^\infty(X, G).$$

We obtain the dynamical system $(\Omega, \mu, C^\infty(X, G))$. Instead of the group $C^\infty(X, G)$ we could have taken one of its covering groups; later we will use this possibility.

As before, take a finite-dimensional vector space V with the antisymmetric form τ, a Hilbert space H^0, and a representation of CCR of the form $\Phi\colon V \to U(H^0)$. Choose a closed V-valued $(n-2)$-form φ^{n-2} on X and define the mapping $t\colon E^1 \times E^1 \to V$ by the formula

$$t(\alpha_1, \alpha_2) = \int_X (\alpha_1 \wedge \alpha_2) \wedge \varphi^{n-2}, \qquad \alpha_1, \alpha_2 \in E^1.$$

Define the mapping $s\colon G \times \Omega \to V$ by the formula

$$s(u, \omega) = t(\dot{u}, \omega) = \int_X (\dot{u} \wedge \omega) \wedge \varphi^{n-2} \tag{18.1}$$

(the form $(\dot{u} \wedge \omega) \wedge \varphi^{n-2}$ is defined in a natural way). One can easily verify that

$$s(u_1 u_2, \omega) = s(u_1, \omega) + s(u_2, \omega u_1) + t\left((u_1^{-1})^{\cdot}, \dot{u}_2\right). \tag{18.2}$$

We see that the function (18.2) is not a cocycle on $C^\infty(X, G)$, with the summand $t((u_1^{-1})^{\cdot}), \dot{u}_2)$ in (18.2) being an obstacle. To correct the situation, take the covering $\widehat{C}^\infty(X, G) \to C^\infty(X, G)$ from 17.3 and lift the function (18.1) to $\widehat{C}^\infty(X, G)$. Repeating the arguments from 17.3, we see that the cocycle $(\widehat{u}_1, \widehat{u}_2) \mapsto t((u_1^{-1})^{\cdot} \dot{u}_2)$, $u_i = p_0(\widehat{u}_i)$, $\widehat{u}_i \in \widehat{C}^\infty(X, G)$, is the coboundary of the function $M\colon \widehat{C}^\infty(X, G) \to V$ given by the formula

$$M(\widehat{u}) = \int_X \beta^2 \wedge \varphi^{n-2}.$$

The form β^2 is defined by formula (17.7). We obtain the cocycle $\widehat{c}\colon \widehat{C}^\infty(X, G) \times \Omega \to V$ given by

$$\widehat{c}(\widehat{u}, \omega) = s(u, \omega) - M(\widehat{u}), \qquad \widehat{u} \in \widehat{C}^\infty(X, G), \quad u = p_0(\widehat{u}).$$

Now we apply the construction from 18.1, taking for A the space of all functions $\psi\colon \Omega \to \mathbb{R}$ of the form

$$\omega \mapsto c_0 + \langle \alpha_0, \omega \rangle + (K_0 \omega, \omega),$$

where $\alpha_0 \in E^1$, K_0 is an integral operator in E^1 with a smooth kernel, the quadratic function $\omega \mapsto (K_0\omega, \omega)$ is defined as in 1.3. It is important to note that the space A is naturally isomorphic to the space A_0 consisting of all functions on E^1 of the form

$$\alpha \mapsto c_0 + (\alpha_0, \alpha) + (K_0\alpha, \alpha), \qquad \alpha \in E^1 = E^1(X, \mathfrak{g}).$$

We obtain the group $A_0 \overset{f}{\times} \widehat{C}^\infty(X,G)$ and a unitary representation T of this group. Denoting by $(a, \widehat{u})$ an element of the group $A_0 \overset{f}{\times} \widehat{C}^\infty(X,G)$ (here $a \in A_0$, $\widehat{u} \in \widehat{C}^\infty(X,G)$), we obtain the following formula for the representation T:

$$(T(a,\widehat{u})f)(\omega) = \sqrt{\frac{d\mu(\omega u)}{d\mu(\omega)}}\,\Phi(t(\dot{u},\omega) + M(\widehat{u}))e^{ia(\omega)}f(\omega u), \qquad u = p_0(\widehat{u}). \tag{18.3}$$

Let us consider irreducibility conditions for the representation T. Recall that we have constructed the mapping $\theta\colon \Gamma \to H^2(X,\mathbb{R})$. Denote by $[\varphi^{n-2}] \in H^{n-2}(X,V)$ the cohomology class of the form φ^{n-2}. Using the pairing $H^2(X,\mathbb{R}) \times H^{n-2}(X,V) \to V$ obtained by integrating the exterior product of forms, we obtain the mapping $b\colon \Gamma \to V$ given by the formula $b(\gamma) = \langle \theta(\gamma), [\varphi^{n-2}]\rangle$.

THEOREM 18.1. *Let the restriction of the representation $v \mapsto \Phi(v)$ to the subgroup $b(\Gamma)$ be irreducible. Then T is irreducible.*

PROOF. Let N be a bounded operator in the space $L_2(\Omega, \mu, H^0) = H$ commuting with the operators of the representation T. Since N commutes with the operators of the form $T(a,e)$, $a \in A_0$, it follows that $(N\psi)(\omega) = N(\omega)\psi(\omega)$, where $\{N(\omega)\}$ is a weakly-measurable family of unitary operators in H^0. The operators $T(0,\gamma)$, $\gamma \in \ker p_0$, are given by the formula

$$f(\omega) \mapsto \Phi(b(\gamma_1))f(\omega), \qquad \gamma_1 \in \Gamma. \tag{18.4}$$

Since the operator N commutes with the operators (18.4) and the family $\{\Phi(b(\gamma)), \gamma \in \Gamma\}$ is irreducible, we see that $N(\omega) = \lambda(\omega)I$, where $\lambda(\omega)$ is a measurable function. Finally, the ergodicity of the action $\omega \mapsto \omega u$ implies that $\lambda(\omega) = \lambda_0 =$ const for $\omega \in \Omega \bmod 0$. Hence, $N = \lambda_0 I$. Theorem 18.1 is proved.

In conclusion, let us consider the Lie algebra of the group $A_0 \overset{f}{\times} \widehat{C}^\infty(X,G)$. This algebra is an extension of the algebra $C^\infty(X,G)$ by the commutative Lie algebra A. The corresponding 2-cocycle is given by the formula

$$c(\xi_1,\xi_2)(\alpha) = \int_{X\times X} \pi_1^*(d\xi_1 \wedge \alpha) \wedge \pi_2^*(d\xi_2 \wedge \alpha) \wedge \Phi^{2n-4},$$
$$\xi_i \in \overline{\mathfrak{g}} = C^\infty(X,\mathfrak{g}), \quad \alpha \in E^1 = E^1(X,\mathfrak{g});$$

here $\pi_k\colon X \times X \to X$ is the kth projection, Φ^{2n-4} is a closed $(2n-4)$-form on $X \times X$ satisfying the condition $\sigma^*\Phi^{2n-4} = \Phi^{2n-4}$, where σ is the flip transformation $\sigma(x_1,x_2) = (x_2,x_1)$, $(x_1,x_2) \in X \times X$. Let us recall also that the 2-forms $d\xi_i \wedge \alpha$ are defined using the exterior product of real forms on X and the inner product in $\mathfrak{g}$. We do not give explicit formulas for the corresponding group 2-cocycle.

§19. Representations of Mickelsson-Rajeev extensions

In this sections we assume the manifold X to be compact.

19.1. The embedding $C^\infty(X, GL_n) \hookrightarrow GL_p(H)$, $p > \dim X$. The group $GL_p(H)$ corresponding to the decomposition $H = H_+ \oplus H_-$ was defined in 2.3. We will need the following realization of this group. Let $E \to X$ be an Hermitian finite-dimensional complex vector bundle and A_0 a pseudodifferential operator acting in the space of its sections and satisfying the following conditions: (1) the principal symbol $a(x, \xi)$ of A_0 is linear in ξ; (2) the closure A of the operator A_0 in the space of L_2-sections is selfadjoint; (3) the operator A admits a bounded inverse operator. Consider the Hilbert spaces H of L_2-sections of the bundle $E \otimes \mathbb{C}^n$. The group $C^\infty(X, GL_n)$ acts in this space, giving an embedding $C^\infty(X, GL_n) \to GL(H)$. The operator A generates, in an obvious way, an operator $A_1 \colon H \to H$. Denote $J = A_1(A_1^2)^{-1/2}$. Then $J^2 = I$ and $J = J^*$, yielding the decomposition $H = H_+ \oplus H_-$, where $H_\pm = \{h \mid h \in H, Jh = \pm h\}$. We define the group $GL_p(H)$ using this decomposition.

The main fact we need is that under the above embedding $C^\infty(X, GL_n) \hookrightarrow GL(H)$ the subgroup $C^\infty(X, GL_n)$ is a subgroup in $GL_p(H)$ for $p > \dim X$. The outline of the proof is contained in the [**33**]; the complete proof is given in [**60**].

For A_0 we can take the Dirac operator acting in the space of sections of the spinor bundle on X (assuming that X is supplied with a spinor structure); to obtain an invertible operator, it suffices to add to A_0 a finite-dimensional summand.

19.2. Mickelsson–Rajeev extensions. In §2 we have defined an extension of the group $GL_p(H)$ by the abelian group $\mathcal{H}^*(\overline{X})$ of holomorphic functions $f \colon \overline{X} \to \mathbb{C}^*$, $\overline{X} = (GL(H_+) \times GL(H_-)) \setminus GL_p(H)$. The natural embedding $\mathcal{H}^*(\overline{X}) \to \mathcal{H}^*(GL_p(H))$ yields an extension of the group $GL_p(H)$ by $\mathcal{H}^*(GL_p(H))$.

Using the above embedding $C^\infty(X, GL_n) \hookrightarrow GL_p(H)$, we obtain an extension of the group $\overline{G} = C^\infty(X, GL_n)$ by the group $A^* = C^\infty(\overline{G}, \mathbb{C})$. To this extension corresponds the extension of the Lie algebra $\overline{\mathfrak{g}} = C^\infty(X, \mathfrak{gl}_n)$ by the abelian Lie algebra $C^\infty(\overline{G}, \mathbb{C})$. Using formulas from §2, we can find the 2-cocycle defining this extension. The formula for this cocycle is quite complicated. However, in the case $\dim X = 3$, $p = 2$, this cocycle is cohomologous to a very simple cocycle $(\xi, \eta) \mapsto c_0(\xi, \eta)$ of the form

$$c_0(\xi, \eta)(u) = \lambda_0 \int_x \operatorname{Tr}(d\xi \wedge d\eta - d\eta \wedge d\xi) \wedge \dot{u}, \qquad u \in \overline{G}, \quad \lambda_0 = \text{const} \neq 0. \tag{19.1}$$

We give the proof of this statement for the torus $\mathbb{T}^3 = \mathbb{R}^3/\mathbb{Z}^3$. In this case $H = L_2(\mathbb{T}^3, \nu) \otimes \mathbb{C}^2 \otimes \mathbb{C}^n$, where ν is the Lebesgue measure on $\mathbb{T}^3$. The operator J constructed by the prescription from 19.1, is given by the formula

$$J = \sum_{j=1}^{3} D_j \otimes e_j \otimes I, \qquad e_1 = \begin{pmatrix} 1 & 0 \\ 0 & -1 \end{pmatrix}, \quad e_2 = \begin{pmatrix} 0 & 1 \\ 1 & 0 \end{pmatrix}, \quad e_3 = \begin{pmatrix} 0 & i \\ -i & 0 \end{pmatrix}.$$

Here D_j is the operator in $L_2(\mathbb{T}^3, \nu)$ acting on the functions $f_k(x) = \exp(2\pi i(k, x))$, $k \in \mathbb{Z}^n$, $x \in \mathbb{T}^3$, by the formula

$$D_j f_k = d_j(k) f_k, \quad j = 1, 2, 3, \quad d_j(k) = \begin{cases} k_j/|k|, & |k| \neq 0, \\ 1/\sqrt{3}, & |k| = 0. \end{cases}$$

The group $\overline{G}$ acts in $L_2(\mathbb{T}^3,\nu)\otimes\mathbb{C}^3$, hence in H, in an obvious way. The cocycle is given (up to a constant factor) by the formula

$$c(\xi,\eta)=\operatorname{Tr}[[J,\xi],[J,\eta]]\left(P_+u^{-1}[J,u]P_+ + P_-u^{-1}[J,u]P_-\right),\qquad P_\pm=(1/2)(I\pm J).$$

In our further computations we use the obvious equality $\operatorname{Tr} e_ie_je_k=-2i\delta_{ijk}$, where δ_{ijk} is the sign of the permutation (ijk) if i,j,k are pairwise distinct, $\delta_{ijk}=0$ if some of the indices i,j,k coincide. Hence,

$$c(\xi,\eta)=-2i\operatorname{Alt}(D_1\xi D_2\eta)-D_1\eta D_2\xi+\xi D_1\eta D_2-\eta D_1\xi D_2-D_1[\xi,\eta]D_2)u^{-1}[D_3,u],$$

where $u\in\overline{G}$ and Alt is the alternation with respect to the indices 1,2,3. For an operator K acting in the space $L_2(\mathbb{T}^3,\nu,\mathbb{C}^n)$, denote

$$\operatorname{Tr}_0 K=\lim_{R\to\infty}\sum_{|k|<R}\operatorname{Tr}P_kKP_k, \tag{19.2}$$

where P_k is the projection to the subspace generated by the vector-valued functions $\exp(2\pi i(k,x))\cdot v$, $v\in\mathbb{C}^n$, $|k|=\sqrt{(k,k)}$ (of course, the limit (19.2) does not necessarily exist). Then

$$\begin{aligned}&c(\xi,\eta)(u)\\&=2i\operatorname{Alt}\operatorname{Tr}_0((D_1\xi D_2\eta)-D_1\eta D_2\xi+\xi D_1\eta D_2-\eta D_1\xi D_2-D_1[\xi,\eta]D_2)u^{-1}D_3u.\end{aligned}\tag{19.3}$$

Introduce the function

$$f(\xi)(u)=2i\operatorname{Alt}\operatorname{Tr}_0(D_1\xi D_2u^{-1}D_3u),\qquad u\in\overline{G}. \tag{19.4}$$

An easy computation shows that the limit in (19.3) always exists. Let us pass from the cocycle (19.2) to an equivalent cocycle $c+\delta f=c_0$. It is given by the formula

$$\begin{aligned}c_0(\xi,\eta)(u)=-2i\operatorname{Alt}\operatorname{Tr}_0(&\xi D_1\eta D_2u^{-1}D_3u-\eta D_1\xi D_2u^{-1}D_3u\\&-D_1\eta D_2u^{-1}D_3u\xi+D_1\xi D_2u^{-1}D_3u\eta).\end{aligned}\tag{19.5}$$

Expand the functions ξ,η,u^{-1},u in Fourier series with respect to the system f_k, $k\in\mathbb{Z}^3$:

$$\xi=\sum\widehat{\xi}(k)f_k,\quad \eta=\sum\widehat{\eta}(k)f_k,\quad u^{-1}=\sum\widehat{u^{-1}}(k)f_k,\quad u=\sum\widehat{u}(k)f_k.$$

Using (19.1), we obtain after tedious but simple computations the following formula:

$$c_0(\xi,\eta)(u)=2i\sum\operatorname{Tr}(\widehat{\xi}(p)\widehat{\eta}(q)-\widehat{\eta}(q)\widehat{\xi}(p))\widehat{u^{-1}}(r)\widehat{u}(-p-q-r)\psi(p,q,r),$$

where

$$\begin{aligned}\psi(p,q,r)=\lim_{R\to\infty}\operatorname{Alt}\Bigg(&\sum_{\substack{|a-p|\le R\\|a|>R}}d_1(a)d_2(a-q)d_3(a-q-r)\\&-\sum_{\substack{|a|\le R\\|a-p|>R}}d_1(a)d_2(a-q)d_3(a-q-r)\Bigg).\end{aligned}$$

It is clear that the last limit depends only on the asymptotic behavior of the function $a \mapsto d_1(a)d_2(a-q)d_3(a-q-r)$ as $|a| \to \infty$, which is computed in an obvious way. As a result, we get the formula

$$\psi(p,q,r) = \frac{4}{3}\pi \det(p,q,r),$$

where $\det(p,q,r)$ is the determinant of the matrix formed by the rows p,q,r. This gives (19.1).

Return to the group $C^\infty(X,G)$, where G is a compact semisimple Lie group; we assume that $G \subset \mathrm{Mat}_k(\mathbb{C})$ for some k. The above arguments yield an extension of the group $C^\infty(X,G)$ by the group $C^\infty(C^\infty(X,G),\mathbb{C}^*)$. The construction from §2 gives a linear representation of this extension.

It is not known whether or not this extension (the Mickelsson–Rajeev extension) has a locally exact unitary representation.

CHAPTER 4

Certain Results on Diffeomorphism Groups

This chapter contains results about diffeomorphism groups that will be used to construct the representations of these groups; this determines also the selection of results presented here. Most of these results are either well known or can be obtained using some standard technique; they are included for completeness. Exceptions are §21 (the theorem about adjoint elements) and §25, which contains an example of an infinite-dimensional symplectic manifold and the corresponding "geometric quantization."

§20. Diffeomorphism groups

When speaking about smooth manifolds, mappings, etc., we always mean C^∞ smoothness. The number of components of a non-connected manifold is assumed to be finite or countable.

20.1. The group $\mathrm{Diff}(X)$ **and its subgroups.** Let X be a smooth manifold without boundary, $n = \dim X < \infty$. Let us introduce the following notation: $\mathrm{Diff}\, X$ is the group of all diffeomorphisms of X, xg is the image of the point x under the action of the diffeomorphism g, e is the identity diffeomorphism. The group $\mathrm{Diff}(X)$ is endowed with the usual compact-open C^∞ topology. The base of neighborhoods of the unit element e in this topology consists of sets of the following form. Take a set $A \subset X$ diffeomorphic to $\mathbb{R}^n$ (so that in A we have local coordinates $x_1, \dots, x_n$), a compact subset $K \subset A$, and numbers $N > 0$, $\varepsilon > 0$. The corresponding neighborhood of e in $\mathrm{Diff}(X)$ consists of all $g \in \mathrm{Diff}(X)$ such that

$$Kg \subset A, \qquad \left| \left(\frac{\partial}{\partial x_1}\right)^{\alpha_1} \cdots \left(\frac{\partial}{\partial x_n}\right)^{\alpha_n} ((xg)_k - x_k) \right| < \varepsilon$$
$$\text{for all } \alpha_i > 0, \quad \alpha_1 + \cdots + \alpha_n \le N, \qquad x \in K.$$

Let us recall the definition of classical infinite-dimensional subgroups in $\mathrm{Diff}(X)$. By $\mathrm{Diff}(X, v^n)$ we denote the group of diffeomorphisms preserving a given volume form v^n on X. If $n = \dim X$ is even and ω^2 is a closed everywhere nondegenerate 2-form on X, then we have a similar group $\mathrm{Diff}(X, \omega^2)$. The pair (X, ω^2) is called a symplectic manifold, and elements of the group $\mathrm{Diff}(X, \omega^2)$ are called symplectomorphisms. Finally, let $\dim X = 2p + 1$, and let α^1 be an 1-form on X such that the form $\alpha^1 \wedge (d\alpha^1)^p$ is nondegenerate everywhere on X. Then by $\mathrm{Diff}(X, \alpha^1)$ we denote the group of all diffeomorphisms $x \mapsto xg$ such that $g\alpha^1 = f \cdot \alpha^1$, where f is a positive smooth function. (Here $g\alpha$ is the action of the transformation $x \mapsto xg$ on a differential form α.) The pair (X, α^1) is called a contact manifold, and an element $g \in \mathrm{Diff}(X, \alpha^1)$ is said to preserve the contact structure on X.

Let Y be a smooth manifold. A mapping $f\colon Y \to \mathrm{Diff}(X)$ is said to be smooth if the mapping $(x, y) \mapsto xf(y)$ from $X \times Y$ to X is smooth.

20.2. The group $D(X)$ and its subgroups. The support of a diffeomorphism g is the set $\operatorname{supp} g = \overline{\{x \mid x \neq xg\}}$, where the bar denotes the closure of a set. By $D(X)$ we denote the connected component of the identity element in the group of all diffeomorphisms with compact support. More precisely, $D(X)$ consists of diffeomorphisms g satisfying the following condition: there exists a family $\{g_t,\ 0 \leq t \leq 1\} \subset \mathrm{Diff}(X)$, continuous in t, such that $g_0 = e$, $g_1 = g$, and the supports of all g_t belong to a compact set independent of t. Similarly one defines the groups $D(X, v^n)$, $D(X, \omega^2)$, $D(X, \alpha^1)$ as the connected component of the identity in the group of compactly supported diffeomorphisms that preserve (respectively) the volume form v^n, the symplectic form ω^2, or, finally, the contact structure determined by the form α^1. (For example, in the definition of $D(X, v^n)$ we must require that the above family $\{g_t\}$ satisfies the condition $g_t v^n = v^n$ for all $t \in [0, 1]$.)

It is important to note that the family $\{g_t\}$ in the above definitions can be chosen to be smooth in t. One can easily prove this for $D(X)$, but for the other three classes the proof is not as simple. Without entering into details, let us outline the proof for, say, the group $D(X, v^n)$ with X compact. The form v^n determines the orientation on X. Denote by $M(X)$ the space of all positive (with respect to this orientation) volume forms on X with the same integral over X as for the form v^n. Then the bundle $D(X) \to M(X)$ given by $g \mapsto gv^n$ admits a smooth section; this follows from the results of Moser [**61**]. Therefore, there exists a smooth retraction $D(X) \to D(X, v^n)$. This implies the above results about the choice of a smooth family $\{g_t\}$.

Usually the continuity of mappings from the group $D(X)$ (and its subgroups) will be defined not with respect to the above topology, but with respect to the following notion of convergence of sequences from $D(X)$. We write $g_k \to g$ as $k \to \infty$ if the supports of all g_k belong to a compact set independent of k and $xg_k \to xg$ together with all partial derivatives in any local coordinate system. Of course, if X is compact, there is no need to distinguish the convergence and the topology on X.

For any open $Y \subset X$ the groups $D(Y)$, $D(Y, v^n)$, $D(Y, \omega^2)$, $D(Y, \alpha^1)$ can be naturally embedded into $D(X)$, etc. (a diffeomorphism $g \in D(Y)$ is extended by the identity to $X \setminus Y$).

20.3. Lie algebras. The Lie algebra of a group of diffeomorphisms consists, by definition, of vector fields of the form $(d/dt)g(t)\mid_{t=0}$, where $t \mapsto g(t)$, $t \in (-1, 1)$, is a smooth mapping into the above group, $g(0) = e$. The Lie algebras of the groups $\mathrm{Diff}(X)$, $\mathrm{Diff}(X, v^n)$, $\mathrm{Diff}(X, \omega^2)$, $\mathrm{Diff}(X, \alpha^1)$ are denoted by $\mathrm{Vect}(X)$, $\mathrm{Vect}(X, v^n)$, $\mathrm{Vect}(X, \omega^2)$, $\mathrm{Vect}(X, \alpha^1)$. The first of these Lie algebras consists of all vector fields on X, and the other three are specified inside $\mathrm{Vect}(X)$ by the conditions $L_\xi v^n = 0$, $L_\xi \omega^2 = 0$, $L_\xi \alpha^1 = f\alpha^1$, where L_ξ is the derivative along the vector field ξ, $f \in C^\infty(X)$.

The Lie algebras of the groups $D(X)$, $D(X, v^n)$, $D(X, \omega^2)$, $D(X, \alpha^1)$ are denoted by $V(X)$, $V(X, v^n)$, $V(X, \omega^2)$, $V(X, \alpha^1)$. They consist of compactly supported vector fields from the Lie algebras $\mathrm{Vect}(X)$, etc. The notation $\xi_k \to \xi$, where ξ_k, ξ belong to one of these algebras, means that all ξ_k are supported on a compact set independent of k and ξ_k converges to ξ uniformly together with all partial derivatives (in local coordinates) on any compact set in X.

Let us describe the relation of these Lie algebras to differential forms on X. Let $E^k(X)$ be the space of real k-forms on X, d the exterior derivative, $Z^k(X)$ the subspace of closed forms, $B^k(X)$ the subspace of exact forms. The spaces of compactly supported forms are denoted by E_c^k, Z_c^k, B_c^k. For a vector field ξ, by i_ξ we denote the inner multiplication with differential forms.

Turning back to Lie algebras of vector fields, we see that the mapping $\xi \mapsto i_\xi v^n$ establishes isomorphisms of linear spaces

$$\text{Vect}(X, v^n) \simeq Z^{n-1}(X), \qquad V(X, v^n) \simeq Z_c^{n-1}(X); \tag{20.1}$$

the mapping $\xi \mapsto i_\xi \omega^2$ establishes isomorphisms of linear spaces

$$\text{Vect}(X, \omega^2) \simeq Z^1(X), \qquad V(X, \omega^2) \simeq Z_c^1(X); \tag{20.2}$$

finally, $\xi \mapsto i_\xi \alpha^1$ yields isomorphisms of linear spaces

$$\text{Vect}(X, \alpha^1) \simeq C^\infty(X), \qquad V(X, \alpha^1) \simeq C_0^\infty(X). \tag{20.3}$$

20.4. Groups associated with the universal covering. Let $p\colon \widetilde{X} \to X$ be the universal covering space, Γ the group of deck transformations of $\widetilde{X}$. The action of $\gamma \in \Gamma$ on $\widetilde{X}$ is denoted by $\gamma\xi$, $\xi \in \widetilde{X}$. Any diffeomorphism $\xi \mapsto \xi\sigma$, $\xi \in \widetilde{X}$, commuting with Γ necessarily maps any fiber $p^{-1}(x)$, $x \in X$, onto a fiber $p^{-1}(x')$, $x' \in X$, and hence generates a diffeomorphism $x \mapsto xg = x'$.

For any connected group $G = G(X) \subset \text{Diff}(X)$ denote by $\widehat{G} = \widehat{G}(X)$ the connected component of the identity in the group of those diffeomorphisms of the manifold $\widehat{X}$ that commute with Γ and generate on X one of the transformations from the group G. We obtain a surjective homomorphism $p_1\colon \widehat{G} \to G$. Denote $\Gamma_0 = \Gamma_0^G = \Gamma \cap \widehat{G}$. Then Γ_0 is a central subgroup in both $\widehat{G}$ and Γ, and $G = \widehat{G}/\Gamma_0$. In particular, we obtain the groups $\widehat{D}(X)$, $\widehat{D}(X, v^n)$, $\widehat{D}(X, \omega^2)$, $\widehat{D}(X, \alpha^1)$, and the corresponding groups $\Gamma_0 = \Gamma_0^{D(X)}$, etc. If the manifold X is not compact, then the group Γ_0 is trivial for any connected subgroup $G \subset D(X)$, so that $\widehat{G} \simeq G$.

In the remaining part of this chapter we consider in more details each of the groups $D(X)$, $D(X, v^n)$, $D(X, \omega^2)$.

§21. The group $D(X)$; theorem on adjoint elements

The main result of this section is Theorem 21.1, which is essential in the study of representations of the group $D(X)$. In 21.1 we present simple properties of this group. Everywhere in this section X denotes a connected C^∞ manifold without boundary.

21.1. Product decomposition of elements of the group $D(X)$. Let Y, Z be open subsets of X such that $\overline{Y} \subset Z$, $Z \simeq \mathbb{R}^n$.

PROPOSITION 21.1. *There exist a neighborhood of unity $U \subset D(X)$ (in the compact-open C^∞ topology) and continuous mappings $\varphi\colon U \to D(Z)$, $\psi\colon U \to D(X \setminus \overline{Y})$ such that (1) $g = \psi(g)\varphi(g)$ for all $g \in U$, (2) if $xg = x$ for some $x \in X$, $g \in U$, then $x\varphi(g) = x\psi(g) = x$.*

PROOF. Let us choose a diffeomorphism $Z \simeq \mathbb{R}^n$. Using this diffeomorphism, transfer the linear space structure from $\mathbb{R}^n$ to Z. Take a region $K \simeq \mathbb{R}^n$ such that $\overline{K} \subset Z$, $\overline{Y} \subset K$. Let $w \in C_0^\infty(K)$ be such that $w(x) = 1$ for $x \in Y$. To a diffeomorphism $g \in D(X)$ such that $\overline{Kg} \subset Z$ we associate the transformation $\varphi(g)\colon X \to X$ defined by the formulas $x\varphi(g) = x + w(x)(xg - x)$ for $x \in K$, $x\varphi(g) = x$ for $x \notin K$. If g is sufficiently close to e on $\overline{K}$, then $\varphi(g)$ is a diffeomorphism. Denote $\psi(g) = g(\varphi(g))^{-1}$. Then $(\varphi(g), \psi(g))$ is the desired pair.

To formulate the next proposition, let us choose a locally finite covering $X = \bigcup Y_i$, $Y_i \simeq \mathbb{R}^n$.

PROPOSITION 21.2. *Any element $g \in D(X)$ can be represented in the form $g = g_1 \dots g_m$, where each factor g_k belongs to one of the subgroups $D(Y_i)$.*

The proof is obtained by consecutive applications of Proposition 21.1 to elements $D(X)$ sufficiently close to the unity $e \in D(X)$.

21.2. Theorem about adjoint elements: formulation. Let $Y \subset X$ be an open set with a finite number of connected components, and $D(Y) \subset D(X)$ the corresponding subgroup.

DEFINITION 21.1. An element $g \in D(X)$ is said to be adjoint to $D(Y)$ if it can be represented in the form

$$g = \lim_{k\to\infty} u_k w_k v_k,$$

where $u_k \in D(Y)$, $v_k \in D(Y)$, $\lim_{k\to\infty} w_k = e$.

The set of all elements adjoint to $D(Y)$ is denoted by $D(X;Y)$. Clearly, $D(Y) \subset D(X;Y)$. The next theorem shows that sometimes the set $D(X;Y)$ can be unexpectedly large.

THEOREM 21.1. *Let L be a submanifold of dimension $n-1$ (possibly, nonconnected) in X. Then the set $D(X; X \setminus L)$ generates the entire group $D(X)$.*

Let us consider a special case when X is the ball

$$B^n = \{x = (x_1, \dots, x_n) \in \mathbb{R}^n \mid x_1^2 + \dots + x_n^2 < 1\}$$

and L is the $(n-1)$-dimensional ball

$$B^{n-1} = \{x = (x_1, \dots, x_{n-1}, 0) \mid x_1^2 + \dots + x_{n-1}^2 < 1\}.$$

In this case the following stronger result holds.

THEOREM 21.1(a). *Any element $g \in D(B^n)$ can be represented in the form*

$$g = uvu^{-1}, \qquad u \in D(B^n; B^n \setminus B^{n-1}), \quad v \in D(B^n \setminus B^{n-1}). \tag{21.1}$$

21.3. Proofs. Choose a number $\delta \in (0,1)$ and a function $p \in C_0^\infty(-1,1)$ satisfying the following conditions:

$$\begin{gathered} p(s) = 0 \quad \text{for all } s \in (-1,-\delta) \cup (\delta, 1), \\ p(s) = p_0 = \text{const} \quad \text{for all } s \in (-\delta/2, \delta/2), \\ 0 < p(s) < \sqrt{\delta^2 - s^2} \quad \text{for all } s \in (-\delta,\delta), \\ |p^{(m)}(s)| \le A(m,\varepsilon)(p(s))^\varepsilon \quad \text{for } 0 < m \in \mathbb{Z},\ \varepsilon \in (0,1), \end{gathered} \tag{21.2}$$

where $A(m,\varepsilon)$ does not depend on s. A point $x = (\xi_1,\dots,\xi_n) \in \mathbb{R}^n$ will be denoted by $x = (x_{n-1},\xi_n)$, where $x_{n-1} = (\xi_1,\dots,\xi_{n-1}) \in \mathbb{R}^{n-1}$. Let us introduce the sets

$$\begin{aligned} K &= \{x = (x_{n-1},\xi_n) \in B^n \mid -p(|x_{n-1}|) \le \xi_n \le 0\},\\ K_- &= \{x = (x_{n-1},\xi_n) \in B^n \mid \xi_n \le -p(|x_{n-1}|)\},\\ K_+ &= \{x = (x_{n-1},\xi_n) \in B^n \mid \xi_n \ge 0\}, \end{aligned}$$

where $|\cdot|$ denotes the length of the vector.

LEMMA 21.1. *There exists a sequence of functions $\varphi_k \in C^\infty(\mathbb{R}^n)$ with the following properties*: (1) $0 \le \varphi_k(x) \le 1$ *for all* $x \in B^n$; (2) $\varphi_k(x) = 1$ *in some neighborhood of the set* $K_+ \cup K_-$ *depending on* k; (3) *the sequence of functions* $\psi_k \in C_0^\infty(B^{n-1})$ *defined by the formula*

$$\psi_k(x_{n-1}) = \int_{-p(|x_{n-1}|)}^{0} \varphi_k(x_{n-1},\xi_n)\,d\xi_n$$

tends to 0 *as* $k \to \infty$ *in the topology of the space* $C^\infty(B^{n-1})$.

Here by a neighborhood of a set $E \subset \mathbb{R}^n$ we mean an open set containing the closure $\overline{E}$.

PROOF. Let $h(s)$, $s \in \mathbb{R}$, satisfy the conditions $h \in C^\infty(\mathbb{R})$, $h(s) \ge 0$ for all s, $h(s) = 1$ for $s < 1$, $h(s) = 0$ for $s > 2$. Denote

$$\begin{aligned} \varphi_k(x) &= h(-k^2\xi_n(\xi_n + p(|x_{n-1}|))), \qquad x = (x_{n-1},\xi_n) \in B^n, \quad k = 1,2,\dots,\\ \varphi_k(x) &= 1, \qquad x \notin B^n. \end{aligned}$$

Let us prove that they satisfy the conditions of the lemma. Conditions (1) and (2) are obviously satisfied. To verify (3), we remark that it is easy to see that

$$\psi_k(x_{n-1}) = k^{-1}U(kp(|x_{n-1}|)),$$

where

$$U(t) = \frac{1}{t}\int_0^{t^2/4} h(z)\left(\frac{1}{4} - \frac{z}{t^2}\right)^{-1/2} dz. \tag{21.3}$$

From the properties of the function $p(s)$ it follows that all functions ψ_k are constant in a fixed (independent of k) neighborhood of the point $(0,\dots,0) \in B^{n-1}$. Therefore, it remains to prove that

$$\left(\frac{d}{ds}\right)^N \left(\frac{1}{k}U(kp(s))\right) \to 0, \qquad k \to \infty, \quad s \in [0,1], \tag{21.4}$$

for any $N = 1,2\dots$ (uniform convergence). The function in (21.4) is a linear combination of the functions of the form

$$U^{(m)}(kp(s))k^{m-1}p^{(i_1)}(s)\cdots p^{(i_r)}(s), \qquad i_1 + \cdots + i_r = m, \quad m \le N,$$

with coefficients that do not depend on k. Formula (21.3) easily implies that $U(t) = t$ for $t < 2$ and

$$U(t) \sim \frac{A_1}{t} + \frac{A_2}{t^2} + \cdots, \qquad t \to \infty,$$

and this asymptotic expansion can be differentiated any number of times. Using the property of the function $p(s)$ for $1 - \frac{1}{m} < \varepsilon < 1$, we obtain the inequality

$$\left|U^{(m)}(kp(s))k^{m-1}p^{(i_1)}(s)\cdots p^{(i_r)}(s)\right| \le \text{const}\left|U^{(m)}(kp(s))k^{m-1}(p(s))^{m\varepsilon}\right|$$
$$\le \text{const}\, k^{m-1-\varepsilon m}\max_{t\ge 0}\left|U^{(m)}(t)t^{m\varepsilon}\right| = \text{const}\, k^{m-1-m\varepsilon} \to 0, \qquad k\to\infty,$$

and (21.4) follows.

LEMMA 21.2. *There exist diffeomorphisms $g_k, h_k \in D(B^n)$, $k = 1, 2, \dots$, satisfying the following conditions: (1) $xg_k = xh_k$ in a neighborhood of the set K_+ (depending on k), (2) $xg_k = xg_1$ in a neighborhood of the set K_- (depending on k), (3) $K_-g_k = K \cup K_-$ for each k, (4) $h_k \to e$ as $k \to \infty$. (Recall that e is the identity diffeomorphism).*

PROOF. Take the functions φ_k, $k = 1, 2, \dots$, constructed in Lemma 21.1. In addition, let $l \in C^\infty[0,\infty)$ be a function such that $l(s) = 1$ for $0 \le s \le \delta$, $l(s) = 0$ for $s \ge (1+\delta)/2$, $0 \le l(s) \le 1$, $l'(s) \le 0$ for all s. Define the mappings $x \mapsto xg_k$, $x \mapsto xh_k$ of the ball B^n to $\mathbb{R}^n$ by the following formulas (where, as before, a point $x \in \mathbb{R}^n$ is written in the form $x = (x_{n-1}, \xi_n)$, $x_{n-1} \in \mathbb{R}^{n-1}$, $\xi_n \in \mathbb{R}$):

$$xg_k = (x_{n-1}, \xi_n'), \quad \xi_n' = \xi_n + l(|x|)\Big(\int_{-p(|x_{n-1}|)}^{\xi_n} \varphi_k(x_{n-1}, \xi)\, d\xi - \xi_n\Big),$$
$$xh_k = (x_{n-1}, \xi_n''), \quad \xi_n'' = \xi_n + l(|x|)\int_{-p(|x_{n-1}|)}^{0} \varphi_k(x_{n-1}, \xi)\, d\xi.$$

Properties of the functions φ_k listed in Lemma 21.1 imply that if k is sufficiently large, then the transformations g_k and h_k are diffeomorphisms. Furthermore, they satisfy the conditions of Lemma 21.2, and the proof is completed.

LEMMA 21.3. $g_1 \in D(B^n; B^n \setminus B^{n-1})$.

To prove this lemma, let us write $g_1 = u_k h_k v_k$, where $u_k = g_k h_k^{-1}$, $v_k = g_k^{-1}g_1$. Lemma 21.2 implies that u_k, v_k are identical transformations in a neighborhood of the ball B^{n-1} and $h_k \to e$ as $k \to \infty$. The lemma is proved.

Now we turn to the proof of Theorem 21.1(a). Take an element $g \in D(B^n)$ and a curve $\{g(t), 0 \le t \le 1\} \subset D(B^n)$, $g(0) = e$, $g(1) = g$. All elements $g(t)$ are supported in the region $|x| < \varepsilon$ for some $\varepsilon \in (0,1)$. Choose (in the notation of Lemma 21.1) a function $p(s)$ such that the corresponding set K_- lies in the region $|x| > \varepsilon$, $\xi_n < 0$. Applying Lemma 21.3, we obtain an element $g_1 \in D(B^n)$ such that $K_-g_1 = K_- \cup K$ (by Lemma 21.2) and $g_1 \in D(B^n; B^{n-1})$ (be Lemma 21.3). Let $v(t) = g_1^{-1}g(t)g_1$. Then all elements $v(t)$ are identical diffeomorphisms in a neighborhood of the semiball $K_- \cup K$, hence belong to the subgroup $D(B^n \setminus B^{n-1})$. Therefore, $g(t) = g_1 v(t) g_1^{-1} \in D(B^n; B^{n-1})$. In particular, $g(1) = g \in D(B^n; B^{n-1})$, and Theorem 21.1(a) is proved.

Now let us prove Theorem 21.1. Cover X by the sets $X_i \simeq \mathbb{R}^n$, $i = 1, 2, \dots$, such that each set $\overline{X_i}$ is either disjoint with L or intersects L in such a way that there exists a diffeomorphism of X_i with the ball $B^n = \{x \in \mathbb{R}^n \mid |x| < 1\}$ that maps $X_i \cap L$ to the ball $B^{n-1} = \{x \in \mathbb{R}^{n-1} \mid x = (x_1, \dots, x_{n-1}, 0) \in B^n\}$.

By Proposition 21.2, any element $g \in D(X)$ can be decomposed into the product of diffeomorphisms g_i, each belonging to one of the groups $D(X_j)$. By Theorem 21.1(a), $g_i \in D(X; X \setminus L)$ for each i. The proof of Theorem 21.1 is completed.

§22. Groups of volume preserving diffeomorphisms

The groups $D(X, v^n)$ and $\widehat{D}(X, v^n)$ were defined in 20.2 and 20.4. Let us introduce also $D_0(X, v^n) \subset D(X, v^n)$ as the subgroup generated by all subgroups $D(U, v^n)$, $U \subset X$, $U \simeq \mathbb{R}^n$. Clearly, $D_0(X, v^n)$ is a normal subgroup in $D(X, v^n)$. In this section we describe certain mappings related to these groups. The missing proofs can be found in [**1, 18, 48**]. Everywhere in this section X is a connected manifold of dimension $n \geq 3$ with the volume form v^n.

22.1. The Lie algebra $\mathfrak{g} = V(X, v^n)$. Recall that this Lie algebra consists of compactly supported vector fields ξ satisfying the condition $L_\xi v^n = 0$. Let $\mathfrak{g}_0 = V_0(X, v^n)$ be the ideal consisting of all vector fields that can be represented as a finite sum of fields each supported in some region $U \simeq \mathbb{R}^n$. Let is define certain mappings.

(a) The isomorphism (20.1) maps $\mathfrak{g}_0$ into the space of exact forms. This yields the mapping

$$V(X, v^n)/V_0(X, v^n) \to H_c^{n-1}(X, \mathbb{R}),$$

where $H_c^{n-1}(X, \mathbb{R})$ is the $(n-1)$-cohomology with compact support. It turns out that this mapping is an isomorphism of Lie algebras if we treat $H_c^{n-1}(X, \mathbb{R})$ as an abelian Lie algebra.

Using the form v^n, we get the pairing $H^1(X, \mathbb{R}) \times H_c^{n-1}(X, \mathbb{R}) \to \mathbb{R}$, hence the isomorphisms $H_c^{n-1}(X, \mathbb{R}) \simeq (H^1(X, \mathbb{R}))' \simeq H_1(X, \mathbb{R})$. Therefore, we get the isomorphism

$$V(X, v^n)/V_0(X, v^n) \to H_1(X, \mathbb{R}). \tag{22.1}$$

(b) Let X be a compact manifold. The space

$$\overset{\circ}{C}{}^\infty(X) = \left\{ f \in C^\infty(X, \mathbb{R}) \,\middle|\, \int_X f v^n = 0 \right\} \tag{22.2}$$

is a $\mathfrak{g}$-module. To a closed 1-form a^1 we associate the mapping $\mathfrak{g} \to \overset{\circ}{C}{}^\infty(X)$ of the form

$$\xi \mapsto i_\xi a^1 - \int_X i_\xi a^1 v^n.$$

This mapping is a 1-cocycle of the $\mathfrak{g}$-module $\overset{\circ}{C}{}^\infty(X)$. If a^1 is an exact form, then this cocycle is trivial. This yields the mapping

$$H^1(X, \mathbb{R}) \to H^1(V(X, v^n), \overset{\circ}{C}{}^\infty(X)).$$

One can prove that this mapping is an isomorphism.

(c) Let the manifold X be noncompact. To any closed 1-form a^1 we associate the mapping $\mathfrak{g} \to C_0^\infty(X)$ of the form $\xi \mapsto i_\xi a^1$. This yields the embedding

$$H^1(X, \mathbb{R}) \to H^1(\mathfrak{g}, C^\infty(X)).$$

For the Lie algebra $\mathfrak{g}_0$ we have the exact sequence

$$0 \to C_0^\infty(X) \to C^\infty(X)/\mathbb{R} \xrightarrow{\alpha} H^1(\mathfrak{g}_0, C_0^\infty(X)) \to H^1(\mathfrak{g}_0, C^\infty(X)) \to 0,$$

where the mapping α associates to a function $f \in C^\infty(X)$ the cocycle $\xi \to L_\xi f$.

22.2. The groups $D(X, v^n)$ and $\widehat{D}(X, v^n)$. Let us construct the mappings of the groups corresponding to the mappings of Lie algebras from 22.1.

First, let the manifold X be compact and

$$\int_X v^n = 1.$$

(a) To a closed 1-form a^1 on X and any $\sigma \in \widehat{D}(X, v^n)$ let us associate the function $f_\sigma^{a^1} \in C^\infty(X)$ as follows:

$$f_\sigma^{a^1}(x) = \int_{p(l)} a^1,$$

where l is a path on $\widetilde{X}$ from a point $\xi \in p^{-1}(x)$ to the point $\xi\sigma$ (the integral does not depend on the choice of ξ and l). Clearly,

$$f_{\sigma_1\sigma_2}^{a^1}(x) = f_{\sigma_1}^{a^1}(x) + f_{\sigma_2}^{a^1}(xg_1), \qquad g_1 = p(\sigma_1),$$

i.e., the mapping $\sigma \mapsto f_\sigma^{a^1}$ is a cocycle of the $\widehat{D}(X, v^n)$-module $C^\infty(X)$. If a^1 is an exact form, then this cocycle is trivial. We obtain the mapping

$$H^1(X, \mathbb{R}) \to H^1(\widehat{D}(X, v^n), C^\infty(X)), \tag{22.3}$$

which turns out to be an isomorphism

(b) Set

$$\int_X f_\sigma^{a^1} v^n = \varphi_\sigma^{a^1}, \qquad \sigma \in \widehat{D}(X, v^n).$$

The mapping $\sigma \mapsto \varphi_\sigma^{a^1}$ is a homomorphism from $\widehat{D}(X, v^n)$ to $\mathbb{R}$; it is linear in a^1 and vanishes on exact forms a^1. We obtain the group homomorphism

$$\widehat{D}(X, v^n) \to H_1(X, \mathbb{R}), \tag{22.4}$$

where we have used the isomorphism $(H^1(X, \mathbb{R}))' \simeq H_1(X, \mathbb{R})$.

(c) The restriction of the mapping (22.4) to the subgroup $\Gamma_0 = \widehat{D}(X, v^n) \cap \Gamma$ coincides with the restriction to Γ_0 of the standard mapping $h \colon \Gamma \to H_1(X, \mathbb{R})$. Hence, (22.4) yields a homomorphism $D(X, v^n) \to H_1(X, \mathbb{R})/h(\Gamma_0)$. This homomorphism vanishes on the subgroup $D_0(X, v^n)$ introduced at the beginning of this subsection. Therefore, we get a homomorphism

$$D(X, v^n)/D_0(X, v^n) \to H_1(X, v^n)/h(\Gamma_0).$$

This homomorphism turns out to be an isomorphism.

(d) The function $f_\sigma^{a^1} - \varphi_\sigma^{a^1}$ is invariant when σ is replaced by $\sigma\gamma$, $\gamma \in \Gamma_0$. Therefore, it defines a function $\psi_g^{a^1}$, $g \in D(X, v^n)$. The mapping $g \mapsto \psi_g^{a^1}$ is a cocycle of the $D(X, v^n)$-module $\overset{\circ}{C}{}^\infty(X)$ (see (22.2)). We obtain a mapping

$$H^1(X, \mathbb{R}) \to H^1(D(X, v^n), \overset{\circ}{C}{}^\infty(X)).$$

This mapping is an isomorphism. Note that the mapping $g \mapsto \psi_g^{a^1}$ can be defined directly from the conditions $d\psi_g^{a^1} = ga^1 - a^1$, $\psi_g^{a^1} \in \overset{\circ}{C}{}^\infty(X)$.

Now let X be a noncompact manifold. To a closed 1-form a^1 we associate a cocycle of the $D_0(X, v^n)$-module $C_0^\infty(X, \mathbb{R})$ defined by the conditions $d\psi_g^{a^1} = ga^1 - a^1$, $\psi_g^{a^1} \in C_0^\infty(X)$. This yields an embedding $H^1(X, \mathbb{R}) \to H^1(D_0(X, v^n), C_0^\infty(X))$. For the group $D_0(X, v^n)$ we have the following exact sequence:

$$0 \to C_0^\infty(X) \to C^\infty(X)/\mathbb{R} \xrightarrow{\alpha} H^1(D_0(X, v^n), C_0^\infty(X)) \\ \to H^1(D_0(X, v^n), C^\infty(X)) \to 0,$$

where the mapping α associates to a function $f \in C^\infty(X)$ the cocycle $g \mapsto \varphi_g$, $\varphi_g(x) = f(xg) - f(x)$.

22.3. The action on sections of a bundle. In this subsection we generalize the mapping (22.3). Let π be a unitary representation of a group Γ in a Hilbert space V. It generates a vector bundle $E \to X$ with the fiber V. Let us recall how it is constructed. The group Γ acts on $V \times \widetilde{X}$ by the formula $\gamma(v, \xi) = (\pi(\gamma)v, \gamma\xi)$. Denote by $\Gamma(v, \xi)$ the orbit of an element (v, ξ) and by E the space of all orbits. Define the projection $p_0 \colon E \to X$ by the formula $p_0(\Gamma(v, \xi)) = p(\xi)$, $\xi \in \widetilde{X}$. For any $x \in X$, $\xi \in p^{-1}(x)$ we have the embedding $V \to E$ given by the formula $v \mapsto \Gamma(v, \xi)$. Hence, we have constructed the bundle $E \to X$.

Let $S(X, \pi)$ be the space of smooth compactly supported sections $f \colon X \to E$ of this bundle. This space can be identified with the space of smooth vector-valued functions $f \colon \widetilde{X} \to V$ satisfying the condition

$$f(\gamma\xi) = \pi(\gamma) f(\xi), \qquad \gamma \in \Gamma, \quad \xi \in \widetilde{X},$$

and vanishing outside some set of the form $p^{-1}(K)$, where K is a compact subset of X. The group $\widehat{D}(X, v^n)$ acts on $S(X, \pi)$ by the formula

$$(\rho(\sigma)f)(\xi) = f(\xi\sigma), \qquad f \in S(X, \pi), \quad \sigma \in \widehat{D}(X, v^n), \quad \xi \in \widetilde{X}. \tag{22.5}$$

Let us construct certain cocycles of this action. First of all, introduce the space $E^k(X, \pi)$ of exterior k-forms taking values in the fibers of the bundle $E \to X$. These forms can be identified with V-valued forms a^k on $\widetilde{X}$ satisfying the condition $\gamma a^k = \pi(\gamma)a^k$, $\gamma \in \Gamma$. This condition is preserved under the usual exterior derivation of V-valued forms. Therefore, we get the mappings $d_k \colon E^k(X, \pi) \to E^{k+1}(X, \pi)$ and the cohomology spaces $H^k(X, \pi) = Z^k(X, \pi)/B^k(X, \pi)$, where $Z^k(X, \pi) = \ker d_k$, $B^k(X, \pi) = \operatorname{im} d_{k-1}$.

Take a form $a^1 \in Z^1(X, \pi)$ and define a mapping $\sigma \mapsto f_\sigma$ from $\widehat{D}(X, v^n)$ to $E^0(X, \pi) = S(X, \pi)$ by the formula

$$f_\sigma(\xi) = \int_l a^1, \tag{22.6}$$

where l is a path from ξ to $\xi\sigma$ in $\widetilde{X}$. The mapping $\sigma \mapsto f_\sigma^{a^1}$, $\sigma \in \widehat{D}(X, v^n)$, is a cocycle of the mapping (22.5). Hence, to any closed form a^1 we associate a cocycle of the action (22.5). This yields the embedding

$$H^1(X, \pi) \to H^1(\widehat{D}(X, v^n), S(X, \pi)).$$

In the case of the one-dimensional trivial representation π this embedding was described in 22.2. If the manifold X is compact, then we have the isomorphisms

$$H^1(X, \pi) \simeq H^1(\widehat{D}(X, v^n), S(X, \pi)) \simeq H^1(\Gamma, \pi).$$

The last isomorphism is given by the formula $f_\sigma \mapsto f_\sigma(\xi_0)$, $\sigma \in \Gamma$, where $\sigma \mapsto f_\sigma$ is a cocycle of the action (22.5) and ξ_0 is an arbitrary point in $\widetilde{X}$.

§23. Groups of volume-preserving diffeomorphisms (continuation); embedding into a semidirect product

We keep the notation of §22. Denote by μ and $\widetilde{\mu}$ the measures on X and $\widetilde{X}$ generated by volume forms v^n and $\widetilde{v^n} = p^*(v^n)$. Let us choose a region $\widetilde{Y} \subset \widetilde{X}$ satisfying the following conditions: (1) $\widetilde{Y} \simeq \mathbb{R}^n$; (2) the sets $\gamma\widetilde{Y}$, $\gamma \in \Gamma$, are pairwise disjoint and $\widetilde{\mu}(\widetilde{X} \setminus \bigcup_\gamma \gamma\widetilde{Y}) = 0$; (3) any compact subset $K \subset \widetilde{X}$ has a nonempty intersection with only a finite number of sets $\gamma\widetilde{Y}$, $\gamma \in \Gamma$. Using the region $\widetilde{Y}$ we construct an embedding

$$\widehat{D}(X, v^n) \hookrightarrow \Gamma^X \cdot M(X, \mu) \tag{23.1}$$

(the latter group was defined in §13). Let us embed the group $\Gamma_0 = \Gamma \cap \widehat{D}(X, v^n)$ into Γ^X as the group of constant functions $X \to \Gamma_0$. Then (23.1) gives the embedding

$$D(X, v^n) \hookrightarrow \Gamma^X/\Gamma_0 \cdot M(X, \mu). \tag{23.2}$$

23.1. Construction of embeddings (23.1) and (23.2). Choose a Riemannian metric on X, lift it to $\widetilde{X}$, and define the distance $\rho(\xi_1, \xi_2)$, $\xi_1, \xi_2 \in \widetilde{X}$, as the infinum of lengths of curves joining ξ_1 with ξ_2. For any measurable transformation $\xi \mapsto \xi\sigma$, $\xi \in \widetilde{X}$, denote $l(\sigma) = \operatorname{ess\,sup} \rho(\xi, \xi\sigma)$, $\xi \in \widetilde{X}$. Finally, denote by Σ the group of all measurable bijections $\xi \mapsto \xi\sigma$, $\xi \in \widetilde{X}$ (more precisely, classes of almost everywhere coinciding bijections), that commute with Γ, preserve the measure $\widetilde{\mu}$, and satisfy the condition $l(\sigma) < \infty$. We construct an isomorphism

$$\Sigma \simeq \Gamma^X \cdot M(X, \mu) \tag{23.3}$$

as follows. To an element $\sigma \in \Sigma$ we associate elements $f_\sigma \in \Gamma^X$, $\overline{\sigma} \in M(X, \mu)$ that are uniquely defined by the formulas

$$\xi\sigma = f_\sigma(x)\xi_1, \qquad \xi, \xi_1 \in \widetilde{Y}, \quad x = p(\xi), \quad f_\sigma(x) \in \Gamma, \quad p(\xi_1) = x\overline{\sigma}.$$

It is clear that the mapping $\sigma \mapsto (f_\sigma, \overline{\sigma})$ yields the isomorphism (23.3). It also yields an isomorphism $\Sigma/\Gamma_0 \simeq \Gamma^X/\Gamma_0 \cdot M(X, \mu)$.

Clearly, we have $\widetilde{D}(X, v^n) \subset \Sigma$, completing the construction of the embedding (23.1), hence also of the embedding (23.2).

Let us note an important property of embeddings (23.1) and (23.2). We start by introducing the notion of convergence of sequences in Σ. The notation $\sigma_k \to \sigma$, $\sigma, \sigma_k \in \Sigma$, means that $\widetilde{\mu}(E\sigma \triangle E\sigma_k) \to 0$ for any subset $E \subset \widetilde{X}$ of finite $\widetilde{\mu}$-measure, and $\sup_k l(\sigma_k) < \infty$. Under the isomorphism (23.3) this convergence becomes the convergence in the group $\Gamma^X \cdot M(X, \mu)$ (see §13).

PROPOSITION 23.1. *The group $\widehat{D}(X, v^n)$ is dense in Σ (with respect to the above convergence).*

PROOF. Let W be the set of limits of sequences of elements from the group $\widehat{D}(X, v^n)$. We must prove that $W = \Sigma$. We will use the region $\widetilde{Y} \subset \widetilde{X}$ constructed at the beginning of this section. The isomorphism (23.3) maps the groups $M(X, \mu)$ and Γ^X to certain subgroups Σ_0 and Σ_1 of Σ. The group Σ_0 consists of the transformations $\sigma \in \Sigma$ preserving $\widetilde{Y}$, whereas Σ_1 consists of all $\sigma \in \Sigma$ preserving each fiber $p^{-1}(x)$, $x \in X$. It is known that the group $D(Y, v^n)$ is dense in $M(Y, \mu)$, $Y = p(\widetilde{Y})$, with respect to weak convergence. Hence, $\Sigma_0 \subset W$.

It remains to prove that $\Sigma_1 \subset W$. Take a point $x_0 \in X$, a sufficiently small neighborhood B of x_0 (the choice of B will be specified later), and an element $\gamma \in \Gamma$. Consider the function $f \in \Gamma^X$ such that $f(x) = \gamma$ for $x \in B$, $f(x) = e_\Gamma$ for $x \notin B$; here e_Γ is the unit element of the group Γ. The isomorphism (23.3) brings this function to a transformation σ, which is identical outside $p^{-1}(B)$ and maps any point $\xi \in \widetilde{Y} \cap p^{-1}(B)$ to the point $\gamma\xi$. Let us prove that $\sigma \in W$ provided that B is a sufficiently small neighborhood with smooth boundary. Take points $\xi_0 \in \widetilde{Y} \cap p^{-1}(x_0)$ and $\gamma\xi_0$, and consider a smooth curve $\widetilde{L}$ joining ξ_0 with $\gamma\xi_0$. Let $L = p(\widetilde{L})$, and assume that $\widetilde{L}$ is chosen in such a way that L is diffeomorphic to a circle; this is possible, since we assumed that $n = \dim X > 2$. Take a small number ε and consider the tubular neighborhood L_ε consisting of all points with the distance to L smaller than ε. We can assume that $L_\varepsilon \simeq S^1 \times B^{n-1}$, $B \cup L_\varepsilon \simeq S^1 \times B^{n-1}$, where B^{n-1} is a ball in $\mathbb{R}^{n-1}$. Let us take also a ball $B_\varepsilon \subset X$ such that $\overline{B_\varepsilon} \subset B$, $\mu(B \setminus B_\varepsilon) < \varepsilon$. There exists a family $\{g_\varepsilon(t), 0 \le t \le 1\} \subset D(B \cup L_\varepsilon, v^n)$ that is continuous in t and satisfies the following condition: for any point $x \in B_\varepsilon$ the curve $t \to xg_\varepsilon(t)$ is closed and homotopic to the oriented curve L, and the lengths of all curves $t \to xg_\varepsilon(t)$ are uniformly bounded. The family $\{g_\varepsilon(t)\}$ can be uniquely lifted to $\widetilde{X}$ so that the resulting family $\{\widetilde{g}_\varepsilon(t)\}$ is continuous and all diffeomorphisms $\widetilde{g}_\varepsilon(t)$ are identical outside $p^{-1}(L_\varepsilon \cup B)$. Finally, substitute for ε the numbers $\varepsilon_k \to 0$. Then the sequence $\widetilde{g}_{\varepsilon_k}(1)$ converges in Σ to the element σ specified at the beginning of this paragraph.

Hence $\sigma \in W$. It is clear that elements of this form generate a dense subgroup in Σ_1. Therefore $\Sigma_1 \subset W$, and the proposition is proved.

It is clear how to define the convergence in the group Σ/Γ_0. Proposition 23.1 implies that the group $D(X, v^n)$ is dense in $\Sigma/\Gamma_0 \simeq \Gamma^X/\Gamma_0 \cdot M(X, \mu)$.

23.2. Extension of cocycles from the groups $D(X, v^n)$ and $\widehat{D}(X, v^n)$. Let us show that the cocycles on the groups $D(X, v^n)$ and $\widehat{D}(X, v^n)$ defined in 22.2 and 22.3 can be extended to $\Gamma^X/\Gamma_0 \cdot M(X, \mu)$ and $\Gamma^X \cdot M(X, \mu)$ (we use the embeddings (23.1) and (23.2)). We assume X to be compact.

(a) First, let us consider the natural representation ρ of the group $M(X, \mu)$ in $L_2(X, \mu)$. Extend it to $\Gamma^X \cdot M(X, \mu)$ assuming that the group Γ^X acts trivially in the space $L_2(X, \mu)$. For a homomorphism $\alpha\colon \Gamma \to \mathbb{R}$, define a cocycle of the representation ρ by the formula $r_\sigma^\alpha(x) = \alpha(f(x))$, where $\sigma = (f, \overline{\sigma}) \in \Gamma^X \cdot M(X, \mu)$. Restricting this cocycle to the subgroup $\widehat{D}(X, v^n)$, we obtain a cocycle on this subgroup, $\sigma \mapsto r_\sigma^\alpha$.

On the other hand, in 22.2 we have constructed the cocycle $\sigma \mapsto f_\sigma^{a^1}$ starting from a closed 1-form $a^1 \in Z^1(X)$. Let us show that the cocycles $\sigma \mapsto r_\sigma^\alpha$ and $\sigma \mapsto f_\sigma^{a^1}$, $\sigma \in \widehat{D}(X, v^n)$, are equivalent (cohomologous) if and only if the homomorphism $\alpha\colon \Gamma \to \mathbb{R}$ and the form $a^1 \in Z^1(X)$ are related in a natural way, that is, if for any point $\xi \in \widetilde{X}$ and $\gamma \in \Gamma$ we have

$$\int_{(\xi, \gamma\xi)} p^*(a^1) = \alpha(\gamma) \tag{23.4}$$

(here (ξ_1, ξ_2) denotes a path in $\widetilde{X}$ between the points ξ_1 and ξ_2). Indeed, according to (23.4), we have

$$r_\sigma^\alpha(x) = \alpha(f_\sigma(x)) = \int_{(\xi_1, \xi\sigma)} p^*(a^1), \qquad \sigma \in \widehat{D}(X, v^n)$$

$$\xi \in \widetilde{Y}, \quad \xi_1 \in \widetilde{Y}, \quad p(\xi) = x, \quad p(\xi_1) = x\overline{\sigma}.$$

Take a path (ξ_1, ξ) in $\widetilde{Y}$ and its projection $(x\overline{\sigma}, x)$ to X. Then

$$r_\sigma^\alpha(x) = \int_{(\xi_1, \xi)} p^*(a^1) + \int_{(\xi, \gamma\xi)} p^*(a^1) = \int_{(x\overline{\sigma}, x)} a^1 + f_\sigma^{a^1}(X).$$

In $Y = p(\widetilde{Y})$, the form a^1 can be written as $a^1 = du$, $u \in C^\infty(Y) \cap L_\infty(Y)$. Therefore, the first summand on the right-hand side of the above formula is a trivial cocycle of the form $\sigma \mapsto u(x\overline{\sigma}) - u(x)$. Therefore, the cocycles $\sigma \mapsto r_\sigma^\alpha$ and $\sigma \mapsto f_\sigma^{a^1}$ are equivalent.

(b) Let ρ_0 be the natural representation of the group $M(X, \mu)$ in the space $L_2^0(X, \mu)$ (the orthogonal complement to constants in $L_2(X, \mu)$); as before, extend ρ_0 to $\Gamma^X \cdot M(X, \mu)$ assuming that Γ^X acts trivially. The function $\sigma \mapsto b_\sigma^\alpha$ of the form

$$b_\sigma^\alpha = r_\sigma^\alpha - \int_X r_\sigma^\alpha(x)\, d\mu, \qquad \sigma \in \Gamma^X \cdot M(X, \mu), \tag{23.5}$$

is a cocycle of the representation ρ_0. It is clear that the function (23.5) does not change when we replace σ with $\sigma\gamma$, $\gamma \in \Gamma_0$. Therefore, we obtain a cocycle on $\Gamma^X / \Gamma_0 \cdot M(X, \mu)$. By the above, it is equivalent to the cocycle $\psi_\sigma^{a^1}$ constructed in 22.2 using a form $a^1 \in Z^1(X)$.

(c) Let us show that the cocycles on $\widehat{D}(X, v^n)$ constructed in 22.3 can also be extended to $\Gamma^X \cdot M(X, \mu)$. Take a unitary representation π of the group Γ in a Hilbert space V and a cocycle $r\colon \Gamma \to V$ of this representation. We obtain a unitary representation of the group $\Gamma^X \cdot M(X, \mu)$ in the Hilbert space $H = L_2(X, \mu, V)$ of vector-valued functions by the formula

$$(T(f, \overline{\sigma})\psi)(x) = \pi(f(x))\psi(x\overline{\sigma}), \qquad \sigma = (f, \overline{\sigma}) \in \Gamma^X \cdot M(X, \mu), \quad \psi \in H.$$

The function $\sigma \mapsto r(f(x)) = c_\sigma(x)$ is a cocycle of this representation. Restricting this cocycle to the subgroup $\widehat{D}(X, v^n)$ embedded into $\Gamma^X \cdot M(X, \mu)$ according to (23.1), we obtain a cocycle of the $\widehat{D}(X, \mu)$-module H.

On the other hand, by (22.3), the representation π of the group Γ and the cocycle $r\colon \Gamma \to V$ generate a closed 1-form $a^1 \in Z^1(X, \pi)$ and a cocycle $\sigma \mapsto f_\sigma$. The cocycles $\sigma \mapsto c_\sigma$ and $\sigma \mapsto f_\sigma$ become equivalent if we embed the space $S(X, \pi)$ into H; to do that it suffices to choose, for each $x \in X$, an isomorphism $p^{-1}(x) \simeq V$ with a Borel dependence on x.

23.3. On double cosets with respect to the subgroup $D(Y, v^n)$**.** Take a region $\widetilde{Y} \subset \widetilde{X}$ as at the beginning of this section, and let $Y = p(\widetilde{Y})$. The group $D(Y, v^n)$ is naturally embedded into $\widehat{D}(X, v^n)$. Recall that back in §13 we constructed a two-side $M(X, \mu)$-invariant mapping $q\colon \Gamma^X \cdot M(X, \mu) \to P(\Gamma)$ (if $\mu(X) = 1$) or $q\colon \Gamma^X \cdot M(X, \mu) \to P_1(\Gamma)$ (if $\mu(X) = \infty$). Restricting q to the subgroup $\widehat{D}(X, v^n)$ embedded into $\Gamma^X \cdot M(X,\mu)$ according to 23.1, we obtain a two-side $D(Y, v^n)$-invariant mapping

$$q\colon \widehat{D}(X,v^n) \to P(\Gamma) \quad \text{if } \mu(X)=1,$$
$$q\colon \widehat{D}(X,v^n) \to P_1(\Gamma) \quad \text{if } \mu(X)=\infty.$$

The mapping q can be written in the form

$$(q(\sigma))(\gamma) = \widetilde{\mu}(\gamma\widetilde{Y} \cap \widetilde{Y}\sigma), \qquad \sigma \in \widehat{D}(X,v^n), \quad \gamma \in \Gamma.$$

Now let us turn to the group $D(X,v^n) \simeq \widehat{D}(X,v^n)/\Gamma_0$. The group Γ_0 acts on $P(\Gamma)$ by the translations $p(\gamma) \mapsto p(\gamma\gamma_0)$. Denoting by $P(\Gamma)/\Gamma_0$ the space of orbits of this action, we obtain the mapping $q_0\colon D(X,\mu) \to P(\Gamma)/\Gamma_0$, which is two-side invariant under the subgroup $D(Y,,v^n)$.

It might be interesting to describe all continuous two-side $D(Y,v^n)$-invariant mappings of the group $D(X,v^n)$ to a topological space (for example, $\mathbb{R}$).

§24. Groups of symplectomorphisms

In this section we consider a connected symplectic manifold (X,ω^2) of dimension $2m$. The groups $\mathrm{Diff}(X,\omega^2)$ and $D(X,\omega^2)$ were defined in 20.2. Let us introduce also the subgroup $D_0(X,\omega^2)$ generated by all subgroups $D(U,\omega^2)$, $U \subset X$, $U \simeq \mathbb{R}^{2m}$. Consider the mappings related to these groups. The proof, omitted here, can be found in [**44, 45, 65**].

24.1. Lie algebras. According to 20.3, to the groups $\mathrm{Diff}(X,\omega^2)$ and $D(X,\omega^2)$ there correspond the Lie algebras $\mathrm{Vect}(X,\omega^2)$ and $\mathfrak{g} = V(X,\omega^2)$. By $\mathfrak{g}_0 = V_0(X,\omega^2)$ we denote the Lie algebra of vector fields that can be represented as finite sums of fields from $V(X,\omega^2)$ supported in regions of the form $U \simeq \mathbb{R}^{2m}$. Consider the following mappings.

(a) The correspondence $\xi \mapsto i_\xi\omega^2$ yields exact sequences of Lie algebras

$$0 \to V_0(X,\omega^2) \to V(X,\omega^2) \to H^1(X,\mathbb{R}) \to 0 \quad \text{if } X \text{ is compact,} \tag{24.1}$$
$$0 \to V_0(X,\omega^2) \to V(X,\omega^2) \to H^1_c(X,\mathbb{R}) \to 0 \quad \text{if } X \text{ is noncompact,} \tag{24.2}$$

Here $H^1(X,\mathbb{R})$ and $H^1_c(X,\mathbb{R})$ are considered as commutative Lie algebras.

(b) Define a mapping

$$C^\infty(X) \to \mathrm{Vect}(X,\omega^2) \tag{24.3}$$

that associates to a function $f \in C^\infty(X,\mathbb{R})$ the vector field $\xi = \xi_f$ satisfying the condition $i_\xi\omega^2 = df$. Define the commutation operation (Poisson bracket) in the space $C^\infty(X,\mathbb{R})$ by the formula $[f_1,f_2] = L_{\xi_1}f_2$, $\xi_1 = \xi_{f_1}$. This operation makes $C^\infty(X,\mathbb{R})$ a Lie algebra, and the mapping (24.3) becomes a homomorphism of Lie algebras. Let $\mathrm{Vect}_0(X,\omega^2)$ be the image of the space $C^\infty(X,\omega)$ under the mapping (24.3). Then (24.3) yields an exact sequence

$$0 \to \mathbb{R} \to C^\infty(X,\mathbb{R}) \to \mathrm{Vect}_0(X,\omega^2) \to 0. \tag{24.4}$$

Vector fields from $\mathrm{Vect}(X,\omega^2)$ are called Hamiltonian, and those from $\mathrm{Vect}_0(X,\omega^2)$, strictly Hamiltonian.

(c) If the manifold X is noncompact, then (24.3) yields an isomorphism of Lie algebras

$$C_0^\infty(X,\mathbb{R}) \simeq V_0(X,\omega^2). \tag{24.5}$$

If X is compact, we obtain an isomorphism

$$\overset{\circ}{C}{}^\infty(X,\mathbb{R}) \simeq V_0(X,\omega^2), \tag{24.6}$$

where

$$\overset{\circ}{C}{}^\infty(X,\mathbb{R}) = \left\{ f \in C^\infty(X,\mathbb{R}) \,\middle|\, \int_X f\omega^{2m} = 0 \right\}.$$

We will write the isomorphisms (24.5), (24.6) in the form $f \mapsto \xi_f$, $\xi \mapsto f_\xi$ (here ξ is a vector field and f is a function).

(d) If X is a noncompact manifold, then the mapping

$$K_0(\xi) = \int_X f_\xi \omega^{2m}, \qquad \xi \in V_0(X,\omega^2), \tag{24.7}$$

is a homomorphism of Lie algebras (i.e., $K_0([\xi_1,\xi_2]) = 0$). The Lie algebra $\ker K_0$ is simple.

If the form ω^2 is exact ($\omega^2 = d\alpha^1$ for some form α^1), then the homomorphism (24.7) can be extended to $V(X,\omega^2)$ by the formula

$$K_0(\xi) = -\frac{1}{2m+1} \int_X L_\xi \alpha^1 \wedge \alpha^1 \wedge \omega^{2m-2}.$$

(e) Choose a point $x_0 \in X$ and introduce the subalgebra $V_0^{x_0}(X,\omega^2) = \{\xi \in V_0(X,\omega^2) \mid \xi(x_0) = 0\}$. The functional

$$K_0^{x_0}(\xi) = f_\xi(x_0), \qquad \xi \in V_0^{x_0}(X,\omega^2), \tag{24.8}$$

is a homomorphism of Lie algebras. Any homomorphism $V_0^{x_0}(X,\omega^2) \to \mathbb{R}$ is of the form $\lambda K_0^{x_0}$ if X is compact, or $\lambda K_0^{x_0} + \alpha K_0$ if X is noncompact.

24.2. The group $G = D(X,\omega^2)$. We want to lift ("integrate") the homomorphisms of Lie algebras from the previous subsection to homomorphisms of the corresponding groups.

(a) Let X be a compact manifold. Consider the universal covering $p\colon \widetilde{G} \to G$. The mapping $V(X,\omega^2) \to H^1(X,\mathbb{R})$ from (24.1) generates on $\widetilde{G}$ a two-side invariant 1-form with values in $H^1(X,\mathbb{R})$. This form is exact, and the corresponding primitive, normalized by the condition that it vanishes at the unit element of the group $\widetilde{G}$, is a homomorphism

$$\widetilde{S}\colon \widetilde{G} \to H^1(X,\mathbb{R}). \tag{24.9}$$

Let us describe this homomorphism more precisely. An element $\widetilde{g} \in \widetilde{G}$ is determined by an element $g \in G$ and the homotopy class of a path $\{g(t),\, 0 \le t \le 1\} \subset G$, $g(0) = e$, $g(1) = g$. The curve $\{g(t)\}$ can be assumed to be smooth. Take an integer-valued smooth singular 1-cycle γ on X and denote by $\gamma\{g(t)\}$ the 2-chain obtain by moving γ according to the formula $x \mapsto xg(t)$, $0 \le t \le 1$. Then

$$\int_{\gamma\{g(t)\}} \omega^2 \tag{24.10}$$

depends only on the homology class of the cycle γ. Therefore, (24.10) determines an element of the space $H^1(X, \mathbb{R})$, and the homomorphism (24.9) is constructed. The differential of this homomorphism at the point $\widetilde{e} \in \widetilde{G}$ coincides with the mapping from (24.1). Let us "descend" the mapping (24.9) from the group $\widetilde{G}$ to G. Denote by $\operatorname{Per} \omega^2$ the set of integrals of the form ω^2 over all integral 2-cycles on X, and by H_0 the subgroup of all elements from $H^1(X, \mathbb{R})$ whose periods (i.e., integrals of the corresponding 1-forms over 1-cycles) belong to $\operatorname{Per} \omega^2$. It is clear that $\widetilde{S}(\ker p) \subset H_0$. This yields a homomorphism

(24.11) $$S \colon G \to H^1(X, \mathbb{R})/H_0.$$

It turns out that $\ker S = G_0 = D_0(X, \omega^2)$.

If X is noncompact, then a similar construction yields a mapping

(24.11a) $$S \colon G \to H^1_c(X, \mathbb{R})/H_0,$$

where H_0 is defined as before. In this case we also have $\ker S = D_0(X, \omega^2)$.

(b) If X is a noncompact manifold, then (24.7) can be lifted to the universal covering $\widetilde{G_0} = \widetilde{D_0}(X, \omega^2)$ of the group $G_0 = D_0(X, \omega^2)$. This yields a homomorphism

(24.11b) $$K \colon G_0 \to \mathbb{R}/\Lambda,$$

where Λ is the group of periods of the 1-form on G_0 obtained from the mapping (24.7).

(c) The homomorphisms (24.11) and (24.11a) are particularly simple in the case $\omega^2 = d\alpha^1$. In this case the form $g\alpha^1 - \alpha^1$ is closed for any $g \in D(X, \omega^2)$, and we obtain a homomorphism

(24.12) $$S \colon D(X, \omega^2) \to H^1_c(X, \mathbb{R}).$$

Next, if $g \in D_0(X, \omega^2) = \ker S$, then the form $g\alpha^1 - \alpha$ is exact; moreover, $g\alpha^1 - \alpha^1 = df_g$, $f_g \in C_0^\infty(X, \mathbb{R})$. The mapping $g \mapsto f_g$ is a cocycle of the natural action of the group $D_0(X, \omega^2)$ in the space $C_0^\infty(X, \mathbb{R})$. Finally, define a group homomorphism

(24.13) $$K \colon G_0 \to \mathbb{R}$$

by the formula

$$K(g) = \int_X f_g \omega^{2m}.$$

The mappings (24.12) and (24.13) coincide (in the case when the form ω^2 is exact) with the mappings (24.11a) and (24.11b). Furthermore, in this case the homomorphism (24.13) can be extended to $D(X, \omega^2)$ by the formula

$$K(g) = -\frac{1}{2m+1} \int_X g\alpha^1 \wedge \alpha^1 \wedge \omega^{2m-2}.$$

(d) Choose a point $x_0 \in X$, and in each of the groups $G = D(X, \omega^2)$, $\widetilde{G} = \widetilde{D}(X, \omega^2)$ consider the stabilizers

$$G^{x_0} = \{g \in G \mid x_0 g = x_0\},$$
$$\widetilde{G}^{x_0} = \{\widetilde{g} \in \widetilde{G} \mid x_0 \widetilde{g} = x_0\};$$

in the second formula $\widetilde{G}$ acts on X by the formula $x\widetilde{g} = xg$, $g = p(\widetilde{g})$; it is clear that $\widetilde{G}^{x_0} = p^{-1}(G^{x_0})$. Our goal is to "lift" the mapping (24.8) to these groups. We restrict ourselves to the case when X is connected and simply connected. In this case G^{x_0} and $\widetilde{G}^{x_0}$ are connected, $V_0(X, \omega^2) = V(X, \omega^2)$, $D_0(X, \omega^2) = D(X, \omega^2)$.

To achieve our goal, consider the mapping $q\colon G \to X$ given by the formula $q(g) = x_0 g$, and the corresponding 2-form $q^*\omega^2$ on G. Let us prove that this form is exact. Indeed, define a 1-form β on G as follows. For $g \in G$, $\xi \in \mathfrak{g}$ denote by (g, ξ) the tangent vector to the curve $t \mapsto g \exp t\xi$ at $t = 0$. Let

$$\beta(g, \xi) = f_\xi(x_0 g).$$

PROPOSITION 24.1. *We have*

$$d\beta = q^*(\omega^2). \tag{24.14}$$

In particular, the restriction of the form β to G^{x_0} is closed.

PROOF. To any vector field $\xi \in \mathfrak{g}$ we associate the vector field $\overline{\xi}$ on G formed by tangent vectors (g, ξ). Then for any $\xi, \eta \in \mathfrak{g}$ we obtain the following sequence of equalities (in the formulas below $L_{\overline{\xi}}$ denotes the derivation along the vector field $\overline{\xi}$, $i_{\overline{\xi}}$ is the inner multiplication by $\overline{\xi}$, and similarly for L_ξ and i_ξ):

$$\begin{gathered}
(d\beta)(\overline{\xi}, \overline{\eta}) = L_{\overline{\xi}} i_{\overline{\eta}} \beta - L_{\overline{\eta}} i_{\overline{\xi}} \beta - \beta([\overline{\xi}, \overline{\eta}]), \\
(i_{\overline{\eta}}\beta)(g) = f_\eta(x), \quad (i_{\overline{\xi}}\beta)(g) = f_\xi(x), \\
(L_{\overline{\xi}} i_{\overline{\eta}} \beta)(g) = (L_\xi f_\eta)(x), \quad (L_{\overline{\eta}} i_{\overline{\xi}} \beta)(g) = (L_\eta f_\xi)(x), \\
\beta([\overline{\xi}, \overline{\eta}]) = f_{[\xi,\eta]}(x),
\end{gathered}$$

where $x = x_0 g$. Hence

$$(d\beta)(\overline{\xi}, \overline{\eta})(g) = (L_\xi f_\eta - L_\eta f_\xi - f_{[\xi,\eta]})(x) = \omega^2(\xi, \eta)(x) \tag{24.15}$$

(here we use the equalities $\omega^2(\xi, \eta) = L_\xi f_\eta = -L_\eta f_\xi = f_{[\xi,\eta]}$.) Formula (24.15) implies (24.14), and Proposition 24.1 is proved.

Using the projection $p\colon \widetilde{G} \to G$, we get the forms $p^*q^*\omega^2$, $\widetilde{\beta} = p^*\beta$ on $\widetilde{G}$. From (24.14) we obtain that $p^*q^*\omega^2 = d\widetilde{\beta}$.

Now we return to the problem of "lifting" the homomorphism (24.8) to the groups G^{x_0} and $\widetilde{G}^{x_0}$. We begin with $\widetilde{G}^{x_0}$. The homomorphism (24.8) generates a two-side invariant 1-form on $\widetilde{G}^{x_0}$. It coincides with the restriction of the form $\widetilde{\beta}$ to $\widetilde{G}^{x_0}$. Consider the periods of the form $\widetilde{\beta} \mid \widetilde{G}^{x_0}$, i.e., the numbers

$$\int_\gamma \widetilde{\beta},$$

where γ is an one-dimensional integral cycle in $\widetilde{G}^{x_0}$. It is clear that $\gamma = d\sigma$, where σ is a two-dimensional chain in $\widetilde{G}$. Proposition 24.1 implies that

$$\int_\gamma \widetilde{\beta} = \int_\sigma d\widetilde{\beta} = \int_\sigma p^*q^*\omega^2 = \int_{qp(\sigma)} \omega^2. \tag{24.16}$$

Here $qp(\sigma)$ is a two-dimensional cycle in X. Let $\widetilde{\Lambda} = \operatorname{Per}(\widetilde{\beta} \mid \widetilde{G}^{x_0})$ be the set of periods of the forms $\widetilde{\beta} \mid \widetilde{G}^{x_0}$. It follows from (24.16) that $\widetilde{\Lambda} \subset \operatorname{Per} \omega^2$.

Hence, the functional (24.8) can be lifted to $\widetilde{G}^{x_0}$ and yields a homomorphism

$$\widetilde{K}^{x_0}\colon \widetilde{G}^{x_0} \to \mathbb{R}/\widetilde{\Lambda}.$$

Now we turn from $\widetilde{G}^{x_0}$ to G^{x_0}. First, let the manifold X be noncompact. We show that, similarly to the case of the group $\widetilde{G}^{x_0}$, the set of periods of the form β on G^{x_0} is contained in $\operatorname{Per}\omega^2$.

Indeed, let $\gamma = \{g(t),\ 0 \le t \le 1\}$, $g(0) = g(1) = e$, be a closed smooth path in G^{x_0}. Also, let α be a smooth path on X from x_0 to ∞. Then

$$\int_\gamma \beta = \int_0^1 f_{\dot g(t)}(x_0)\,dt = \int_0^1 \left(\int_\alpha i_{\dot g(t)}\omega^2\right) dt = \int_\sigma \omega^2,$$

where

$$\dot g(t) = \frac{d}{dt} g(t)\cdot (g(t))^{-1}$$

and σ is the two-dimensional cocycle obtained by moving α along the family $\{g(t),\ 0 \le t \le 1\}$. Hence,

$$\int_\gamma \beta \in \operatorname{Per}\omega^2,$$

as required. Let us note that the above arguments do not use Proposition 24.1.

Therefore, in this case also the functional (24.8) can be lifted to a homomorphism

$$K^{x_0}\colon G^{x_0} \to \mathbb{R}/\Lambda_0, \qquad \Lambda_0 = \operatorname{Per}(\beta \mid G^{x_0}).$$

Now let X be a compact (and simply connected) manifold. In this case we know nothing about the set $\Lambda_0 = \operatorname{Per}(\beta \mid G^{x_0})$. The reason is, of course, that we do not have sufficient information about the group $\pi_1(G^{x_0})$.

(e) Let $x_0 \neq x_1$ be two points of X. Take the group $G^{x_0,x_1} = G^{x_0} \cap G^{x_1}$. We can define a homomorphism of the universal covering group $\widetilde{G}^{x_0,x_1}$ to $\mathbb{R}$ as follows. Take a smooth path l from x_0 to x_1 and a smooth family $\{g(t),\ 0 \le t \le 1\} \subset G^{x_0,x_1}$, $g(0) = e$, $g(1) = g$. A homotopy class of such families (for a fixed g) defines an element $\widetilde{G}^{x_0,x_1}$. Denote

$$f(\widetilde{g}) = \int_{l\{g(t)\}} \omega^2,$$

where $l\{g(t)\}$ is the two-dimensional cycle obtained by moving l along $\{g(t)\}$. We get a homomorphism. Set $\Lambda_1 = f(\ker p)$, where $p\colon \widetilde{G}^{x_0,x_1} \to G^{x_0,x_1}$ is the universal covering. It is clear that $\Lambda_1 \subset \operatorname{Per}\omega^2$. We obtain a homomorphism $f_0\colon G^{x_0,x_1} \to \mathbb{R}/\Lambda_1$.

24.3. The A. Weinstein mapping. Proposition 24.1 yields an interesting homomorphism

$$\varphi\colon \pi_1(G) \to \mathbb{R}/\operatorname{Per}\omega^2, \qquad G = D(X,\omega^2) \tag{24.17}$$

(provided that X is a compact simply connected manifold); this homomorphism was introduced by Weinstein in [**65**]. To construct φ let us consider a part of the exact homotopy sequence:

$$\cdots \to \pi_2(X) \xrightarrow{i_1} \pi_1(G^{x_0}) \xrightarrow{j} \pi_1(G) \to \{1\}. \tag{24.18}$$

Define the homomorphism $\psi\colon \pi_1(G^{x_0}) \to \mathbb{R}$ by the formula

$$\psi(\gamma) = \int_{\overline{\gamma}} \beta, \qquad \gamma \in \pi_1(G^{x_0}),$$

where $\overline{\gamma}$ is a closed smooth path representing the element γ, and β is the form from Proposition 24.1. Proposition 24.1 easily implies that if $\gamma \in i_1(\pi_2(X))$, i.e., $\gamma = i_1(\sigma)$ for some $\sigma \in \pi_2(X)$, then

$$\psi(\gamma) = \int_{\overline{\gamma}} \beta = \int_{\overline{\sigma}} \omega^2,$$

where $\overline{\sigma}$ is a smooth 2-cycle representing the element σ. Hence, $\psi(\gamma) \in \operatorname{Per}\omega^2$ for $\gamma \in i_1(\pi_2(X))$. But then the exact sequence (24.17) yields a homomorphism of the form (24.18), as desired.

24.4. Bundles with connection; automorphisms preserving the connection. In the next two subsections we briefly discuss relations with geometric quantization theory. For detailed presentations of this theory, see [**24, 25**]. We describe the principal construction under certain simplifying assumptions; this construction will be used in §25 where we will present an example of an infinite-dimensional symplectic manifold.

Consider a connected symplectic manifold (X, ω^2). Let us assume that the following integrality condition is satisfied:

$$[\omega^2] \in i\left(H^2(X, \mathbb{Z})\right),$$

where $[\omega^2]$ is the cohomology class of the form ω^2 and the mapping $i\colon H^2(X,\mathbb{Z}) \to H^2(X,\mathbb{R})$ is generated by the embedding $\mathbb{Z} \to \mathbb{R}$. Denote $\mathbb{T}^1 = \{z \in \mathbb{C} \mid |z| = 1\}$. Then there exist a principal $\mathbb{T}^1$-bundle

$$\xi\colon E \to X \tag{24.19}$$

and a connection form β^1 on X such that $d\beta^1 = \xi^*\omega^2$.

Consider the group of automorphisms of the bundle (24.19) preserving the form β^1. Denote by $A(E, \beta^1)$ the subgroup consisting of all automorphisms that can be connected with the identity automorphism by a smooth curve. Let $\mathfrak{a}(E, \beta^1)$ be the Lie algebra of this group. On the other hand, denote by $\operatorname{Diff}_0(X, \omega)$ the group of diffeomorphisms $g \in \operatorname{Diff}(X, \omega^2)$ satisfying the following condition: there exists a smooth family $\{g(t),\, 0 \le t \le 1\}$ such that $g(0) = e$, $g(1) = g$, and the vector field $\frac{d}{dt} g(t) \cdot (g(t))^{-1}$ belongs to the Lie algebra $\operatorname{Vect}_0(X, \omega^2)$ of strictly Hamiltonian vector fields (see 24.1). Then we have the following exact sequences:

$$1 \longrightarrow \mathbb{T}^1 \longrightarrow A(E, \beta^1) \xrightarrow{\alpha} \operatorname{Diff}(X, \omega^2) \longrightarrow 1, \tag{24.20}$$

$$0 \longrightarrow \mathbb{R} \longrightarrow \mathfrak{a}(E, \beta^1) \xrightarrow{\alpha} \operatorname{Vect}_0(X, \omega^2) \longrightarrow 0, \tag{24.21}$$

The correspondence $\eta \mapsto i_\eta \beta^1$, $\eta \in \mathfrak{a}(E, \beta^1)$, defines an isomorphism between $\mathfrak{a}(E, \beta^1)$ and the Lie algebra $C^\infty(X, \mathbb{R})$ (with the commutator given by the Poisson bracket). Under this identification of Lie algebras $\mathfrak{a}(E, \beta^1)$ and $C^\infty(X, \mathbb{R})$, the sequence (24.21) becomes (24.4).

Let us give exact formulas for the above objects. Let $X = \bigcup X_k$ be a covering of X such that all X_k are open sets and all nonempty sets $X_i \cap X_k$, $X_i \cap X_k \cap X_s$

are connected. Let us assume that the principal $\mathbb{T}^1$-bundle (24.19) is trivial over each X_k, so that we have diffeomorphisms

$$\varphi_k\colon X_k \times \mathbb{T}^1 \to \xi^{-1}(X_k) \tag{24.22}$$

that commute with the action of the group $\mathbb{T}^1$. On the intersections $X_k \cap X_j$ these diffeomorphisms are related as follows:

$$\varphi_k(x, z_1) = \varphi_i(x, z_2) \Leftrightarrow \frac{z_1}{z_2} = u_{kj}(x),$$
$$x \in X_j \cap X_j, \quad z_1, z_2 \in \mathbb{T}^1.$$

Here $u_{kj}\colon X_k \cap X_j \to \mathbb{T}^1$ are smooth functions. We require that these functions can be represented in the following form:

$$u_{kj}(x) = \exp\{2\pi i f_{kj}(x)\}, \qquad x \in X_j \cap X_k,$$

where $f_{kj}\colon X_k \cap X_j \to \mathbb{R}$ are smooth functions. (All these conditions are automatically satisfied if the sets X_i, $X_k \cap X_j$, $X_k \cap X_j \cap X_s$ are contractible.) If the set $X_k \cap X_j \cap X_s$ (with pairwise distinct k, l, s) is nonempty, then the function $f_{kj} + f_{js} + f_{sk}$ is an integer c_{kjs}, and the family $\{c_{kjs}\}$ is a two-dimensional cocycle for the covering $\{X_k\}$. Under the identification (24.22), the form β^1 becomes a form on $X_k \times \mathbb{T}^1$ that can be written as follows:

$$\alpha_k^1 - \frac{1}{2\pi i}\frac{dz}{dz}, \qquad z = e^{is} \in \mathbb{T},$$

where α_k^1 is a form on X_k satisfying the condition $d\alpha_k^1 = \omega^2$.

Now we consider the group $A(E, \beta^1)$. To describe certain elements of this group, take an open connected subset $Y \subset X$ such that the principal $\mathbb{T}^1$-bundle is trivial over Y. Construct an embedding

$$j_Y\colon D_0(Y, \omega) \to A(E, \beta^1), \tag{24.23}$$

such that $\alpha \circ j_Y$ is identical on $D_0(Y, \omega^2)$ (where $\alpha\colon A(E, \beta^1) \to \mathrm{Diff}_0(X, \omega)$ is taken from (24.20)). To this end, choose a diffeomorphism

$$\varphi\colon Y \times \mathbb{T}^1 \to p^{-1}(Y) \tag{24.24}$$

commuting with the action of the group $\mathbb{T}^1$. Let the connection form β^1, on being transferred to $Y \times \mathbb{T}^1$, take the form

$$\alpha_Y^1 - \frac{1}{2\pi i}\frac{dz}{z}, \qquad z = e^{is} \in \mathbb{T}^1,$$

where α_Y^1 is a form on Y; this form satisfies the condition $d\alpha_Y^1 = \omega^2$. To an arbitrary $g \in D_0(Y, \omega^2)$ we associate an element $\widetilde{g} \in A(E, \beta^1)$ such that $\widetilde{g}$ is identical outside $\xi^{-1}(Y)$, while its action on $\xi^{-1}(Y)$ after the identification (24.22) becomes the action

$$(x, z)\widetilde{g} = (xg, e^{i\psi(x)}), \qquad \psi \in C_0^\infty(Y), \quad d\psi = 2\pi(g\alpha_Y^1 - \alpha_Y^1).$$

We get the desired embedding (24.23). Denote by $\check{D}(E, \beta^1)$ the group of transformations of the manifold E generated by all $\widetilde{g}$, where $g \in D_0(Y, \omega^2)$ and Y is an arbitrary open connected region over which the bundle (24.19) is trivial. We have the surjective homomorphism $\check{p}\colon \check{D}(E, \beta^1) \to D_0(X, \omega)$. The action of an element of the group $\ker \check{p}$ on E coincides with the action of a number from $\mathbb{T}^1$ on E.

It is clear that for a noncompact X the mapping $\check{p}$ is an isomorphism. If X is compact, then $\check{p}$ is a covering over $D_0(X,\omega)$. This can be proved using the fact that the corresponding extension of Lie algebras splits.

§25. An example of an infinite-dimensional symplectic manifold

In this section we consider a connected manifold X with the volume form v^n, $n \geq 3$. We will describe an infinite-dimensional symplectic manifold, related to X, with the transitive action of the group $D(X,v^n)$. (We do not dwell on general definitions related to infinite-dimensional symplectic manifolds.) The construction from 24.4 provides a central extension of the group $D_0(X,v^n)$; we recall (see §22) that this group is generated by all subgroups of the form $D(U,v^n)$, $U \subset X$, $U \simeq \mathbb{R}^n$.

Let us recall some notation. (1) $E^k = E^k(X)$ is the space of all real exterior k-forms on the manifold X; (2) $Z^k = Z^k(X)$, $B^k = B^k(X)$ are the space of closed form and the space of exact forms; (3) $E^k_c = E^k_c(X)$, $Z^k_c = Z^k_c(X)$, $B^k_c = B^k_c(X)$ are the spaces of the forms with compact support; (4) $\mathrm{Vect}(X,v^n)$ is the Lie algebra of vector fields preserving the volume form v^n; $V(X,v^n)$ is the subalgebra of compactly supported fields; $V_0(X,v^n)$ is the subalgebra of vector fields that can be represented as finite sums of fields supported in regions of the form $U \simeq \mathbb{R}^n$. The correspondence $\xi \mapsto i_\xi v^n$, $\xi \in \mathrm{Vect}(X,v^n)$, defines the isomorphisms

$$\begin{aligned} j&\colon \mathrm{Vect}(X,v^n) \to Z^{n-1}(X),\\ j&\colon V(X,v^n) \to Z^{n-1}_c(X),\\ j&\colon V_0(X,v^n) \to B^{n-1}(X). \end{aligned} \tag{25.1}$$

In 25.1–25.3 the manifold X is supposed to be compact; the case of a noncompact X is simpler, and is briefly discussed in 25.7.

25.1. The Lie algebra E^{n-1}/B^{n-2}. The exterior derivative $d\colon E^{n-2} \to B^{n-2}$ yields the mapping $\bar{d}\colon E^{n-2}/B^{n-2} \to B^{n-1}$ with $\ker \bar{d} = H^{n-2}(X,\mathbb{R})$. We get an exact sequence of linear spaces

$$0 \to H^{n-2}(X,\mathbb{R}) \to E^{n-2}/B^{n-2} \xrightarrow{j^{-1}\circ\bar{d}} V_0(X,v^n) \to 0. \tag{25.2}$$

The goal of this subsection is to define a Lie algebra structure on E^{n-2}/B^{n-2} in such a way that (25.2) becomes a central extension of Lie algebras. First, we construct a bilinear mapping

$$E^{n-2} \times E^{n-2} \to E^{n-2}/B^{n-2} \tag{25.3}$$

as follows. Take a region $U \subset X$, $U \simeq \mathbb{R}^n$. To forms $\alpha \in E^{n-2}$, $\beta \in E^{n-2}_c(U)$ we associate the element $\gamma + B^{n-2} \in E^{n-2}/B^{n-2}$, where the form $\gamma \in E^{n-2}_c(U)$ is defined by the formula $j[j^{-1}d\alpha, j^{-1}d\beta] = d\gamma$. This defines a mapping (25.3) for pairs (α,β) with $\alpha \in E^{n-2}$, $\beta \in E^{n-2}_c(U)$. To define this mapping for all pairs $(\alpha,\beta) \in E^{n-2} \times E^{n-2}$ it suffices to represent β as the sum of terms supported in the regions $U \simeq \mathbb{R}^n$; it is easy to verify that the resulting mapping is well defined. The constructed mapping is skew-symmetric and vanishes on $E^{n-2} \times B^{n-2}$. Therefore, we get a mapping

$$E^{n-2}/B^{n-2} \times E^{n-2}/B^{n-2} \to E^{n-2}/B^{n-2}.$$

One can easily verify that it makes the space E^{n-2}/B^{n-2} a Lie algebra, and we get a central extension of the form (25.2).

This construction is somewhat similar to the extension of the Lie algebra of all those vector fields on a symplectic manifold (X, ω^2) that preserve the form ω^2. The Lie bracket in the space E^{n-2}/B^{n-2} is an analog of the Poisson bracket in $C^\infty(X)$.

Consider the 2-cocycle corresponding to the extension (25.2):

$$c\colon \mathfrak{g}_0 \times \mathfrak{g}_0 \to H^{n-2}(X, \mathbb{R}), \qquad \mathfrak{g}_0 = V_0(X, \mathbb{R}^n). \tag{25.4}$$

To define this cocycle let us choose a continuous linear mapping $J\colon B^{n-1} \to E^{n-2}$ such that the composition $d \circ J$ is the identity on B^{n-1}; it is known that such an operator exists. We obtain a natural mapping $\overline{J}\colon B^{n-1} \to E^{n-2}/B^{n-2}$. Now we split the exact sequence of linear spaces (not of Lie algebras!) (25.2) using the mapping $S = \overline{J} \circ j\colon \mathfrak{g}_0 \to E^{n-2}/B^{n-2}$. This mapping generates the required 2-cocycle (25.4) by the formula $c(\xi.\eta) = S([\xi, \eta]) - [S(\xi), S(\eta)]$. We omit the simple proof of the fact that $c(\xi, \eta)$ is the cohomology class of the closed 2-form

$$(J \circ d - I) i_\xi i_\eta v^n,$$

where I is the identity operator in E^{n-2}.

Concluding this subsection, we note that any element $\gamma \in H_{n-2}(X, \mathbb{R})$ leads to a one-dimensional central extension

$$0 \to \mathbb{R} \to (E^{n-2}/B^{n-2})_\gamma \to V_0(X, v^n) \to 0 \tag{25.5}$$

that arises from the mapping $H^{n-2}(X, \mathbb{R}) \to \mathbb{R}$ given by the pairing $H^{n-2}(X, \mathbb{R}) \times H_{n-2}(X, \mathbb{R}) \to \mathbb{R}$.

The goal of the subsequent exposition is to describe the extension of the group $D_0(X, v^n)$ corresponding to the extension (25.2) of Lie algebras. This extension is related to a symplectic manifold $M_{n-2}(X)$ that will be described below.

25.2. The manifold $M_{n-2}(X)$. Recall that X is a compact manifold with the volume form v^n. Choose a compact connected oriented $(n-2)$-dimensional submanifold $s_0 = s_0^{n-2} \subset X$. Denote by $M_{n-2} = M_{n-2}(X)$ the set of all oriented submanifolds $s \subset X$ that can be obtained from s_0 by the action of an element of the group $D_0(X, v^n)$ (instead of $D_0(X, v^n)$ one can take $D(X, v^n)$ or $D(X)$, but we will not discuss this here). Make M_{n-2} an infinite-dimensional manifold. For that, introduce on X a Riemannian metric. Take a submanifold $s \in M_{n-2}$. A sufficiently narrow ε-neighborhood $V(s, \varepsilon)$ of s is diffeomorphic to the space $N(s)$ of the normal bundle $N \to s$; the fiber of this bundle is the two-dimensional real linear space. Each section of this bundle yields (using the diffeomorphism $V(s, \varepsilon) \simeq N(S)$) a point in M_{n-2}. The set of all such points is declared to be a chart of the manifold M_{n-2}. Hence, M_{n-2} becomes a manifold modeled by the linear topological space of sections of the bundle $N \to s$.

A tangent vector to M_{n-2} at the point s is an element of the linear space $\operatorname{Vect}_s(X)/\operatorname{Vect}(s)$, where $\operatorname{Vect}_s(X)$ is the space of vector fields on X at points of s and $\operatorname{Vect}(s)$ is the space of vector fields on X that are tangent to the submanifold s. Denote the tangent space to M_{n-2} at the point s by $T_s M_{n-2}$, or simply by T_s. In each space T_s we introduce a bilinear form w_s^2 of the form

$$w_s^2(\overline{\xi}, \overline{\eta}) = \int_s i_\xi i_\eta v^n, \qquad \overline{\xi}, \overline{\eta} \in T_s, \tag{25.6}$$

where $\xi, \eta \in \mathrm{Vect}_s(X)$, $\overline{\xi} = \xi + \mathrm{Vect}(s)$, $\overline{\eta} = \eta + \mathrm{Vect}(s)$. One can easily see that if $\xi \in \mathrm{Vect}(s)$, the integral (25.6) vanishes; therefore w_s^2 is well defined. Therefore, we get a skew-symmetric 2-form w^2 on M_{n-2}.

Let us prove that this form is closed. We first note that any n-form on a connected noncompact manifold of dimension n is exact. Take a point $x_0 \in X$. By the above remark, there exists an $(n-1)$-form $v_{x_0}^{n-1}$ on the set $X \setminus \{x_0\}$ such that $dv_{x_0^{n-1}} = v^n$ on this set. Denote by $M_{n-1}^{x_0}$ the subset of all $s \in M_{n-2}$ that do not contain the point x_0. Define the form $w_{x_0}^1$ on $M_{n-1}^{x_0}$ by the formula

$$w_{x_0}^1(\overline{\xi}) = \int_s i_\xi v_{x_0}^{n-1}, \qquad \overline{\xi} \in T_s, \quad \overline{\xi} = \xi + \mathrm{Vect}(s), \quad \xi \in \mathrm{Vect}_s(X).$$

Let us prove that

$$dw_{x_0}^1 = w^2 \tag{25.7}$$

on the manifold $M_{n-2}^{x_0}$. Indeed, for any $\xi, \eta \in \mathrm{Vect}_s(X)$ we have

$$\begin{aligned} dw_{x_0}^1(\overline{\xi}, \overline{\eta}) &= L_{\overline{\xi}} w_{x_0}^1(\overline{\eta}) - L_{\overline{\eta}} w_{x_0}^1(\overline{\xi}) - w_{x_0}^1([\overline{\xi}, \overline{\eta}]) \\ &= \int_s (L_\xi i_\eta - L_\eta i_\xi - i_{[\xi,\eta]}) v_{x_0}^{n-1}. \end{aligned}$$

On the other hand, $L_\xi i_\eta - L_\eta i_\xi - i_{[\xi,\eta]} = i_\eta i_\xi s - d i_\eta i_\xi$ (as a handbook for similar computations one can use the book [**11**], and, in particular, Chapter IV there). Therefore,

$$dw_{x_0}^1(\overline{\xi}, \overline{\eta}) = \int_s (i_\eta i_\xi s - d i_\eta i_\xi) v_{x_0}^{n-1} = \int_s i_\eta i_\xi v^n.$$

This proves (25.7). Hence, the form w^2 is exact on any submanifold $M_{n-2}^{x_0}$. Therefore, it is closed. Formula (25.6) implies that it is nondegenerate. Hence, we obtain a symplectic manifold (M_{n-2}, w^2).

25.3. The principal $\mathbb{T}^1$-bundle over M_{n-2}; the connection. From now on we assume that X is oriented and

$$\int_X v^n = 1. \tag{25.8}$$

Let us construct a principal $\mathbb{T}^1$-bundle

$$\xi \colon E \to M_{n-2} \tag{25.9}$$

(recall that $\mathbb{T}^1 = \{z \in \mathbb{C} \mid |z| = 1\}$). In the previous subsection we defined, for any point $x \in X$, the set $M_{n-1}^x = \{s \in M_{n-2} \mid x \notin s\}$ and the $(n-1)$-form v_x^{n-1} on $X \setminus \{x\}$ such that $dv_x^{n-1} = v^n$ on $X \setminus \{x\}$. The bundle (25.9) we want to construct will be trivial over each M_{n-2}^x. Therefore, to describe the bundle (25.9) we must define the transition functions that relate trivializations of the bundle over the sets $M_{n-2}^{x_1}$, $M_{n-2}^{x_2}$ for two distinct points $x_1, x_2 \in X$. It suffices to consider pairs of sets taken from a family $\{M_{n-2}^x\}$ that cover the manifold M_{n-2}.

To achieve our goal, let us select a region $U \simeq \mathbb{R}^n$ in X. For a point $x \in U$ let us choose an $(n-1)$-form v_x^{n-1} on $X \setminus \{x\}$ so that the following conditions are satisfied: (1) $dv_X^{n-1} = v^n$ on $X \setminus \{x\}$; (2) for any points $x_1 \neq x_2$ in X the support of the form $v_{x_1}^{n-1} - v_{x_2}^{n-1}$ is a compact subset of U. To construct such a family $\{v_x^{n-1}\}$, choose a point x_0 and a form $v_{x_0}^{n-1}$ on $X \setminus \{x_0\}$ such that $dv_{x_0}^{n-1} = v^n$ on $X \setminus \{x_0\}$. Then for any point $x \in U$ we choose a diffeomorphism $g_x \in D(U, v^n)$ such that

$xg_x = x_0$ and set $v_x^{n-1} = g_x v_{x_0}^{n-1}$. This completes the construction of the required family of $(n-1)$-forms $\{v_x^{n-1},\ x \in U\}$.

For two points $x_1 \neq x_2$ in U the form $v_{x_1}^{n-1} - v_{x_2}^{n-1}$ is supported in U and closed in $X \setminus \{x_1, x_2\}$. If we joint the points x_1 and x_2 by a simple curve l lying inside U, then the form will be exact on $X \setminus l$. Therefore, there exists an $(n-2)$-form $v_{x_1,x_2,l}^{n-2}$ on $X \setminus l$ such that

$$dv_{x_1,x_2,l}^{n-2} = v_{x_1}^{n-1} - v_{x_2}^{n-1} \tag{25.10}$$

on $X \setminus l$; the form $v_{x_1,x_2,l}^{n-2}$ can be chosen in such a way that it vanishes outside a compact subset of U. Below we fix the forms $v_{x_1,x_2,l}^{n-2}$ (one for each pair of points $x_1 \neq x_2$ in X and a curve $l \subset U$ joining these points).

Now we can construct the transition function for the pair of sets $M_{n-2}^{x_1}$, $M_{n-2}^{x_2}$, where $x_1, x_2 \in U$, $x_1 \neq x_2$. Take a point $s \in M_{n-2}^{x_1} \cap M_{n-2}^{x_2}$ (so that $x_1 \notin s$, $x_2 \notin s$). Let l be a simple curve that joins x_1 with x_2, lies in U, and is disjoint from s. Set

$$f_{x_1,x_2}(s) = \exp\left(2\pi i \int_s v_{x_1,x_2,l}^{n-2}\right), \qquad s \in M_{n-2}^{x_1} \cap M_{n-2}^{x_2}. \tag{25.11}$$

Let us show that this number does not depend on the choice of the curve l. Indeed, take two curves l_1, l_2 in U that join x_1 with x_2 and are disjoint from s. Let us prove that

$$\int_s \left(v_{x_1,x_2,l_1}^{n-2} - v_{x_1,x_2,l_2}^{n-2}\right) \tag{25.12}$$

is an integer. The form under the integral sign is closed on $U \setminus (l_1 \cup l_2)$ and supported in U. Therefore, we can reduce the integral (25.12) to the integral over an integral $(n-2)$-dimensional cycle lying in $U \setminus (l_1 \cup l_2)$. Deforming this cycle, we can assume that it lies in a small ε-neighborhood of a point $y_0 \in l_1$, $y_0 \neq x_1, x_2$ (here the ε-neighborhood is defined using some Riemannian metric on X). Denote this cycle by $\sigma_{n-2}^{\varepsilon}$. We can assume that this cycle is the boundary of an $(n-1)$-dimensional chain τ_{n-1}^{ε} in $X \setminus l_1$. Let $\varepsilon \to 0$. Then the cycle $\sigma_{n-2}^{\varepsilon}$ tends to the point y_0; we can assume that the chain τ_{n-1}^{ε} tends to an $(n-1)$-dimensional cycle τ_{n-1}^{0}. Since the form v_{x_1,x_2,l_2}^{n-2} is smooth near the point y_0, the integral (25.12) can be reduced to the form

$$\begin{aligned}\int_{\sigma_{n-2}^{\varepsilon}} v_{x_1,x_2,l_1}^{n-2} + o(1) &= \int_{\tau_{n-1}^{\varepsilon}} dv_{x_1,x_2,l_1}^{n-2} + o(1)\\ &= \int_{\tau_{n-1}^{\varepsilon}} (v_{x_1}^{n-1} - v_{x_2}^{n-1}) + o(1) = \int_{\tau_{n-1}^{0}} (v_{x_1}^{n-1} - v_{x_2}^{n-1}) + o(1),\end{aligned}$$

where $o(1)$ denotes terms that tend to 0 as $\varepsilon \to 0$. The cycle τ_{n-1}^{0} is homologous to the sum of a cycle of the form $m\partial B_\delta$, where $m \in \mathbb{Z}$ and B_δ is the oriented ball of radius δ centered at the point x_1, and a similar cycle centered at x_2. As $\delta \to 0$, we have

$$\begin{aligned}m\int_{\partial B_\delta} (v_{x_1}^{n-1} - v_{x_2}^{n-1}) + o(1) &= m\int_{\partial B_\delta} v_{x_1}^{n-1} + o(1) = m\int_{X \setminus B_\delta} dv_{x_1}^{n-1} + o(1)\\ &= m\int_X v^n + o(1) = m + o(1)\end{aligned}$$

(here we use the equality (25.8)). But the integral (25.12) is invariant under these transformations, so that it equals the integer m, as required.

Therefore, the function $s \mapsto f_{x_1,x_2}(s)$, $s \in M^{x_1}_{n-2} \cap M^{x_2}_{n-2}$, is well defined by formula (25.11). Similar arguments show that for three points $x_1, x_2, x_3 \in U$ the number

$$\int_s \left(v^{n-2}_{x_1,x_2,l_1} + v^{n-2}_{x_2,x_3,l_2} + v^{n-2}_{x_3,x_1,l_3} \right)$$

is an integer. Therefore,

$$f_{x_1,x_2}(s) f_{x_2,x_3}(s) f_{x_3,x_1}(s) = 1 \quad \text{for any } s \in M^{x_1}_{n-2} \cap M^{x_2}_{n-2} \cap M^{x_3}_{n-2},$$

and the family of functions $\{f_{x_1,x_2}(s)\}$ defines a principal $\mathbb{T}^1$-bundle (25.9). The space E is obtained from the disjoint union of the sets $M^x_{n-2} \times \mathbb{T}^1$, $x \in U$, using the identification

$$(s,u) \equiv (s,v) \Leftrightarrow \frac{u}{v} = f_{x_1,x_2}(s), \qquad s \in M^{x_1}_{n-2} \cap M^{x_2}_{n-2}. \tag{25.13}$$

Introduce a connection form β^1 on E, defining it on each set $\xi^{-1}(M^x_{n-2}) \simeq M^x_{n-2} \times \mathbb{T}^1$ by the formula

$$w^1_x - \frac{1}{2\pi i} \frac{du}{u}, \qquad u \in \mathbb{T}^1,$$

where the form w^1_x was defined earlier.

25.4. The group of connection preserving automorphisms. First of all, we choose, for any region $V \simeq \mathbb{R}^n$ in X, a continuous linear mapping $J_V \colon B^{n-1}_c(V) \to E^{n-2}_c(V)$ such that $d \circ J_V$ is the identity operator in $B^{n-1}_c(V)$.

Take a region $V \simeq \mathbb{R}^n$ in X such that the set $U \setminus \overline{V}$ is nonempty and arcwise connected; here U is the region indicated in 25.13. For any point $x \in U \setminus \overline{V}$ and any $g \in D(V, v^n)$ define the transformation $\widetilde{g}_x$ of the set $\xi^{-1}(M^x_{n-2}) \subset E$ as follows. Identify the set $\xi^{-1}(M^x_{n-2})$ with $M^x_{n-1} \times \mathbb{T}^1$ as in 25.3. Set

$$(s,u)\widetilde{g}_x = \left(sg, u \exp\left\{ 2\pi i \int_s J_V(g v^{n-1}_x - v^{n-1}_x) \right\} \right), \qquad s \in M^x_{n-2}, \quad u \in \mathbb{T}^1. \tag{25.14}$$

Using the above identification, extend the transformation $\widetilde{g}_x$ to $\xi^{-1}(M^x_{n-2})$ (keeping the previous notation $\widetilde{g}_x$). Formula (25.14) easily implies that the transformations $\widetilde{g}_{x_1}$ and $\widetilde{g}_{x_2}$ act in the same way on the set $\xi^{-1}(M^{x_1}_{n-2}) \cap \xi^{-1}(M^{x_2}_{n-2})$, where $x_1, x_2 \in U \setminus \overline{V}$ (here we use the assumption that the set $U \setminus \overline{V}$ is arcwise connected). However, it is clear that the sets M^x_{n-2}, $x \in U \setminus \overline{V}$, cover the manifold M_{n-2}. Therefore, the transformations $\widetilde{g}_x$, $x \in U \setminus \overline{V}$, define the transformation $\widetilde{g}$ of the manifold E. It is clear that $\widetilde{g}$ is an automorphism of the bundle (25.9), and the corresponding transformation of the base X of this bundle coincides with the transformation $s \mapsto sg$, $s \in M_{n-2}$, $g \in D(V, v^n)$. Finally, simple computations show that $\widetilde{g}$ preserves the connection form β^1.

Up to now the region V was fixed. Now we consider all regions $V \simeq \mathbb{R}^n$ such that $U \setminus \overline{V}$ is nonempty and arcwise connected. Introduce the group of all automorphisms of the bundle (25.9) generated by all transformations $\widetilde{g}$, $g \in D(V, v^n)$, constructed above, and transformations of the group $\mathbb{T}^1$ acting on E. Denote the resulting group by $A^{s_0}(E, \beta^1) = A^{s_0}$, where s_0 is the $(n-2)$-dimensional oriented submanifold in X with which the entire construction started. The group A^{s_0} preserves the connection form β^1. There is a surjective homomorphism $A^{s_0} \to D_0(X, v^n)$ with the kernel $\mathbb{T}^1$.

Denote by $\mathfrak{a}^{s_0}(E, \beta^1)$ the Lie algebra of the group $A^{s_0}(E, \beta^1)$. Let α_0 be the element of the homology space $H_{n-2}(X, \mathbb{R}^n)$ corresponding to the manifold s_0. Let us construct the isomorphism

$$\mathfrak{a}^{s_0}(E, \beta^1) \simeq (E^{n-2}/B^{n-2})_{\alpha_0}, \tag{25.15}$$

where the Lie algebra on the right-hand side was defined in 25.1. To a form $a \in E^{n-2}$ we associate the vector field $\xi \in \mathfrak{a}^{s_0}(E, \beta^1)$ by the formula

$$\int_s a = i_\xi \beta^1,$$

where β^1 is the connection form defined in 25.3. The correspondence $a \mapsto \xi$ easily gives the isomorphism (25.15).

25.5. The group $A(X, \beta^1)$ corresponding to the algebra E^{n-2}/B^{n-2}. Up until now the submanifold s_0 has been fixed. Suppose we have a family of submanifolds $\{s_j,\ 1 \le j \le m\}$ such that the corresponding homology classes form a basis of the space $H_{n-2}(X, \mathbb{R})$. We obtain central extensions

$$1 \to \mathbb{T}^1 \to A^{s_j}(E, \beta^1) \to D_0(X, v^n) \to 1.$$

The well-known operation of summation of central extensions yields a central extension of the group $D_0(X, v^n)$ by the torus $\mathbb{T}^m$. The latter torus can be identified with the group $H^{n-2}(X, \mathbb{R})/L$, where L consists of the elements of the space $H^{n-2}(X, \mathbb{R})$ for which the integrals of the corresponding $(n-2)$-forms over all cycles s_j are integers. As a result, we obtain the central extension

$$1 \to H^{n-2}(X, \mathbb{R})/L \to A(X, v^n) \xrightarrow{\pi} D_0(X, v^n) \to 1 \tag{25.16}$$

corresponding to the extension (25.2) of Lie algebras. This achieves the main goal of this section.

The mapping $\pi\colon A(X, v^n) \to D_0(X, v^n)$ admits a continuous section in a neighborhood W of the unit element $e \in D_0(X, v^n)$. This allows us to construct the local 2-cocycle $(h, g) \mapsto a(h, g)$, $h, g \in W$, corresponding to the extension (25.16); this cocycle takes values in $H^{n-2}(X, \mathbb{R})$. We give, without proof, an explicit formula for this 2-cocycle (under the appropriate choice of the above section).

Let us choose a continuous linear mapping $J\colon B^{n-1}(X) \to E^{n-2}(X)$ such that $d \circ J$ is the identity mapping of $B^{n-1}(X)$. Choose a point $x_0 \in X$ and a neighborhood U of x_0. There is an $(n-1)$-form v^{n-1} on $X \setminus \{x_0\}$ such that $dv^{n-1} = v^n$ for $x \ne x_0$. If $g \in D_0(X, v^n)$ is sufficiently close to the identity, then the form $gv^{n-1} - v^{n-1}$ is exact outside U and there exists an exact form v_g^{n-1} on all of X that coincides with $gv^{n-1} - v^{n-1}$ outside U. Consider the $(n-2)$-form

$$(J \circ h - h \circ J)v_g^{n-1}, \tag{25.17}$$

where h, g are taken from the above neighborhood of the identity $W \subset D_0(X, v^n)$. This form is closed, and the corresponding cohomology class $a(h, g) \in H^{n-2}(X, \mathbb{R})$ yields a mapping $W \times W \to H^{n-2}(X, \mathbb{R})$. This mapping is the desired local cocycle corresponding to the extension (25.16). We must emphasize that it cannot be extended to the entire group $D_0(X, v^n)$ (even if we allow discontinuous extensions).

It is only the factorization with respect to the subgroup L that allows us to construct the extension (25.16).

25.6. The group $A(X, v^n)$ and the inductive limit of the groups $\{D(U, v^n), U \subset X, U \simeq \mathbb{R}^n\}$. In this subsection we consider (without proofs) yet another problem leading to the group $A(X, v^n)$. First we recall the notion of the inductive limit of a family of groups.

Let there be given a family of groups $\{G_\alpha\}$, and for any α, β a family $f^\gamma_{\alpha\beta}\colon G_\alpha \to G_\beta$ (for some α, β this family can be empty). The inductive limit of the family $\{G_\alpha, f^\gamma_{\alpha\beta}\}$ is the group A with homomorphisms $u_\alpha\colon G_\alpha \to A$ such that (1) $u_\beta \circ f^\gamma_{\alpha\beta} = u_\alpha$ for all α, β, γ for which a homomorphism $f^\gamma_{\alpha\beta}$ is defined; (2) for any group A^0 and homomorphisms $u^0_\alpha\colon G_\alpha \to A^0$ satisfying the condition similar to (1) there exists a unique homomorphism $\psi\colon A \to A^0$ such that $u^0_\alpha = \psi \circ u_\alpha$ for each α.

Now let us return to a connected manifold X with a volume form v^n, and consider the family of all groups $\{D(U, v^n), U \subset X, U \simeq \mathbb{R}^n\}$ and inclusion homomorphisms $f_{UV}\colon D(U, v^n) \to D(V, v^n)$ for $U \subset V$. The problem is to find the inductive limit of the family $\{D(U, v^n), f_{UV}\}$. The answer is as follows. We must consider all connected coverings of the group $A(X, v^n)$ trivial over each subgroup of the form $D(U, v^n)$, $U \simeq \mathbb{R}^n$. Among all these coverings there is a universal one (covering all others). This is the required group A (with obvious homomorphisms $D(U, v^n) \to A$); see **[18]**.

A similar problem can be solved for the family of Lie algebras $\{V(U, v^n), U \subset X, U \simeq \mathbb{R}^n\}$. The solution is given by the Lie algebra E^{n-2}/B^{n-2} introduced in 25.1, together with obvious inclusions $V(U, v^n) \to E^{n-2}/B^{n-2}$.

25.7. The group $A(X, v^n)$ for a noncompact manifold X. In this case everything is simpler. The form v^n is exact on M, so that $v^n = dv^{n-1}$ for some form v^{n-1}. The closed $(n-1)$-form (25.17) is defined for all $h, g \in D_0(X, v^n)$. Taking the corresponding element from $H^{n-2}(X, \mathbb{R})$, we obtain a 2-cocycle on $D_0(X, v^n)$ with values in $H^{n-2}(X, \mathbb{R})$. This results in a central extension of the group $D_0(X, v^n)$ by $H^{n-2}(X, \mathbb{R})$.

§26. The action of diffeomorphism groups on the configuration spaces $\Omega_m(X)$; "small" subgroups

In this section by $G = G(X)$ we denote one of the groups $D(X)$, $D(X, v^n)$, $D(X, \omega^2)$, $D(X, \alpha^1)$, where X is a smooth manifold with $\dim X = n < \infty$. We recall that for any region $Y \subset X$ the group $G(Y)$ can be naturally considered as a subgroup of $G(X)$.

The set $B \subset X$ is called a ball if there exists a diffeomorphism of a neighborhood of the closure of B to $\mathbb{R}^n$ that transforms the set B itself into the ball $B^n = \{x \in \mathbb{R}^n \mid |x| < 1\}$.

26.1. "Small" subgroups. By $G_* = G_*(X)$ we denote the subgroup of G defined as follows: (1) if $G \neq D(X, \omega^2)$, then G_* is generated by all subgroups of the form $G(B)$, where B is a ball; (2) if $G = D(X, \omega^2)$, then G_* is generated by all subgroups of the form $\{g \in G(B, \omega^2) \mid K(g) = 0\}$. (The mapping $K\colon D(B, \omega^2) \to \mathbb{R}$ was defined in 24.2.)

The subgroup $G_* = G_*(X)$ is called the "small" subgroup of the group $G = G(X)$. One can easily see that any smooth homomorphism of G_* to a finite-dimensional Lie group is trivial.

26.2. The manifold $\Omega_m(X)$. By $\Omega_m = \Omega_m(X)$ we denote the set of all subsets $\omega \subset X$ consisting of m pairwise distinct points, $m < \infty$. One can easily make Ω_m a manifold. The group $\mathrm{Diff}(X)$ acts transitively on Ω_m by diffeomorphisms $\omega \mapsto \omega g$. Choose a point $\omega_0 = (x_1^0, \dots, x_m^0) \in \Omega_m$. The goal of this section is to study the stabilizer

$$G^{\omega_0} = \{g \in G \mid \omega_0 g = \omega_0\}.$$

Till the end of this section we assume that $\dim X = n \geq 3$.

26.3. Homomorphisms of the group G^{ω_0}. In the next chapter we will consider unitary representations of the group G induced from representations of the group G^{ω_0}. It is natural to restrict ourselves to those representations of G^{ω_0} that can be factored through a homomorphism of G to a finite-dimensional (not necessarily connected) Lie group. A natural problem here is to describe all such homomorphisms. We start with the description of two such homomorphisms. Choose a point $x_0 \in X$ and an integer $r \geq 0$, and denote by J_r the group of r-jets of diffeomorphisms $g \in G^{x_0}$. Also, choose diffeomorphisms from G that map points $x_i^0 \in \omega_0$, $1 \leq i \leq m$, to the point x_0; this allows us to identify a neighborhood of each point x_i^0 with a neighborhood of the point x_0. This yields a natural surjective homomorphism

$$\gamma_r \colon G^{\omega_0} \longrightarrow J_r^m \cdot S_m, \tag{26.1}$$

where S_m is the symmetric group, $J_r^m = J_r \times \cdots \times J_r$ (m factors), $J_r^m \cdot S_m$ is the semidirect product (S_m acts on J_r^m by permutations of factors). To construct another homomorphism, let us consider the universal covering $p \colon \widetilde{X} \to X$ and the group of deck transformations Γ. Let us recall that in 20.4 we constructed the groups $\widehat{G}$, $\Gamma_0 = \Gamma \cap \widehat{G}$ and the central extension

$$1 \longrightarrow \Gamma_0 \longrightarrow \widehat{G} \xrightarrow{p_1} G \longrightarrow 1.$$

Denote $\widehat{G}^{\omega_0} = p_1^{-1}(G^{\omega_0})$. The group $\widehat{G}^{\omega_0}$ acts transitively on the set $p^{-1}(\omega_0)$ by transformations that commute with Γ. Choose points $\xi_i^0 \in p^{-1}(x_i^0)$, $1 \leq i \leq m$. To an element $\sigma \in \widehat{G}^{\omega_0}$ we associate the m-tuple $(\gamma_1, \dots, \gamma_m) \in \Gamma^m$ and the element $s \in S_m$ according to the formula

$$\xi_i^0 \sigma = \gamma_i \xi_{is}^0.$$

The mapping $\sigma \mapsto (\gamma_1, \dots, \gamma_m)s$ is a surjective homomorphism of the group $\widehat{G}^{\omega_0}$ to the semidirect product $\Gamma^m \cdot S_m$ (with S_m acting on Γ^m by permutations). On the subgroup $\Gamma_0 = \Gamma \cap \widehat{G}$ this homomorphism is reduced to $\gamma_0 \mapsto (\gamma_0, \dots, \gamma_0)$. This results in the homomorphism

$$\beta \colon G^{\omega_0} \to (\Gamma^m/\Gamma_0) \cdot S_m; \tag{26.2}$$

here Γ_0 is embedded into Γ^m by the formula $\gamma_0 \mapsto (\gamma_0, \dots, \gamma_0)$. Combining (26.1) and (26.2), we obtain a surjective homomorphism

$$\alpha_r \colon G^{\omega_0} \to (J_r^m \times \Gamma^m/\Gamma_0) \cdot S_m. \tag{26.3}$$

Now we return to the problem of describing all homomorphisms of G^{ω_0} to finite-dimensional Lie groups. For an integer $r \geq 0$ we define a normal subgroup $G_r^{\omega_0}$

such that $G^{\omega_0} \supset G_0^{\omega_0} \supset G_1^{\omega_0} \supset \cdots$ and each smooth homomorphism from G^{ω_0} to a finite-dimensional Lie group can be factored through the canonical homomorphism

$$\tau_r \colon G^{\omega_0} \to K_r^{\omega_0} = G^{\omega_0}/G_r^{\omega_0}$$

for some $r \geq 0$. The group $G_r^{\omega_0}$ we define similarly to the definition of the "small" subgroup from 26.1. If $G \neq D(X, \omega^2)$, then by $G_r^{\omega_0}$ we denote the subgroup of G^{ω_0} generated by all subgroups of the form $G(B) \cap \ker \gamma_r$, where B is a ball in X. If $G = D(X, \omega^2)$, the group $G_r^{\omega_0}$ is generated by all subgroups of the from

$$G(B) \cap \ker \gamma_r \cap \ker K \cap \left(\bigcap_{x_i^0 \in B \cap \omega_0} \ker K^{x_i^0} \right).$$

It is clear that $G_r^{\omega_0}$ is a normal subgroup of G^{ω_0}. Consider the group $K_r^{\omega_0} = G^{\omega_0}/G_r^{\omega_0}$ and the canonical homomorphism $\tau_r \colon G^{\omega_0} \to K_r^{\omega_0}$. The next proposition shows that the homomorphisms τ_r, $r \geq 0$, solve our problem.

PROPOSITION 26.1. *Let L be a finite-dimensional Lie group. Any smooth homomorphism $G^{\omega_0} \to L$ can be represented in the form $\varphi \circ \tau_r$ for some $r \geq 0$, where $\varphi \colon K_r^{\omega_0} \to L$ is a homomorphism of Lie groups.*

As we have already noted, the homomorphism $G^{\omega_0} \to L$ is trivial on each subgroup $G_*(B)$, where B is a ball disjoint from ω_0. Therefore, to prove Proposition 26.1 it suffices to prove the following result, which will be also useful later (see 30.1).

LEMMA 26.1. *Let there be given diffeomorphisms $g_r \in G_r^{\omega_0}$, $r = 0, 1, \cdots$. Then g_r can be represented in the form $g_r = u_r v_r$, $r = 0, 1, \cdots$, where $u_r \in G_*(X \setminus \omega_0)$ for all $r \geq 0$, and $v_r \to e$ as $r \to \infty$.*

To prove the lemma, let us consider a ball $B \subset X$ and a point $x_0 \in B$. Let h_r^0 be jets of diffeomorphisms from G leaving the point x_0 fixed and having at this point tangency of order r with the identity diffeomorphism. We show that the jet h_r^0 can be extended to a diffeomorphism $h_r \in G_r^{x_0}(B)$ (recall that $G_r^{x_0}(B)$ is the special case of the group $G^{\omega_0}(X)$ when $X = B$, $\omega_0 = \{x_0\}$). This is obvious for $G = D(X)$; therefore, it suffices to consider, for example, the group $G = D(X, \omega^2)$.

There exists a smooth family of jets of diffeomorphisms $\{h_r^0(t),\ 0 \leq t \leq 1\} \subset G = D(X, \omega^2)$ that leave the point x_0 fixed, have at this point tangency of order r with the identity diffeomorphism, and satisfy the conditions $h_r^0(0) = e$ and $h_r^0(1) = h_r^0$. Let $\xi_r^0(t)$, $0 \leq t \leq 1$, be the jets of vector fields from $\mathfrak{g} = V(X, \omega^2)$ that are tangent to the family $\{h_r^0(t)\}$. The field $\xi_r^0(t)$ has a zero of order r at the point x_0. One can easily extend the jets $\xi_r^0(t)$ to vector fields $\xi_r(t) \in V(B, \omega^2)$ in such a way that the functionals K_0 and $K_0^{x_0}$ vanish on $\xi_r(t)$ and $\xi_r(t) \to 0$ as $r \to \infty$ uniformly with respect to t in the C^∞-topology of the space $V(B, \omega^2)$. (Here we pass from jets of vector fields to jets of 1-forms using the correspondence $\xi \mapsto i_\xi \omega^2$, $\xi \in V(B, \omega^2)$.) Finally, using $\xi_r(t)$ we construct a family of homomorphisms $h_r(t)$ tangent to the fields $\xi_r(t)$ for $0 \leq t \leq 1$. The diffeomorphisms $h_r(1)$, $r = 1, 2, \ldots$, satisfy the desired conditions. For other groups the construction is similar.

Lemma 26.1 follows from this result.

26.4. Remarks about the structure of the group $K_r^{\omega_0}$, $r \geq 0$. It is clear that the homomorphism (26.3) is trivial on the group $G_r^{\omega_0}$. Therefore, we get a surjective homomorphism

$$\overline{\alpha}_r\colon K_r^{\omega_0} \to (J_r^m \times \Gamma^m/\Gamma_0)\cdot S_m, \qquad r \geq 0.$$

We consider the group $\ker\overline{\alpha}_r$ for $G = D(X), D(X, v^n), D(X, \omega^2)$ and $r \geq 1$.

Let $G = D(X)$. We prove that

$$\ker\overline{\alpha}_r = \mathbb{Z}_2^p$$

for some $p \leq m$. Consider the following family of normal subgroups of G^{ω_0}:

$$G_r^{\omega_0} \subset \ker\alpha_r \subset \ker\beta \subset G^{\omega_0}.$$

Note that if H is a group with normal subgroups $B \subset A \subset H$, then there are natural mappings $H/B \to H/A$, $\ker(H/B \to H/A) \simeq A/B$. Hence,

$$\ker\overline{\alpha}_r = \ker\left(G^{\omega_0}/G_r^{\omega_0} \to G^{\omega_0}/\ker\alpha_r\right) \simeq \ker\left(\ker\beta/G_r^{\omega_0} \to \ker\beta/\ker\alpha_r\right).$$

The group $\ker\beta$ is the connected component of the unit element in the group G^{ω_0}, $\ker\beta/\ker\alpha_r \simeq J_r^m$. The mapping $\ker\beta/G^{\omega_0} \to \ker\beta/\ker\alpha_r$ is a connected covering. Since $\pi_1(J_r^m) \simeq \mathbb{Z}_2^m$, $\ker\overline{\alpha}_r \simeq \mathbb{Z}_2^p$ for some $p \leq m$.

EXAMPLE. Let $G = D(\mathbb{R}^n)$ and $\omega_0 = \{0_n\}$, where $0_n = (0, \ldots, 0)$. Let us show that

$$K_r^{\omega_0} \simeq \widetilde{GL}_+(n,\mathbb{R})\cdot J_r^1, \qquad J_r^1 = \{g \in J_r \mid (dg)(0_n) = I\}, \tag{26.4}$$

where $\widetilde{GL}_+(n,\mathbb{R})$ is the two-sheet (universal) covering group of $GL_+(n,\mathbb{R})$, and $\widetilde{GL}_+(n,\mathbb{R})\cdot J_r^1$ is the semidirect product. Therefore, $\ker\alpha_r = \mathbb{Z}_2$.

It suffices to verify that the natural homomorphism $G^{\omega_0} \to GL(n,\mathbb{R})$ generates the zero homomorphism of fundamental groups. Take a path $g_t \in G^{\omega_0}$, $0 \leq t \leq 1$, $g_0 = e$, $g_1 = e$, and the corresponding path $h_t \in GL_+(n,\mathbb{R})$. Choose a curve $s \mapsto x(s) \in \mathbb{R}^n$, $s \in [0,\infty)$, such that $x(0) = 0_n$, $x(s) \to \infty$ as $s \to \infty$. Taking the Jacobi matrix $\dot{g}_t$ of the mapping g_t at the point $x(s)$, we obtain the elements $\dot{g}_t(s) \in GL_+(n,\mathbb{R})$. It is clear that $\dot{g}_t(0) = h_t$, and $\dot{g}_t(s) = I_n$ for a sufficiently large s; here I_n is the identity matrix. Therefore, the path $\{h_t\}$ can be contracted to a point. Therefore, the homomorphism $G^{\omega_0} \to GL_+(n,\mathbb{R})$ yields the zero mapping of fundamental groups, and (26.4) follows.

If $\dim X = 2$, then the group $K_r^{\omega_0}$ admits a natural homomorphism to the braid group $B_m(X)$, $m = |\omega_0|$ (see, for example, [**49**]).

Let $G = D(X,\omega^2)$ with a simply connected X. If X is a compact manifold, then $\ker\overline{\alpha}_r \simeq \mathbb{R}/\Lambda_0 \times (\mathbb{R}/\Lambda_1)^{m-1}$, where the groups Λ_0, Λ_1 were defined in 24.2. If X is a noncompact manifold, then $\ker\overline{\alpha}_r \simeq (\mathbb{R}/\Lambda_0)^m$. The case of a non-simply-connected X requires mappings from 24.2; we will not dwell on this.

CHAPTER 5

Representations of Diffeomorphism Groups Associated with Configuration Spaces

This chapter opens with the description of representations of the group $D(X)$ that are spherical with respect to a subgroup concentrated in a small coordinate neighborhood. We show that such representations can be naturally realized using the space of closed subsets (configurations) in X. Next, we study representations of diffeomorphism groups considered in Chapter 4; these representations are related with the spaces of finite or countable configurations. A special role is played by the Poisson measure on the space of countable configurations.

§27. Class O representations of the group $D(X)$; the commutative operator algebra generated by a representation

About the results of this section, see [**15**]. Everywhere by X we denote a connected C^∞-manifold of dimension $n < \infty$. Recall that a subset $B \subset X$ is called a ball if there is a diffeomorphism of a neighborhood of its closure $\overline{B}$ to $\mathbb{R}^n$ that maps B to the unit ball.

27.1. Class O representations. Consider a unitary representation $g \mapsto T(g)$, $g \in D(X)$, in a Hilbert space H. For a ball $B \subset X$, consider the subspace

$$\{h \in H \mid T(g)h = h \text{ for all } g \in D(B)\}. \tag{27.1}$$

DEFINITION 27.1. A representation T is said to be of class O if the closure of the linear span of all subspaces of the form (27.1) coincides with H.

An equivalent condition is that there is a ball $B \subset X$ such that the smallest closed subspace in H invariant under T and containing the subspace (27.1), equals H.

The goal of this section is to associate to a class O representation a certain von Neumann algebra; the main result of the section is that this algebra turns out to be commutative.

27.2. Projections P_F and the algebra A. For a closed subset $F \subset X$ consider the subgroup $D(X \setminus F)$ and define the subspace

$$H_F = \{h \in H \mid T(g)h = h \text{ for all } g \in D(X \setminus F)\}. \tag{27.2}$$

Denote by P_F the orthogonal projection onto H_F. Finally, let A be the von Neumann algebra generated by the projections P_F, where F runs over all closed subsets of X. The next theorem describes some properties of the projections P_F and the algebra A. Below, F denotes a closed subset in X.

THEOREM 27.1. *(1) If $g \in D(X)$, then*

$$P_{Fg} = T(g^{-1}) P_F T(g). \tag{27.3}$$

(2) If $F_1 \supset F_2 \supset \cdots$, $F = \bigcap F_k$, then $P_{F_1} \geq P_{F_2} \geq \cdots$, $P_F = \prod_{k=1}^{\infty} P_{F_k}$ (the product converges strongly).

(3) For any closed sets F_1, F_2 we have

$$P_{F_1} \cdot P_{F_2} = P_{F_1 \cap F_2}. \tag{27.4}$$

(4) The algebra A is commutative.

PROOF. Property (1) is obvious. Property (2) follows from the fact that $D(X \setminus F_1) \subset D(X \setminus F_2) \subset \cdots$, $D(X \setminus F) = \bigcup_k D(X \setminus F_k)$. Property (4) follows from (3). Therefore, it remains to prove (3).

LEMMA 27.1. *Let $F \subset X$ be a closed subset, L a submanifold of dimension $n-1$ (not necessarily connected). Then $P_{F \cup L} = P_F$.*

PROOF OF THE LEMMA. Let $h \in H_{F \cup L}$. We prove that $h \in H_F$. Consider the group $D(X \setminus F)$ and the subset $D(X; X \setminus (F \cup L)) \subset D(X \setminus F)$ defined as in Definition 21.1. Let g be an element of this subset. By Definition 21.1, $g = \lim_{k \to \infty} u_k w_k v_k$, where $u_k, v_k \in D(X \setminus (F \cup L))$, $w_k \to e$ as $k \to \infty$. Therefore,

$$\begin{aligned}(T(g)h, h) &= \lim_{k \to \infty} (T(u_k w_k v_k)h, h)) = \lim_{k \to \infty} (T(w_k)T(v_k)h, T(u_k^{-1})h) \\ &= \lim_{k \to \infty} (T(w_k)h, h) = (h, h).\end{aligned}$$

Hence, $(T(g)h, h) = (h, h)$, so that $T(g)h = h$ for $g \in D(X; X \setminus (F \cup L))$. But by Theorem 21.1, the latter set generates the entire group $D(X \setminus F)$. Hence, $T(g)h = h$ for all $g \in D(X \setminus F)$. Therefore, $h \in H_F$.

We have proved that $H_{F \cup L} \subset H_F$. On the other hand, $H_F \subset H_{F \cup L}$ (by part (2) of Theorem 27.1). Therefore, $H_F = H_{F \cup L}$. Lemma 27.1 is proved.

LEMMA 27.2. *Let F and K be two closed subsets of X, $K_1 = \overline{X \setminus K}$. Assume that the set $L = \partial K$ is a submanifold of dimension $n-1$ (not necessarily connected). Then $P_F = P_{F \cup K} \cdot P_{F \cup K_1}$.*

PROOF OF THE LEMMA. Denote $K^0 = K \setminus L$, $K_1^0 = K_1 \setminus L$. Let A and A' be the von Neumann algebras generated by the operators $T(g)$, $g \in D(K^0 \setminus F)$, and $T(g)$, $g \in D(K_1^0 \setminus F)$, respectively. The subgroups $D(K^0 \setminus F)$ and $D(K_1^0 \setminus F)$ are supported on disjoint sets, hence commute. Therefore, the algebras A and A' commute as well. The projections $P_{F \cup K}$ and $P_{F \cup K_1}$ lie in the centers of the corresponding algebras, hence also commute. Next, the group $D(X \setminus (F \cup L))$ is a direct product of the subgroups $D(K^0 \setminus F)$ and $D(K_1^0 \setminus F)$, hence $H_{F \cup L} = H_{F \cup K} \cap H_{F \cup K_1}$. But by Lemma 27.1, $H_{F \cup L} = H_F$. Hence, $H_F = H_{F \cup K} \cap H_{F \cup K_1}$. Since $P_{F \cup K}$ and $P_{F \cup K_1}$ commute, we have $P_F = P_{F \cup K} \cdot P_{F \cup K_1}$. Lemma 27.2 is proved.

Now we prove (27.4). First, let F_1 be the set K indicated in Lemma 27.2. Then by this lemma,

$$\begin{aligned}P_{K \cap F_2} &= P_{(K \cap F_2) \cup K} P_{(K \cap F_2) \cup K_1} = P_K P_{F_2 \cup K_1}, \\ P_{F_2} &= P_{F_2 \cup K} P_{F_2 \cup K_1}.\end{aligned} \tag{27.5}$$

Multiply the second equality on the left by P_K and notice that $P_K P_{F_2 \cup K} = P_K$ (by part (2) of Theorem 27.1). We obtain

$$P_K P_{F_2} = P_K P_{F_2 \cup K} P_{F_2 \cup K_1} = P_K P_{F_2 \cup K_1}.$$

Comparing with (27.5), we get (27.4) with K taken for F_1.

Now let F_1 be an arbitrary closed set. Then (27.4) follows from the above special case, part (2) of the theorem, and the following lemma.

LEMMA 27.3. *Let F be a closed subset of X. Then there exist closed subsets F_1, $i = 1, 2, \ldots$, such that $F_1 \supset F_2 \supset \cdots$, $F = \bigcap F_i$, and for each i the boundary ∂F_i is a (not necessarily connected) manifold of dimension $n-1$, $n = \dim X$.*

PROOF OF THE LEMMA. There exists a function $f \in C^\infty(X)$ such that $f(x) = 0$ for $x \in F$, $f(x) > 0$ for $x \notin F$. By Sard's lemma, for almost all $y \in \mathbb{R}$ either the set $F_y = \{x \mid f(x) < y\}$ is empty, or its boundary is an $(n-1)$-dimensional submanifold. In this set of numbers y take the sequence $y_i \searrow 0$. The sets F_{y_i}, $i \geq 1$, satisfy the required conditions. Lemma 27.3 and Theorem 27.1 are proved.

§28. Description of class O representations ([15])

28.1. Construction of representations. Formulations of theorems. We introduce an important dynamical system $(\Omega, \mu, D(X))$ as follows. Let $\Omega = \Omega(X)$ be the collection of all nonempty closed subsets of X. For any $F \subset X$ denote

$$\Omega_F = \{\omega \in \Omega \mid \omega \subset F\}.$$

Denote by $B = B^F$ the smallest σ-algebra containing the sets Ω^F, where F runs over all open subsets $F \subset X$. Let us emphasize that if in this definition we allow F to run over all closed (or all Borel) subsets of X, we obtain the same σ-algebra B. In the constructions below it will be convenient to take for F all closed sets.

The group $D(X)$ acts naturally on Ω, preserving the σ-algebra B; we will denote this action by $\omega \mapsto \omega g$. Let μ be a measure on Ω that is quasi-invariant with respect to this action and satisfies the continuity conditions from 1.4. We obtain the dynamical system $(\Omega, \mu, D(X))$.

Now we pass to representations. Let H^0 be a Hilbert space and $a(g, \omega)$, $g \in D(X)$, $\omega \in \Omega \bmod 0$, a cocycle of the dynamical system $(\Omega, \mu, D(X))$ taking values in the group of unitary operators $U(H^0)$; we assume the usual measurability and continuity conditions in the definition of a cocycle (see 1.4) to be satisfied. We get the unitary representation $g \mapsto T(g)$, $g \in D(X)$, in the space $H = L_2(\Omega, \mu, H^0)$ given by the formula

$$(T(g)f)(\omega) = \sqrt{j(g,\omega)}\,a(g,\omega)f(g\omega), \qquad j(g,\omega) = d\mu(\omega g)/d\mu(\omega). \tag{28.1}$$

Let us impose two conditions on the cocycle $a(g, \omega)$. To describe these conditions, take a ball $B \subset X$ and consider the sets

$$\Omega_{X \setminus \overline{B}} = \{\omega \in \Omega \mid \omega \subset X \setminus \overline{B}\},$$
$$\Omega^B = \{\omega \in \Omega \mid B \subset \omega\}.$$

The group $D(B)$ acts trivially on both these sets. Now we describe the conditions the cocycle should satisfy.

(α) For any ball $B \subset X$

$$a(g, \omega) = I, \qquad g \in D(B), \quad \omega \in \Omega_{X \setminus B} \bmod 0.$$

The second condition is imposed only in the case $\mu(\Omega^B) > 0$; in this case we get the representation of the group $D(B)$ in the space $L_2(\Omega^B, \mu)$ by the formula

$$f(\omega) \mapsto a(g, \omega) f(\omega), \qquad f \in L_2(\Omega^B, \mu), \quad g \in D(B). \tag{28.2}$$

The second condition is formulated as follows.

(β) The representation (28.2) does not have nonzero invariant vectors.

Representations of the form (28.1) for which the corresponding cocycle satisfies the conditions (α) and (β) are called *elementary*; $\dim H^0$ is called the *rank* of an elementary representation. One can easily see that any elementary representation belongs to the class O (see §27); for more details, see 28.2.

Let us note that for the majority of "natural" dynamical systems $(\Omega, \mu, D(X))$ almost all (with respect to the measure μ) configurations $\omega \in \Omega$ have no inner points. Therefore, $\mu(\Omega^B) = 0$ and condition (β) becomes redundant.

To analyze the irreducibility of representations, we need two additional conditions (they are not included into the definition of an elementary representation):

(γ) The dynamical system $(\Omega, \mu, D(X))$ is ergodic.

(δ) The cocycle $a(g, \omega)$ is irreducible (see 1.4).

Let us formulate the main results of this section. Take two elementary representations T_i constructed from measures μ_i on Ω, Hilbert spaces H_i^0, and cocycles $a_i(g, \omega)$, $i = 1, 2$. Let us recall that in 1.4 we introduced the space $L(a_1, a_2)$ of intertwining operators for a pair of cocycles, the space $L(T_1, T_2)$ of intertwining operators for a pair of representations, and the natural mapping

$$L(a_1, a_2) \to L(T_1, T_2). \tag{28.3}$$

THEOREM 28.1. (1) *If $\mu_1 \perp \mu_2$, then $T_1 \perp T_2$, (i.e., $L(T_1, T_2) = \{0\}$).*

(2) *If $\mu \sim \mu_2$, then the mapping* (28.3) *is an isomorphism.*

COROLLARY 28.1. *An elementary representation of the form* (28.1) *is irreducible if and only if the conditions* (γ) *and* (δ) *(and, of course, the conditions* (α) *and* (β) *entering the definition of an elementary representation) are satisfied.*

COROLLARY 28.2. *Two elementary representations T_i, $i = 1, 2$, are equivalent if and only if the measures μ_1, μ_2 and the cocycles a_1, a_2 are equivalent.*

These results are rather simple. The following theorem is much deeper.

THEOREM 28.2. *Each representation of class O is equivalent to a sum of elementary representations with pairwise distinct ranks.*

COROLLARY 28.3. *Each irreducible class O representation of the group $D(X)$ is equivalent to an elementary representation satisfying the conditions* (γ) *and* (δ) *(and, of course, conditions* (α) *and* (β)).

Now we pass to proofs.

28.2. Properties of elementary representations. In the next lemma B denotes a ball in X, $\Omega(B)$ the space of nonempty subsets $\omega \subset B$ that are closed in B.

LEMMA 28.1. *Let ν be a finite measure on $\Omega(B)$ invariant under $D(B)$. Then it is concentrated on a single element $\omega_0 = B$ of $\Omega(B)$, i.e., $\nu(\Omega(B) \setminus \{\omega_0\}) = 0$.*

PROOF. Take balls B_1, B_2 in B such that $\overline{B_1} \subset B_2$, $\overline{B_2} \subset B$. There exists a diffeomorphism $g \in D(B)$ mapping B_1 to B_2. Since the measure ν is invariant under g, we have $\nu(\Omega(B \backslash \overline{B}_1)) = \nu(\Omega(B \backslash \overline{B}_2))$. On the other hand, $\Omega(B \backslash \overline{B}_2) \subset \Omega(B \backslash \overline{B}_1)$. Therefore, the sets $\Omega(B \backslash \overline{B}_1)$ and $\Omega(B \backslash \overline{B}_2)$ differ by a set of ν-measure zero. Hence, the latter assertion holds for any two balls B_1, B_2 in B such that $\overline{B_1} \subset B_2$, $\overline{B_2} \subset B$. Fix the ball B_1 and start to decrease the ball B_2, replacing it by balls $B_3, B_4, \ldots$, such that $B_2 \subset B_3 \subset \ldots$, $\bigcap B_k = B$. This yields the equality $\nu(\Omega(B \setminus B_1)) = 0$, where B_1 is a ball in B such that $\overline{B_1} \subset B$. Hence, $\nu(\Omega(B) \setminus \{\omega_0\}) = 0$. The lemma is proved.

Now we fix an elementary representation of the form (28.1). Below we will use the notation H_F, P_F introduced in 27.2.

LEMMA 28.2. *Let F be a closed subset of X. Then the subspace*

$$H_F = \{f \in H \mid T(g)f = f \text{ for each } g \in D(X \setminus F)\} \tag{28.4}$$

coincides with the subspace of all $f \in H$ such that $f(\omega) = 0$ for $\omega \in \Omega \setminus \Omega_F \bmod 0$.

Lemma 28.2 means that the operator P_F (projection to H_F) coincides with the operator of multiplication by the characteristic function of the set Ω_F.

PROOF. Property (α) in 28.1 implies that any function $f \in H$ vanishing outside of Ω_F belongs to H_F. Let us prove the converse result. Take a function $f \in H_F$ and prove that it vanishes if $\omega \in \Omega \setminus \Omega_F \bmod 0$. Take a ball B such that $\overline{B} \subset X \setminus F$ and set

$$E^B = \{\omega \in \Omega \mid \omega \cap \overline{B} \neq \varnothing\}.$$

Let us prove that the restriction of f to E^B vanishes everywhere. Since B is an arbitrary ball, this implies the lemma.

Define the measure τ on Ω by the formula $d\tau = |f(\omega)|^2_{H^0} d\mu$; this measure is finite since $f \in L_2(\Omega, \mu, H^0)$. Furthermore, the invariance of the function f under the subgroup $D(B)$ implies the invariance of the measure τ under this subgroup. Denote by η the restriction of the measure τ to E^B. The mapping $E^B \to \Omega(B)$ given by $\omega \mapsto \omega \cap B$ transfers the measure η to $\Omega(B)$; the resulting measure is finite and invariant under $D(B)$. By Lemma 28.1, it is concentrated on the configuration $\omega_0 = B$ from $\Omega(B)$. This means that the measure τ is concentrated on the subset $\Omega^B \subset E^B$. Hence, the function $f|_{E^B}$ is concentrated on Ω^B. In this case, the invariance of $f|_{E^B}$ under $D(B)$ is written in the form

$$a(g, \omega) f(\omega) = f(\omega), \qquad g \in D(B), \quad \omega \in \Omega^B \bmod 0.$$

By the condition (β) from the definition of an elementary representation, we have $f|_{E^B} = 0$. The lemma is proved.

In particular, this lemma implies that an elementary representation belongs to class O.

28.3. Proof of Theorem 28.1. Consider an intertwining operator M. We have $MP_F^1 = P_F^2 M$, where P_F^k is the projection to the subspace

$$H_F^k = \{f \in H_k = L_2(\Omega, \mu_k, H_k^0) \mid T_k(g)f = f \text{ for any } g \in D(X \setminus F)\}.$$

The description of subspaces H_F^k from Lemma 28.2 implies that

$$MQ_1(E) = Q_2(E)M, \tag{28.5}$$

where E is an arbitrary Borel set from Ω, $Q_k(E)$ the operator of multiplication by the characteristic function of the set E; this operator acts in the space $H_k = L_2(\Omega, \mu_k, H_k^0)$. This easily implies that if $\mu_1 \perp \mu_2$, then $M = 0$, i.e., $T_1 \perp T_2$. Therefore, the first statement of Theorem 28.1 is proved.

Now let $\mu_1 \sim \mu_2$. Then (28.5) implies that M is given by the formula

$$(Mf)(\omega) = M(\omega)f(\omega), \qquad \omega \in \Omega \bmod 0,$$

where $\omega \mapsto M(\omega)$ is a weakly measurable family of linear mappings from H_1^0 to H_2^0. Finally, the fact that M intertwines the representations T_1 and T_2 implies that $M \in L(a_1, a_2)$. Theorem 28.1 is proved.

28.4. Proof of Theorem 28.2. We consider the case of a compact manifold X (for a noncompact X the proof is similar). Consider a representation V of class O, the corresponding projections P_F, and the von Neumann algebra A generated by these projections. By Theorem 27.1, the algebra A is commutative; it is clear also that $I \in A$. Denote by M the spectrum (the space of maximal ideals) of the algebra A. Then A can be identified with the Banach algebra $C(M)$ and the space H with the direct integral $\int_M H(m)\,d\tau = \mathcal{H}$, where τ is a Borel measure on M, $\{H(m)\}$ is the direct integral of Hilbert spaces (see [**26**, Chapter IV, §17]). To any Borel subset $E \subset M$ we associate the operator Q_E in $\mathcal{H}$ acting as the multiplication by the characteristic function χ_E of the set E. Let us discuss in more details the passage from A and H to $C(M)$ and $\mathcal{H} = \int_M H(m)\,d\tau$. Any element $g \in D(X)$ determines an automorphism of the algebra A by the formula $B \mapsto T(g)BT(g^{-1})$, $B \in A$. Under the isomorphism $A \simeq C(M)$, this automorphism becomes the automorphism of the algebra $C(M)$ given by the formula $f(m) \mapsto f(mg)$, $f \in C(M)$, where $m \mapsto mg$, $m \in M$ is a homeomorphism of the space M.

To a closed subset $F \subset X$ there corresponds an open and closed set $F^* \subset M$ such that under the isomorphism $A \simeq C(M)$ the operator $P_F \in A$ becomes the characteristic function χ_{F^*} of the set F^*. Theorem 27.1 implies that the correspondence $F \mapsto F^*$ has the following properties:

(A_1) if $F_1 \supset F_2 \supset \cdots$, $F = \bigcap_k F_k$, then $F_1^* \supset F_2^* \supset \cdots$, $F^* = \bigcap_k F_k^*$;
(A_2) $(\varnothing)^* = \varnothing$;
(A_3) $(F_1 \cap F_2)^* = F_1^* \cap F_2^*$;
(A_4) $(Fg)^* = F^*g$ for $g \in D(X)$.

Under the isomorphism of Hilbert spaces $H \simeq \int_M H(m)\,d\tau$ the representation V becomes the representation U in the space $\int_M H(m)\,d\tau$, and the operator P_F becomes the operator Q_{F^*}.

LEMMA 28.3. *We have $U(g^{-1})Q_E U(g) = Q_{Eg}$ for any Borel set $E \subset M$ and any $g \in D(X)$.*

PROOF. The collection of all Borel sets $E \subset M$ satisfying this relation is closed under Borel operations. For sets of the form $E = F^*$, where F is a closed subset of X, Lemma 28.3 follows from Theorem 27.1. But the sets of the form F^* generate the entire σ-algebra of Borel sets in M, and Lemma 28.3 is proved.

LEMMA 28.4. *The measure τ is quasi-invariant with respect to $D(X)$.*

PROOF. If E is a Borel subset of M and $\tau(E) = 0$, then $Q_E = 0$. Lemma 28.3 implies that $Q_{Eg} = 0$, whence $\tau(Eg) = 0$. The lemma is proved.

LEMMA 28.5. *For each $k \in \{1, 2, \dots, \infty\}$ the set*

$$M_k = \{m \in M \mid \dim H(m) = k\}$$

is invariant mod 0 *under each $g \in D(X)$ (i.e., $\tau(M_k \triangle M_k g) = 0$ for $g \in D(X)$).*

PROOF. Choose $k \in \{1, 2, \dots, \infty\}$ and consider the set

$$N_k = \{m \in M \mid \dim H(m) \le k\}.$$

Then there exists a collection $\{f_j\} \subset Q(N_k)\mathcal{H}$ of cardinality k such that the vectors $\{Q_E f_j,\ E \subset N_k \text{ a Borel subset}\}$ span the subspace $Q_{N_k}\mathcal{H}$.

Lemma 28.3 implies that the vectors $\{Q_L T(g) f_j,\ L \subset N_k g \text{ is a Borel subset}\}$ span the subspace $Q_{N_k g}$. This means that $\dim H(m) \le k$ for $m \in N_k g$. Therefore, $\tau(N_k g \setminus N_k) = 0$. Similarly, $\tau(N_k \setminus N_k g) = 0$. Therefore, $\tau(N_k g \triangle N_k) = 0$. Lemma 28.5 is proved

Lemma 28.5 gives a decomposition of the space $\mathcal{H}$ into $D(X)$-invariant subspaces, with the representation in each of these subspaces having the described form with $\dim H(m) = \text{const}$ for $m \in M$. Therefore in the considerations below we can assume that $\dim H(m) = \text{const}$. This assumption leads to an isomorphism of Hilbert spaces

$$\mathcal{H} \simeq L_2(M, \tau, H^0), \tag{28.6}$$

where H^0 is a fixed Hilbert space. Hence the initial representation is realized in the space $L_2(M, \tau, H^0)$ using the dynamical system $(M, \tau, D(X))$. At the next step we pass from the space M to the space Ω consisting of all closed subsets of the manifold X (see 28.1).

Let us construct a mapping $\Phi\colon M \to \Omega$ as follows. To a point $m \in M$ we associate the subset $\omega = \bigcap_\alpha F_\alpha \subset X$, where F_α runs over all closed subsets of X such that $m \in F_\alpha^*$. The set ω is nonempty: indeed, otherwise we could find a finite family $\{F_k,\ 1 \le k \le r\} \subset \Omega(X)$ such that $\bigcap_{k=1}^r F_k = \varnothing$. By properties ($A_2$), ($A_3$) above, $\bigcap_{k=1}^r (F_k)^* = \varnothing$, contradicting the assumption $m \in \bigcap_{k=1}^r (F_k)^*$. The mapping $\Phi\colon M \to \Omega$ is constructed. Let us consider its properties.

LEMMA 28.6. *If $F = \overline{F} \subset X$, then $\Phi^{-1}(\Omega_F) = F^*$.*

PROOF. Take a point $m \in F^*$. Then $\Phi(m) = \bigcap F_\alpha$, where $F_\alpha = \overline{F_\alpha} \subset X$, $m \in F_\alpha^*$. In particular, we can take F as one of the sets F_α. This gives $\Phi(m) \subset F$, i.e., $\Phi(m) \in \Omega_F$. We proved that $\Phi(F^*) \subset \Omega_F$.

Now let $m \in M$ and $\Phi(m) \subset F$. Let us prove that $m \in F^*$. Take a metric $d(x_1, x_2)$ on X and for each $\varepsilon > 0$ consider the set $F_\varepsilon = \{x \in X \mid d(x, F) \le \varepsilon\}$. By the definition of the mapping Φ, there exist closed sets $X_1 \supset X_2 \supset \dots$ in X such that $X_k \subset F_\varepsilon$, $m \in \bigcap X_k^*$. Since $\omega \subset F$, we have $X_k \subset F_\varepsilon$ for sufficiently large k. But then $X_k^* \subset F_\varepsilon^*$, hence $m \in F_\varepsilon^*$. Therefore,

$$m \in \bigcap_{n=1}^{\infty} F_{1/n}^* = \Big(\bigcap_{n=1}^{\infty} F_{1/n}\Big)^* = F,$$

as required. Lemma 28.6 is proved.

This lemma implies, in particular, that the mapping $\Phi\colon M \to \Omega$ is Borel (with respect to Borel structures on M and Ω). Define the measure μ on Ω (more precisely, on the σ-algebra B^Ω) by the formula $\mu(E) = \tau(\Phi^{-1}(E))$, $E \in B^\Omega$). Lemma 28.6 implies that the mapping Φ is an isomorphism of spaces with measure (Ω, μ) and (M, τ). Therefore, we get an isomorphism of Hilbert spaces

$$\Phi_*\colon L_2(\Omega, \mu, H^0) \to L_2(M, \tau, H^0)$$

given by the formula $(\Phi_* f)(m) = f(\Phi(m))$, $f \in L_2(\Omega, \mu, H^0)$. Combining this isomorphism with the isomorphism (28.6), we obtain an isomorphism of Hilbert spaces

$$H \simeq L_2(\Omega, \mu, H^0).$$

This allow us to consider operators acting in H as operators in $L_2(\Omega, \mu, H^0)$. In particular, the representation V becomes a representation T in $L_2(\Omega, \mu, H^0)$. By Lemma 28.6, the operator P_F becomes the operator of multiplication by the characteristic function of the set Ω_F (here $F = \overline{F} \subset X$). Therefore, the linear spaces

$$\{f \in L_2(\Omega, \mu, H^0) \mid T(g)f = f \text{ for all } g \in D(X \setminus F)\}, \tag{28.7}$$

$$\{f \in L_2(\Omega, \mu, H^0) \mid f(\omega) = 0 \text{ for } \omega \in \Omega \setminus \Omega_F \bmod 0\} \tag{28.8}$$

coincide.

Next, Lemma 28.3 implies that for any Borel subset $L \subset \Omega$ we have

$$T(g^{-1}) Q_L T(g) = Q_{Lg}, \tag{28.9}$$

where Q_L denotes the operator of multiplication by the characteristic function of the set L.

Now we can easily prove that the representation T is of the form (28.1) and the conditions (α), (β) from 28.1 are satisfied.

Introduce the operators

$$(W(g)f)(\omega) = \sqrt{j(g,\omega)}\, f(\omega g), \qquad g \in D(X), \quad f \in L_2(\Omega, \mu, H^0).$$

in the space $L_2(\Omega, \mu, H^0)$. It is clear that for these operators the relation similar to (28.9) holds. This implies that the operators $T(g)W(g^{-1})$ commute with the operators Q_L for any Borel set in Ω. Therefore, $T(g)W(g^{-1})$ is the operator of multiplication by a vector-valued function $a(g, \omega)$. This proves formula (28.1). Properties (α) and (β) in the definition of an elementary representation follow from the coincidence of spaces (28.7) and (28.8). Theorem 28.2 is proved.

§29. On dynamical systems of the form $(\Omega, \mu, D(X))$; the Shavgulidze measure

Results of §28 reduce the classification of class O representations to the description of dynamical systems of the form $(\Omega, \mu, D(X))$ and cocycles satisfying the conditions from 1.4. The last two problems are "wild" and do not allow nice solutions. One can only speak about sufficiently natural examples of such objects. This will be done in §§30–32. In this section we present a beautiful construction suggested by E. T. Shavgulidze [**37**].

Consider the group $D_r^n = \mathrm{Diff}_r([0,1]^n)$ of all C^r-smooth transformations of the cube $[0,1]^n$. The group $D^n = \mathrm{Diff}_\infty^0([0,1]^n)$ of the diffeomorphisms of the cube that are identical on the boundary, acts on D_r^n by left shifts. Shavgulidze constructed a measure on D_r^n that is quasi-invariant under this action (subject to some relations between r and n). We reproduce this construction for $n = 1$.

Let D_2 be the group of all C^2-smooth transformations g of the segment $[0,1]$ that preserve the points 0 and 1 and satisfy the conditions $g'(0) = 1$, $g''(0) = 0$. Let C^0 be the space of continuous functions $f\colon [0,1] \to \mathbb{R}$ such that $f(0) = 0$. The formula

$$g \mapsto \frac{g''}{g'} \tag{29.1}$$

defines a one-to-one mapping from D_2 to C_0. Under this correspondence, the left shift $g \mapsto g_0 g$, $g \in D_2$, $g_0 \in D(0,1)$, becomes the action

$$f \mapsto f + \frac{g_0''}{g_0'} e^{Jf}, \qquad Jf(x) = \int_0^x f(t)\, dt.$$

Consider the Wiener measure in C_0. It is quasi-invariant under the latter action (as follows from general theorems about shifts of Gaussian measures). Passing this measure to D_2 by the mapping (29.1), we obtain a measure on D_2 that is quasi-invariant under the action of the group $D(0,1)$.

The Shavgulidze measure also gives a simple method to construct a measure on $\Omega = \Omega(0,1)$ (the set of closed subsets of the segment $(0,1)$) that is quasi-invariant under $D(0,1)$. It suffices to choose an arbitrary set $\omega_0 \in \Omega$ and to consider the mapping $D_2 \to \Omega$ given by the formula $g \mapsto \omega_0 g^{-1}$. It maps the above measure to the space Ω; the resulting measure on Ω is quasi-invariant under the action of $D(0,1)$.

It is not known whether or not the above actions are ergodic; all the more, there is no information about the unitary representations generated by these actions. Further investigation of the Shavgulidze measure could provide interesting examples of representations of diffeomorphism groups.

§30. Representations associated to spaces of finite configurations

It was shown in §28 that a large class of representations of the group $D(X)$ can be realized using the space of closed subsets (configurations) of the manifold X. Now we study representations associated to the configuration spaces Ω_m consisting of collections of m pairwise distinct points of the manifold X; let us recall that the manifold Ω_m was considered in 26.2. We consider not only the group $D(X)$ itself, but also the subgroups indicated in §20; for the group $D(X)$ itself the main result of this section follows from the results of §28. We will also prove an analog of the Mackey theorem for induced representations of $D(X)$.

Everywhere in this section by X we denote a connected manifold, and by $G = G(X)$ one of the groups $D(X)$, $D(X, v^n)$, $D(X, \omega^2)$, $D(X, \alpha^1)$; let us recall that for an open region $Y \subset X$ the group $G(Y)$ is embedded into the group $G(X)$.

30.1. Induced representations of the group G. Consider a dynamical system (Ω_m, μ_m, G), where μ_m is a fixed measure on Ω_m given in local coordinates by a smooth positive density. For a point $\omega \in \Omega_m$ we define (following §26) the stabilizer $G^\omega = G^\omega(X) = \{g \in G \mid \omega = \omega g\}$; similarly one defines the group $G^\omega(Y)$ for a region $Y \subset X$. Choose a point $\omega_0 \in \Omega_m$ and take a unitary representation V of its stabilizer G^{ω_0} in a Hilbert space H^0. Let $T = \operatorname{Ind} V$ be the induced representation of the group G. To write it down explicitly, choose a Borel mapping $\theta \colon \Omega_m \to G$ satisfying the condition $\omega_0\theta(\omega) = \omega$, $\omega \in \Omega_m$. The representation T can be realized in the space $H = L_2(\Omega_m, \mu_m, H^0)$ by the formula

$$(T(g)f)(\omega) = \sqrt{j(g,\omega)}a(g,\omega)f(\omega g), \qquad g \in G, \quad f \in H, \tag{30.1}$$

$$j(g,\omega) = d\mu_m(\omega g)/d\mu_m(\omega),$$

$$a(g,\omega) = V(\theta(\omega)g\theta(\omega g))^{-1}). \tag{30.2}$$

Formulate the conditions on the representation V; in the formulation of these conditions we use the groups introduced in §26.

(α) The restriction of the representation V to $G_*(X \setminus \{\omega_0\})$ is trivial.

(β) Let $\omega_0 = \{x_i^0,\ 1 \le i \le m\}$. Then there exists a ball $B \subset X$ that contains the points x_i^0, $1 \le i \le m-1$, but not the point x_m^0, such that the restriction of the representation V to $G_*(X \setminus \overline{B}) \cap G^{\omega_0}$ does not have nonzero invariant vectors.

Let us remark that the condition (α) is obligatory, whereas the condition (β) will be imposed only occasionally.

Before turning to induced representations, let us prove an important consequence of the condition (α).

PROPOSITION 30.1. *A representation V of the group G^{ω_0} satisfying the condition (α) is a sum of representations, each of which is trivial on one of the subgroups $G_r^{\omega_0}$, $r \ge 0$.*

PROOF. Let h be a vector in the space of representation V, $|h| = 1$. Lemma 26.1 implies $(V(g_r)h, h) \to 1$ for any sequence $\{g_r \in G_r^{\omega_0}\}$. Hence, $|(V(g)h,h) - 1| < 1/2$ for all $g \in G_r^{\omega_0}$ if r is sufficiently large. But then the representation $g \to V(g)$ has a nonzero invariant vector, and the proposition follows.

Proposition 30.1 allows us to consider only those representations V of the group G^{ω_0} that come from representations of the finite-dimensional (in general, not connected) Lie group $K_r^{\omega_0} = G^{\omega_0}/G_r^{\omega_0}$. As a special case, we consider representations obtained from the representations of the group $(J_r^m \times \Gamma^m/\Gamma_0) \cdot S_m$ using the homomorphism (26.3).

30.2. Intertwining operators; formulations of the theorems. Take two representations $g \mapsto V_i(g)$, $g \in G^{\omega_0}$, $i = 1, 2$, satisfying the condition (α) from the previous subsection. Let $T_i = \operatorname{Ind} V_i$, $i = 1, 2$, be the induced representations. Recall (see 1.4) that there exists a natural mapping of spaces of intertwining operators

$$L(V_1, V_2) \to L(T_1, T_2). \tag{30.3}$$

THEOREM 30.1. *Let one of the following two conditions be satisfied:* (1) *there is no finite positive G-invariant measure on X;* (2) *the representations V_i, $i = 1, 2$, satisfy the condition* (β). *Then the mapping* (30.3) *is an isomorphism of linear spaces.*

COROLLARY 30.1. *If a representation V is irreducible, then the representation* $\operatorname{Ind} V$ *is also irreducible. If representations V_i, $i = 1, 2$ are disjoint, then the representations $T_i = \operatorname{Ind} V_i$ are also disjoint.*

Let us list all the cases when the condition (1) of Theorem 30.1 is satisfied: (1) $G = D(X)$; (2) $G = D(X, \alpha^1)$; (3) $G = D(X, v^n)$, $\int_X v^n = \infty$; (4) $G = D(X, \omega^2)$, $\int_X \omega^{2p} = \infty$, $2p = \dim X$. (Of course in the last two cases the manifold X is assumed to be oriented.)

In the next theorem we consider the spaces Ω_{m_1}, Ω_{m_2} with $m_1 \neq m_2$. Choose points $\omega_i \in \Omega_{m_i}$ and consider representations V_i of the groups G^{ω_i}, $i = 1, 2$, satisfying the condition (α).

THEOREM 30.2. *Let $m_1 \neq m_2$. Let at least one of the conditions* (1), (2) *in Theorem* 30.1 *be satisfied. Then the representations $T_i = \operatorname{Ind} V_i$, $i = 1, 2$, are disjoint, i.e., $L(T_1, T_2) = \{0\}$.*

30.3. Proofs of Theorems 30.1 and 30.2. Take a representation V of the group G^{ω_0} satisfying the condition (α) and consider the representation $T = \operatorname{Ind} V$ of the group G. Let at least one of the conditions (1), (2) of Theorem 30.1 be satisfied (of course, in (2) we must write V instead of V_i). Choose a ball $B \subset X$ and consider the subspace $H_B \subset H = L_2(\Omega_m, \mu_m, H^0)$ of the form

$$H_B = \{f \in H \mid T(g)f = f \text{ for all } g \in G_*(X \setminus B)\}.$$

LEMMA 30.1. *The subspace H_B coincides with the set of functions $f \in H$ satisfying the condition $f(\omega) = 0$ for all $\omega \in \Omega_m \setminus \Omega_m(B)$* mod 0.

PROOF OF THE LEMMA. Since the representation V acts trivially on the subgroup $G_*(X \setminus \{\omega_0\})$, any function $f \in H$ vanishing outside $\Omega_m(B)$ belongs to the subspace H_B. Let us prove the converse. For each $k = 0, 1, \ldots, m$ introduce the set $E_k = \{\omega \in \Omega_m \mid |\omega \cap B| = k\}$. Here $|\omega \cap B|$ is the cardinality of the set $\omega \cap B$. Fix a number $k = 0, 1, \ldots, m-1$ and a function $f \in H_B$, and consider the function $f \mid E_k$. Represent an element $\omega \in E_k$ in the form $\omega = (\alpha, \beta)$, where $\alpha = \omega \cap B$, $\beta = \omega \cap (X \setminus B)$. The invariance of $f \mid E_k$ with respect to the group $G_*(X \setminus \overline{B})$ can be written in the form

$$\begin{aligned} &\sqrt{j(\alpha, \beta g)}a(g, \alpha, \beta)f(g, \alpha, \beta) = f(\alpha, \beta), \\ &g \in G_*(X \setminus \overline{B}), \quad (\alpha, \beta) \in E_k \bmod 0, \\ &j(g, \alpha, \beta) = j(g, \omega) = d\mu_m(\omega g)/d\mu_m(\omega). \end{aligned} \tag{30.4}$$

Let us prove that $f \mid E_k = 0$. Let condition (1) of Theorem 30.1 be satisfied. Formula (30.4) implies that the measure ν on E_k defined by $d\nu = |f(\alpha, \beta)|^2 d\mu$ is invariant under $G_*(X \setminus \overline{B})$. Furthermore, $\nu(E_k) \leq |f| < \infty$. This easily implies the existence of a smooth finite G-invariant measure on X. Condition (1) of Theorem 30.1 gives $\nu = 0$, i.e., $f \mid E_k = 0$ for all $k < m$. Hence $f(\omega) = 0$ for all ω that are not subsets of B. This proves the statement of the lemma if condition (1) of Theorem 30.1 is satisfied.

Now let there exist a positive finite G-invariant smooth measure on X. This means that either $G = D(X, v^n)$, $\int_X v^n = 1$, or $G = D(X, \omega^2)$, $\int_X \omega^{2p} < \infty$. Assume that in the construction of the representation (30.1) we use this G-invariant measure. Formula (30.4) is rewritten as follows:

$$\begin{aligned} &a(g,\alpha,\beta)f(\alpha,\beta g) = f(\alpha,\beta), \qquad g \in G_*(X \setminus B), \quad (\alpha,\beta) \in E_k, \\ &a(g,\alpha,\beta) = V(\theta(\alpha,\beta)g(\theta(\alpha,\beta g))^{-1}). \end{aligned} \tag{30.5}$$

Let us identify the function $f(\alpha, \beta)$ with the vector-valued function $\beta \mapsto f_1(\beta)$, $\beta \in \Omega_{m-k}(X \setminus B)$, with values in the Hilbert space $L_2(\Omega_k(B), \mu_k, H^0)$. Then the equality (30.5) can be rewritten in the form

$$A(g,\beta)f_1(\beta g) = f_1(\beta), \qquad \beta \in \Omega_{m-k}(X \setminus \overline{B}) \bmod 0, \quad g \in G_*(X \setminus \overline{B}).$$

Let us recall that in the realization of the representation $T = \operatorname{Ind} V$ by formula (30.1) we can freely change the point ω_0 and the mapping $\theta\colon \Omega_m \to G$ satisfying the condition $\omega_0\theta(\omega) = \omega$. Using this freedom, we can assume that the mapping $(\alpha, \beta) \mapsto A(g, \beta)$ is strongly continuous whenever g and β run over neighborhoods of elements $e \in G_*(X \setminus \overline{B})$ and $\beta_0 = \omega_0 \cap (X \setminus \overline{B})$ respectively. Let us prove that in a neighborhood of the point β_0 the function f_1 coincides with a strongly continuous function. Choose a family $\{g_t\} \subset G_*(X \setminus \overline{B})$ depending continuously on the point t from a ball in a finite-dimensional Euclidean space and acting simply transitively in a neighborhood of β_0 (one can easily see that such a family exists). Then for any t we have

$$A(g_t,\beta)f_1(\beta g_t) = f_1(\beta) \tag{30.6}$$

for almost all β from the above neighborhood. But then there exists a point β_1 such that for almost all t the equality (30.6) holds for $\beta = \beta_1$. The mapping $t \mapsto (A(g_t, \beta_1))^{-1} f_1(\beta_1)$ yields a strongly continuous vector-valued function that coincides with f_1 almost everywhere in a neighborhood of β_0.

Hence, in formulas (30.6) we can assume that $f \mapsto f_1(\beta)$ is a strongly continuous vector-valued function in a neighborhood of the point β_0. This means that in this formula we can take $\beta = \beta_0$, $g \in G_*(X \setminus B) \cap G^{\omega_0}$. Next, using the freedom in the choice of the mapping θ, we can assume that $\theta(\alpha, \beta) = \theta_0(\alpha)\theta_1(\beta)$, where $\theta_0(\alpha) \in G(B)$, $\theta_1(\beta) \in G(X \setminus \overline{B})$, and the mapping $\beta \mapsto \theta_1(\beta)$ is continuous in a neighborhood of β_0. Therefore, (30.5) yields

$$V(g)f(\alpha_0,\beta_0) = f(\alpha_0,\beta_0), \qquad g \in G_*(X \setminus B) \cap G^{\omega_0}, \quad \alpha \in \Omega_k(B) \bmod 0. \tag{30.7}$$

Let us recall that the representation V satisfies condition (2) of Theorem 30.1. This condition and (30.7) show that $f(\alpha, \beta_0) = 0$ for all $\alpha \in \Omega_k(B) \bmod 0$. This shows that $f \mid E_k = 0$ for $k < m$. Therefore, $f(\omega) = 0$ for almost all $\omega \not\subset B$. Lemma 30.1 is proved.

Below we will need the following notation. For a measurable set $Y \subset X$ consider the set $\Omega_m(Y) = \{\omega \in \Omega_m \mid \omega \subset Y\}$ and denote by $P^m(Y)$ the operator of multiplication (in the space $L_2(\Omega_m, \mu_m, H^0)$) by the characteristic function of the set $\Omega_m(Y)$.

LEMMA 30.2. *Let a bounded operator K acting from $H_1 = L_2(\Omega_{m_1}, \mu_{m_1}, H_1^0)$ to $H_2 = L_2(\Omega_{m_2}, \mu_{m_2}, H_2^0)$ satisfies the condition*

$$KP^{m_1}(B) = P^{m_2}(B)K \tag{30.8}$$

for any ball $B \subset X$. Then:

(1) *if $m_1 = m_2$, then*

$$(Kf)(\omega) = \mathcal{N}(\omega)f(\omega), \tag{30.9}$$

where $\omega \mapsto \mathcal{N}(\omega)$, $\omega \in \Omega_m \bmod 0$, is a weakly measurable mapping from Ω_m to the space of linear operators from H_1^0 to H_0^2, $\operatorname{ess\,sup} \|\mathcal{N}(\omega)\| < \infty$;

(2) *if $m_1 \neq m_2$, then $K = 0$.*

PROOF. Formula (30.8) easily implies that

$$KP^{m_1}(Y) = P^{m_2}(Y)K \tag{30.10}$$

for any measurable set $Y \subset X$.

Let $m_1 = m_2 = m$. In this case it suffices to consider the case $H_1 = H_2 = H$. The operators $P^m(Y)\colon H \to H$ generate the algebra consisting of operators of multiplication by all bounded measurable functions. This implies the lemma for $m_1 = m_2$.

Now let $m_2 < m_1$. Take pairwise disjoint measurable sets $Y_j \subset X$, $1 \le j \le r$. Denote by $Q^{m_i}_{\{Y_j\}}$ the operator of multiplication by the characteristic function of the set $\{\omega \in \Omega_{m_i} \mid \omega \cap Y_j \neq \varnothing \text{ for all } j\}$. One can easily verify the formula (where we use the notation $Y_j^c = X \setminus Y_j$)

$$Q^{m_i}_{\{Y_j\}} = \prod_j (I - P^{m_j}(Y_j^c)) = \sum_s (-1)^s P^{m_j}(Y^c_{j_1} \cap \dots \cap Y^c_{j_s}).$$

Together with (30.10), this formula implies that

$$KQ^{m_1}_{\{Y_j\}} = Q^{m_2}_{\{Y_j\}}K.$$

In particular, let $m_1 = r$. Since $m_2 < m_1$, we have $Q^{m_2}_{\{Y_j\}} = 0$. Therefore,

$$KQ^{m_1}_{\{Y_j\}} = 0.$$

It is clear that the identity operator can be represented as a sum of operators of the form $Q^{m_1}_{\{Y_j\}}$. Therefore, $K = 0$ and Lemma 30.2 is proved.

Now we return to the representations T_1, T_2 indicated in Theorems 30.1 and 30.2. Let $K \subset L(T_1, T_2)$. Lemma 30.1 implies that if B is a ball in X, then

$$K(L_2(\Omega_{m_1}(B), \mu_{m_1}, H_1^0)) \subset L_2(\Omega_{m_2}(B), \mu_2, H_2^0),$$
$$K^*(L_2(\Omega_{m_2}(B), \mu_{m_2}, H_2^0)) \subset L_2(\Omega_{m_1}(B), \mu_1, H_1^0).$$

(Here we used the obvious embedding of $L_2(\Omega_m(B), \mu_m, H^0)$ into $L_2(\Omega_m, \mu, H^0)$.) This implies formula (30.8) for any ball $B \subset X$. Now we apply Lemma 30.2. If $m_1 \neq m_2$, then $K = 0$, and Theorem 30.2 follows. If $m_1 = m_2 = m$, then K is given by (30.9). Since $K \in L(T_1, T_2)$, we have

$$\mathcal{N}(\omega)a_1(g, \omega) = a_2(g, \omega)\mathcal{N}(\omega g), \qquad g \in G, \quad \omega \in \Omega_m \bmod 0. \tag{30.11}$$

Now Theorem 30.1 follows from the lemma below, in which the cocycles $a_i(g,\omega)$, $i=1,2$, are constructed according to formula (30.2) for arbitrary unitary representations $g\mapsto V_i(g)$, $g\in G^{\omega_0}$ (the previous restrictions on V_i are dropped here).

LEMMA 30.3. *Let formula* (30.11) *be satisfied. Then* $\mathcal{N}(\omega)=\mathcal{N}_0$ *for* $\omega\in\Omega_m \bmod 0$, *where* $\mathcal{N}_0\in L(V_1,V_2)$.

PROOF. Using the arguments from the proof of Lemma 30.1, we can assume that the mappings $(g,\omega)\mapsto a_i(g,\omega)$, $\omega\mapsto\mathcal{N}(\omega)$ are strongly continuous in some neighborhoods of the points (e,ω_0) and ω_0 respectively. Therefore, in (30.11) we can take $\omega=\omega_0$, $g\in G^{\omega_0}$. Denote $\mathcal{N}_0=\mathcal{N}(\omega_0)$. From (30.11) one easily sees that $\mathcal{N}_0\in L(V_1,V_2)$ and $\mathcal{N}(\omega)=\mathcal{N}_0$ for $\omega\in\Omega_m \bmod 0$.

Lemma 30.3 and Theorem 30.1 are proved.

30.4. Representations of the group $G=D(X,v^n)$ induced from the symmetric group. Let X be a connected manifold with the volume form v^n, π a unitary representation of the symmetric group S_m in a Hilbert space E^π. Following 30.3, consider the action of the group $G=D(X,v^n)$ in the space Ω_m, the natural homomorphism $\alpha_0\colon G^{\omega_0}\to S_m$, the representation $V=\pi\circ\alpha_0$ of the group G^{ω_0}, and, finally, the induced representation T^π of the group G. The following realization of the representation G^π will be convenient. The form v^n generates G-invariant measures μ and μ^m on X and X^m. Denote by H^π the subspace of vector-valued functions $f\in H=L_2(X^m,\mu^m,E^\pi)$ such that

$$f(x_{1s},\dots,x_{ns})=f(x_1,\dots,x_n),\quad s\in S_n.$$

The representation T^π acts in this space by the formula

$$(T^\pi(g)f)(x_1,\dots,x_n)=f(x_1g,\dots,x_ng).$$

If $\mu(X)=\infty$, then the representations T^π are irreducible and pairwise nonequivalent; this follows from Theorem 30.1.

Now let $\mu(X)=1$, so that the assumptions of Theorem 30.1 are not satisfied. The subspace $H_0^\pi\subset H^\pi$ specified by the condition

$$\int_X f(x_1,\dots,x_m)\,d\mu(x_1)=0\quad\text{for all } (x_2,\dots,x_m)\in X^{m-1} \bmod 0$$

is invariant under T^π. Denote $T_0^\pi=T^\pi\mid H_0^\pi$.

THEOREM 30.3. *The representations* T_0^π *are irreducible and pairwise nonequivalent.*

For the proof, let us choose an integer $r>0$ and consider the space $C_0^\infty(X^r)$ of all smooth compactly supported (in the case of a noncompact X) functions $f(x_1,\dots,x_r)$, $x_i\in X$. Let $W\in(C_0^\infty(X^r))'$ satisfy the following two conditions:

(a) W is invariant under the action $x_i\mapsto x_ig$, $1\le i\le r$, of the group $G=D(X,v^n)$;

(b) for any functions $\varphi_j\in C_0^\infty(X)$, $1\le j\le r$, we have

$$|\langle\varphi_1\otimes\cdots\otimes\varphi_r,W\rangle|\le C|\varphi_1|_{L_2}\cdots|\varphi_r|_{L_2};$$

here $\varphi_1\otimes\cdots\otimes\varphi_r$ is a function of the form $\varphi_1(x_1)\cdots\varphi_r(x_r)$ and $|\varphi|_{L_2}$ is the norm in the space $L_2(X,\mu)$. We give an example of such a functional. As usual, we write

$\langle \psi, W \rangle$ for $\psi \in C_0^\infty(X^r)$ in the form

(30.12) $$\langle \psi, W \rangle = \int_{X^r} w(x_1, \dots, x_r)\psi(x_1, \dots, x_r)\, d\mu(x_1) \dots d\mu(x_r).$$

Here $w(x_1, \dots, x_r)$ is a distribution. Take a family of pairs of numbers $(i_1, j_1), \dots, (i_s, j_s)$ such that all $2s$ numbers $i_1, j_1, \dots, i_s, j_s$ are pairwise distinct and belong to the set $\{1, \dots, r\}$. To such a family we associate the distribution of the form

(30.13) $$\delta(x_{i_1}, x_{j_1}) \cdots \delta(x_{i_s}, x_{j_s}) = w(x_1, \dots, x_r).$$

Here the distribution $\delta(x_1, \dots, x_i)$ is given by the formula

$$\int_{X_i} \delta(x_1, \dots, x_i)\varphi(x_1, \dots, x_i)\, d\mu(x_1) \dots d\mu(x_i) = \int_X \varphi(x, \dots, x)\, d\mu(x).$$

For $s = 0$ we set $w = 1$.

LEMMA 30.4. (1) *Any functional W satisfying the conditions* (a), (b) *is given by formula* (30.12), *where $w(x_1, \dots, x_r)$ is a linear combination of distributions of the form* (30.13).

(2) *Let r_1, r_2 be positive integers, $r_1 + r_2 = r$. Let a functional W satisfy the condition* (a) *and the following condition* (b′):

$$|\langle \varphi_1 \otimes \varphi_2, W \rangle| \le c|\varphi_1|_{L_2} \cdot |\varphi_2|_{L_2} \quad \text{for any } \varphi_i \in C_0^\infty(X^{r_i}),\ i = 1, 2$$

(*the function $\varphi_1 \otimes \varphi_2$ is defined to be $\varphi_1(x_1, \dots, x_{r_1})\varphi_2(x_{r_1+1}, \dots, x_r)$*). *Then the corresponding distribution $w(x_1, \dots, x_r)$ is a linear combination of distributions of the form* (30.13), *where $\{i_1, \dots, i_s\} \subset \{1, \dots, r_1\}$, $\{j_1, \dots, j_s\} \subset \{r_1 + 1, \dots, r\}$.*

PROOF. (1) First, let us assume that the set $\operatorname{supp} w$ lies in the set $X_*^r = \{(x, \dots, x) \in X^r,\ x \in X\}$. Let us prove that if $r > 2$ then $w = 0$, and if $r = 2$ then $w = c\delta(x_1, x_2)$, $c = \text{const}$. We can assume that $X = B^n = \{x \in \mathbb{R}^n,\ |x| < 1\}$, $v^n = dx_1 \wedge \cdots \wedge dx_n$. Condition (a) implies that

$$w(x_1, \dots, x_r) = u(x_1 - x_r, \dots, x_{r-1} - x_r),$$

where the distribution $u(y_1, \dots, y_{r-1})$ is supported at the point $(0, \dots, 0)$. Therefore, $u(y_1, \dots, y_{r-1}) = L\delta(y_1, \dots, y_{r-1})$, where L is a differential operator with constant coefficients. The condition (b) easily implies that w is of order zero, so that

$$\langle \varphi, W \rangle = c \int_X \varphi(x, \dots, x)\, d\mu(x).$$

However, for $c \neq 0$, this functional satisfies the condition (b) only if $r = 2$. This proves our statement.

(2) Let us prove the first part of the lemma by induction on r. For $r = 1$ it is obvious. Let $r \geq 2$. Break the set $\{1, \dots, r,\}$ arbitrarily into two nonempty parts $\{i_1, \dots, i_k\}$, $\{j_1, \dots, j_{r-k}\}$, and restrict $w(x_1, \dots, x_r)$ to the open set

(30.14) $$X_{i_1, \dots, i_k}^r \{(x_1, \dots, x_r) \mid x_{i_\alpha} \neq x_{j_\beta} \text{ for } \alpha = 1, \dots, k,\ \beta = 1, \dots, r - k\}.$$

The inductive hypothesis implies that this restriction is a linear combination of distributions of the form (30.13). We note that for any open region $E \subset X^r$ nonconstant restrictions to E of distributions of the form (30.13) are linearly independent. Next, the union of all regions of the form (30.14) is the set $X^r \setminus X_*^r$. These two facts imply that $w(x_1, \dots, x_r)$ can be represented as a linear combination of distributions of the form (30.13) and a distribution supported on the set X^r_*. Using the previous result, we prove the first part of the lemma.

The second part of the lemma easily follows from the first.

In the next lemma we consider the subspace $L_2^0(X^m, \mu_m) \subset L_2(X^m, \mu_m)$ defined by the conditions

$$\int_X f(x_1, \dots, x_m)\, d\mu(x_i) = 0 \quad \text{for all } i \text{ and all } (x_1, \dots, \check{x}_i, \dots, x_m) \in X^{m-1} \bmod 0.$$

The group $G = D(X, v^n)$ acts in $L_2^0(X^m, \mu_m)$ by the formula $f(x_1 g, \dots, x_m g)$.

LEMMA 30.5. *Let $A\colon L_2^0(X^{m_1}, \mu_{m_1}) \to L_2^0(X^{m_2}, \mu_{m_2})$ be a bounded linear operator commuting with the action of G on X^{m_1} and X^{m_2}. Then*

(1) *if $m_1 \neq m_2$, then $A = 0$;*

(2) *if $m_1 = m_2 = m$, then A is a linear combination of operators of the form $f(x_1, \dots, x_m) \mapsto f(x_{1s}, \dots, x_{ms})$, $s \in S_m$.*

PROOF OF THE LEMMA. Consider a functional $W \in (C_0^\infty(X^{m_1+m_2}))'$ whose value at a function $(\varphi_1 \otimes \varphi_2)(x_1, \dots, x_{\overline{m}})$, $\overline{m} = m_1 + m_2$, is given by the formula

$$\langle \varphi_1 \otimes \varphi_2, W \rangle = (A\varphi_1^0, \overline{\varphi_2^0}), \qquad \varphi_i \in C_0^\infty(X^{m_i}),$$

where φ_i^0 is the orthogonal projection of the function φ_i to $L_2^0(X^{m_i}, \mu_{m_i})$. The second part of Lemma 30.4 implies that if the corresponding distribution $w(x_1, \dots, x_{\overline{m}})$ is nonzero, then $m_1 = m_2 = m$; moreover, $w(x_1, \dots, x_{\overline{m}})$ is a linear combination of distributions of the form (30.13), where $s = m$ and $i_1, \dots, i_m$ is a permutation of the set $\{1, \dots, m\}$. The lemma is proved.

Now we prove Theorem 30.3. Let an operator $A\colon H_0^{\pi_1} \to H_0^{\pi_2}$ intertwines the representations $T_0^{\pi_1}$ and $T_0^{\pi_2}$, where π_1 and π_2 are representations of the groups S_{m_1} and S_{m_2}. Lemma 30.5 implies that if $m_1 \neq m_2$, then $A = 0$. If $m_1 = m_2 = m$, then A is a linear combination of operators $B\colon H_0^{\pi_1} \to H_0^{\pi_2}$ of the form $(Bf)(x_1, \dots, x_m) = B_0(f(x_1, \dots, x_m))$, where B_0 acts from E^{π_1} to E^{π_2} intertwining the representations π_1 and π_2. This implies Theorem 30.3.

30.5. Smooth representations; an analog of the Mackey theorem on induced representations. In this subsection we consider the group $D(X)$ and its action on Ω_m. The well-known Mackey theorem claims that a unitary representation of a locally compact group generated by the action of this group on a space with measure is induced by a representation of the stabilizer of some point in this space. A similar result for the action of the group $D(X)$ on $\Omega_m(X)$ can be obtained under the additional assumption that the corresponding representation is smooth. Let us define the notion of the smoothness of a representation.

Consider a linear representation $g \mapsto T(g)$, $g \in D(X)$, in a Banach space K. A vector $f \in K$ is said to be smooth if for any integer $m > 0$ and any smooth mapping φ of the ball $B^m = \{x \in \mathbb{R}^m \mid |x| < 1\}$ to $D(X)$ the composition $y \mapsto T(\varphi(y))$, $y \in B^m$, has partial derivatives of all orders that are continuous in the strong topology. (We recall that the smoothness of the mapping $\psi\colon B^m \to D(X)$ means the smoothness of the mapping $(y, x) \mapsto x\psi(y)$; see 20.1.) Smooth vectors form a linear space; we call it the Gårding space of the representation T. A representation T is said to be smooth if the Gårding space is dense in K.

Consider a unitary representation of the group $D(X)$ in the Hilbert space $H = L_2(\Omega_m, \mu_m, H^0)$ acting by the formula

$$(T(G)f)(\omega) = \sqrt{j(g,\omega)}a(g,\omega)f(\omega g), \quad f \in H, \quad g \in G, \quad \omega \in \Omega_m \bmod 0. \tag{30.15}$$

Here $(g,\omega) \mapsto a(g,\omega)$ is a cocycle taking values in the group of unitary operators of the space H^0 (see 1.4).

THEOREM 30.4. *Let a representation of the form* (30.15) *be smooth. Then it is induced by a unitary representation of the stabilizer of a point* $\omega_0 \in \Omega_m$.

PROOF. We consider only the case $m = 1$ (the case $m > 1$ brings no essentially new complications). So, we are given a smooth representation T acting in the space $H = L_2(X, \mu, H^0)$ by the formula

$$(T(g)f)(x) = \sqrt{j(g,x)}a(g,x)f(xg). \tag{30.16}$$

We must prove that this representation is induced by a representation of the subgroup $G^{x_0} = \{g \mid x_0 g = x_0\}$. Define the representation ρ of the group $D(X)$ in $L_1(X, \mu)$ by the formula

$$(\rho(g)f)(x) = j(g,x)f(xg).$$

LEMMA 30.6. *Let a function* $\alpha \in L_1(X,\mu)$ *belong to the Gårding space of the representation* ρ. *Then it coincides almost everywhere with a smooth function.*

PROOF. The conditions of the lemma imply that for any ball B all partial derivatives of the function α belong to $L_1(X,\mu)$. This implies the desired result.

Let us return to the representation (30.16) and denote by $\mathcal{L}$ its Gårding space.

LEMMA 30.7. *For any vectors* $f_i \in \mathcal{L}$, $i = 1, 2$, *there exists a* C^∞*-function* $x \mapsto \Phi(f_1, f_2, ; x)$, $x \in X$, *such that*

$$(f_1(x), f_2(x))_{H^0} = \Phi(f_1, f_2; x), \qquad x \in X \bmod 0.$$

PROOF. Consider the mapping

$$\Phi\colon L_2(X,\mu,H^0) \times L_2(X,\mu,H^0) \to L_1(X,\mu)$$

that associates to the functions f_i, $i = 1, 2$, the function $a \mapsto (f_1(x), f_2(x))_{H^0}$. It intertwines the representation $H \otimes H$ and the representation ρ. If $f_i \in \mathcal{L}$, then the vector $\Phi(f_1, f_2)$ is smooth for ρ. Applying Lemma 30.6, we get the desired result.

LEMMA 30.8. *Let* $f_i \in \mathcal{L}$, $i = 1, 2$, $g_k \to e$ *as* $k \to \infty$. *Then* $\Phi(T(g_k)f_1, f_2; x) \to \Phi(f_1, f_2; x)$ *uniformly on any compact subset in* X.

PROOF. Since $g_k \to e$, the above convergence takes place in the sense of $L_1(X,\mu)$. Furthermore, on any compact subset of X the partial derivatives of any order of the above functions are bounded uniformly with respect to k. This yields the desired result.

It is clear that $\Phi(f, f; x) \geq 0$ for all $x \in X$. Choose a point $x_0 \in X$ and define the inner product in the Gårding space $\mathcal{L}$ by the formula $\Phi(f_1, f_2; x_0)$. Denote by

$\overline{\mathcal{L}}$ the completion of $\mathcal{L}$ by the norm $\sqrt{\Phi(f,f;x_0)}$. For a vector $f \in \mathcal{L}$ denote by f^0 its image under the canonical mapping $\mathcal{L} \to \overline{\mathcal{L}}$. Denote the inner product in $\overline{\mathcal{L}}$ by $(\cdot,\cdot)_{\overline{\mathcal{L}}}$.

Let us construct a unitary representation V of the group G^{x_0} in the space $\overline{\mathcal{L}}$ as follows. First of all, note the obvious equality

$$\Phi(T(g)f_1, T(g)f_2; x) = j(g,x)\Phi(f_1, f_2; xg), \qquad f_i \in \mathcal{L}, \quad g \in D(X). \tag{30.17}$$

Define the operators $V(g)$, $g \in G^{x_0}$, in the space $\mathcal{L}$ by the formula

$$V(g)f = (j(g,x))^{-1/2}T(g)f, \qquad f \in \mathcal{L}, \quad g \in G^{x_0}.$$

It follows from formula (30.17) that the operators $V(g)$ preserve the inner product $\Phi(f_1, f_2; x_0)$. Therefore, these operators can be extended to a unitary representation of the group G^{x_0} in the Hilbert space $\mathcal{L}$. The continuity of the dependence of $V(g)$ on g follows from Lemma 30.8.

Using the representation V of the group G^{x_0} in $\overline{\mathcal{L}}$, we construct the induced representation $U = \operatorname{Ind} V$ of the group $D(X)$. It can be realized in the space $H_1 = L_2(X, \mu, \overline{\mathcal{L}})$ by the formula

$$(U(g)\varphi)(x) = \sqrt{j(g,x)}V(\theta(x)g(\theta(xg))^{-1})\varphi(xg), \qquad \varphi \in H_1, \tag{30.18}$$

where $\theta(x) \in D(X)$ is such that $x_0\theta(x) = x$ and $x \mapsto \theta(x)$ is a Borel mapping. Let us prove that the representations (30.16) and (30.18) are equivalent. To a function $f \in \mathcal{L}$ we associate the function $\varphi_f \in H_1$ by the formula $\varphi_f(x) = (j(\theta(x), x_0))^{-1/2}(T(\theta(x))f)^0$; let us recall that $(T(\theta(x))f)^0$ is the image of the element $T(\theta(x))f \in \mathcal{L}$ under the canonical mapping $\mathcal{L} \to \overline{\mathcal{L}}$. The mapping $f \mapsto \varphi_f$ gives a linear mapping $W\colon \mathcal{L} \to H_1$. It preserves the inner product, since on applying (30.17) with $g = \theta(x)$ we get

$$\begin{aligned} |\varphi_f|^2_{H_1} &= \int_X |\varphi_f(x)|^2_{\overline{\mathcal{L}}} d\mu = \int_X (j(\theta(x)), x_0))^{-1}\Phi(T(\theta(x))f, T(\theta(x))f; x_0)\, d\mu \\ &= \int_X \Phi(f,f;x)\, d\mu = |f|^2_H, \qquad H = L_2(X,\mu,H^0), \quad H_1 = L_2(X,\mu,\overline{\mathcal{L}}). \end{aligned}$$

One can easily see that the mapping W intertwines the representations (30.16) and (30.18). Theorem 30.4 is proved.

30.6. Remarks. (1) For $\dim X = 2$ we obtain induced representations of the group G related to the homomorphism of the group $K_r^{\omega_0}$ to the braid group and with unitary representations of the latter group (see [**66, 35**], and also [**15**]); we do not dwell on this topic here.

(2) For 2-transitive diffeomorphism groups the irreducibility criterion for certain induced representations was obtained by P. Chernoff [**46**]. Representations related to subspaces of finite configurations, in particular, irreducibility conditions, were considered in [**8, 12, 15, 46, 56**]. The approach presented in 30.2, 30.3 differs from the methods used in the above papers.

(3) The example from 26.3 allows us to construct induced representations of the group $D(\mathbb{R}^n)$ associated to representations of the two-sheet covering of the group $GL_+(n, \mathbb{R}^n)$; this possibility was first indicated in [**53**].

(4) An analog of the Mackey theorem (30.5) is presented here for the first time.

§31. Representations of the group $D(S^1)$ associated to countable configurations with one limit point ([14])

By $D(S^1)$ we denote the group of orientation-preserving diffeomorphisms of the circle. In this section we consider an example of the elementary representation of the group $D(S^1)$ (elementary representations were defined in 28.1) associated to the space of countable subsets $\omega \subset S^1$ with one limit point; we denote this space by Ω_*. This is a Borel subset of the space $\Omega = \Omega(S^1)$ introduced in 28.1.

31.1. The dynamical system $(\Omega_*, \mu, D(S^1))$ and the corresponding representations of the group $D(S^1)$; formulations of theorems. Let us realize the circle S^1 as the group $\mathbb{R}/2\mathbb{Z}$. Let dx be the Lebesgue measure on S^1. For $x \in S^1$ denote by $|x|$ the length (measure) of the shortest arc joining the points x and 0.

Let us construct a measure on Ω_* that is quasi-invariant with respect to the action $\omega \mapsto \omega g$, $\omega \in \Omega_*$, $g \in D(S^1)$. Here it is convenient to use probabilistic language. Choose numbers $\delta_k > 0$, $k = 1, 2, \ldots$, satisfying the condition

$$\sum_{k=1}^{\infty} \delta_k < \infty \tag{31.1}$$

and consider a sequence of independent random points $x_k \in S^1$, $k = 1, 2, \ldots$, distributed on S^1 with the density (with respect to the Lebesgue measure) $p_k(x)$, where

$$p_1(x) = \frac{1}{2}, \qquad p_k(x) = \frac{\delta_k}{2}|x|^{\delta_k - 1}, \quad k \geq 2, \qquad x \in S^1.$$

Set $y_k = x_1 + \cdots + x_k$; here $+$ is the addition in the group S^1. We get a random sequence

$$y = (y_1, y_2, \ldots), \qquad y_k \in S^1, \tag{31.2}$$

with independent increments $x_k = y_k - y_{-1}$. Therefore, we get a probabilistic measure P in the space of all sequences of the form (31.2). This measure can be explicitly described as follows: if E is an arbitrary Borel set in $(S^1)^k = S^1 \times \cdots \times S^1$ (k times), then the set E^* of sequences (31.2) such that $(y_1, \ldots, y_k) \in E$ has the measure

$$P(E^*) = \int_E p_1(y_1) p_2(y_2 - y_1) \cdots p_k(y_k - y_{k-1})\, dy_1 \cdots dy_k.$$

Condition (31.1) implies that almost every (with respect to the measure P) sequence of the form (31.2) has a limit $\lim y_k$; denote by Y the set of all sequences (31.2) satisfying this condition. To any sequence from Y we associate the closure of the set of all its points. As a result, we get a mapping $q\colon T \to \Omega_*$. This mapping generates the measure $\theta = q_*(P)$ on Ω_* by the formula $\theta(F) = P(q^{-1}(F))$ for a Borel set $F \subset \Omega_*$.

The next theorem describes the main properties of the form θ.

THEOREM 31.1. *The measure θ is quasi-invariant with respect to the action $\omega \mapsto \omega g$ of the group $D(S^1)$ on Ω_* (and satisfies the continuity conditions* (1) *and* (2) *from* 1.4). *The dynamical system $(\Omega_*, \theta, D(S^1))$ is ergodic.*

Now we define a representation of the group $D(S^1)$ in $L_1(\Omega_*, \theta)$ by the formula (where s is a real parameter)

$$(T_s(g)f)(\omega) = (j(g,\omega))^{1/2+is} f(\omega g), \qquad j(g,\omega) = d\theta(\omega g)/d\theta(\omega). \tag{31.3}$$

Theorem 31.1 and results from §28 give the following theorem.

THEOREM 31.2. *The representation* (31.3) *is irreducible* (*for any* $s \in \mathbb{R}$). *The representations* T_{s_1} *and* T_{s_2} *are nonequivalent for* $s_1 \neq s_2$.

The proof of Theorem 31.1 is presented in subsections 31.2 and 31.3.

31.2. Proof of the quasi-invariance of the measure θ. Denote $j(g,x) = d(xg)/dx$, $x \in S^1$, $g \in D(X)$. Consider the infinite product

$$\prod_{k=1}^{\infty} z_k(y), \quad z_k(y) = \frac{p_k(y_k g - y_{k-1} g)}{p_k(y_k - y_{k-1})} j(g, y_k), \qquad y_k \in S^1. \tag{31.4}$$

Let us prove that this product converges for almost all (with respect to the measure P) sequences of the form (31.2) to the derivative $dP(yg)/dP(y)$, $y \in Y$.

LEMMA 31.1. *For any* $g \in D(S^1)$ *and any sequence* (31.2) *we have*

$$\exp\{-A(\delta_k + |x_k|)\} \leq z_k(y) \leq \exp\{A(\delta_k + |x_k|)\} \qquad x_k = y_k - y_{k-1}, \tag{31.5}$$

where $A = A(g)$ *depends only on* g (*more precisely, on* $\max_x(j(g,x), j(g^{-1},x))$).

PROOF. Introduce the function

$$f(s_1, s_2) = \log\left(\frac{|s_1 - s_2|}{|s_1 g - s_2 g|} j(g, s_2)\right), \qquad s_1, s_2 \in S^1, \quad s_! \neq s_2. \tag{31.6}$$

This function is continuous for $s_1 \neq s_2$, and $f(s_1, s_2) \to 0$ as $|s_1 - s_2| \to 0$. Defining f on the diagonal $s_1 = s_2$ by the formula $f(s,s) = 0$, we obtain a function that is continuous on $S^1 \times S^1$, is infinitely differentiable in a neighborhood of the diagonal $s_1 = s_2$ and vanishes at $s_1 = s_2$. But then

$$-A|s_1 - s_2| \leq f(s_1, s_2) \leq A|s_1 - s_2| \tag{31.7}$$

with some constant $A = A(g)$. From (31.4), (31.6) it follows that

$$z_k(y) = \left(\frac{|y_k g - y_{k-1} g|}{|y_k - y_{k-1}|}\right)^{\delta_k} e^{f(y_{k-1}, y_k)}. \tag{31.8}$$

The number $A = A(g)$ can be chosen so that

$$e^{-A} \leq \frac{|s_2 g - s_1 g|}{s_2 - s_1} \leq e^A. \tag{31.9}$$

Formulas (31.7), (31.8), and (31.9) imply (31.5). The lemma is proved.

LEMMA 31.2. *The infinite product (31.4) converges almost everywhere (with respect to the measure P) on Y to an integrable function $z(g, y)$, i.e.,*

$$\prod_{k=1}^{\infty} z_k(y) = z(g, y), \qquad y = (y_1, y_2, \dots) \in Y \mod 0.$$

The functions $y \mapsto z(g, y)$, $y \mapsto (z(g, y))^{-1}$ are integrable with respect to the measure P for each $g \in D(S^1)$.

PROOF. By Lemma 31.1,

$$|\log z_k(y)| \leq A(\delta_k + |x_k|). \tag{31.10}$$

Note that (31.1) implies the convergence of the series $\sum |y_k - y_{k-1}| = \sum |x_k|$ almost everywhere on Y. Together with (31.10), this implies the convergence of the series $\sum \log z_k(y)$. Hence, the product (31.4) convergence almost everywhere on Y to a function $y \mapsto z(g, y)$. Define also the function

$$\varphi(y) = \prod_{k-1}^{\infty} \varphi_k(y), \qquad \varphi_k(y) = e^{A(\delta_k + |x_k|)}, \quad x_k = y_k - y_{k-1},$$

for which we have

$$(\varphi(y))^{-1} \leq z(g, y) \leq \varphi(y). \tag{31.11}$$

The function φ is integrable with respect to the measure P because

$$\int_Y \varphi(y)\, dP(y) = \prod_{k=1}^{\infty} \int_{S^1} e^{A(\delta_k + |x_k|)} p_k(x)\, dx = B, \qquad 0 < B < \infty.$$

Therefore, (31.11) also implies the integrability of $z(g, y)$ and $(z(g, y))^{-1}$. The lemma is proved.

LEMMA 31.3. *For any Borel subset $E \subset Y$ and any $g \in D(S^1)$ we have*

$$P(Eg) = \int_E z(g, y)\, dP(y). \tag{31.12}$$

PROOF. It suffices to verify (31.12) for a cylinder set $E = \{y = (y_1, y_2, \dots) \in Y \mid (y_1, \dots, y_n) \in F\}$, where F is a Borel set in $(S^1)^n = S^1 \times \cdots \times S^1$. Since all the functions $\prod_{k=1}^{m} z_k(y)$ are bounded from above by an integrable function $\varphi(y)$, the Lebesgue theorem shows that

$$\int_E z(g, y)\, dP(y) = \lim_{m \to \infty} \int_F \prod_{k=1}^{m} z_k(y)\, dP(y).$$

Now, for $m > n$ we have

$$\begin{aligned}\int_F \prod_{k=1}^{m} z_k(y)\, dP(y) &= \int_F p_1(y_1 g) \cdots p_n(y_n g - y_{n-1} g) \prod_{k=1}^{n} j(g, y_k)\, dy_k \\ &= \int_{Fg} p_1(y_1) \cdots p_n(y_n - y_{n-1})\, dy_1 \cdots dy_n = P(Eg),\end{aligned}$$

implying (31.12). The lemma is proved.

Lemma 31.3 implies that the measure P is quasi-invariant under the action of the group $D(S^1)$ and

$$dP(yg)/dP(y) = z(g, y), \qquad g \in D(S^1), \quad y \in Y \mod 0.$$

Finally, let us note that the function $g \mapsto z(g, y)$ is continuous for almost all $y = (y_1, y_2, \dots) \in Y$; this again follows from the Lebesgue theorem applied to the series $\sum \log z_k(y)$ on any set of the form $\{g \mid M^{-1} \le j(g, x) \le M\}$ (see Lemma 31.1). Therefore, the action of the group $D(S^1)$ on Y satisfies the continuity conditions (see 1.4) in the definition of the dynamical system.

Finally, carrying the measure P to the set Ω_* by the formula $\theta = q_*(P)$, we obtain a dynamical system $(\Omega_*, \theta, D(S^1))$.

31.3. The ergodicity of the dynamical system $(\Omega_*, \theta, D(S^1))$**.** Let S_∞ be the group of permutations s of the set $\{1, 2, \dots\}$ such that each s is the identity permutation outside a finite set. The group S_∞ acts on Y by the formula $(y_1, y_2, \dots)s = (y_{1s^{-1}}, y_{2s^{-1}}, \dots)$. Hence, we obtain an action of the group $G \times S_\infty$ on Y. Let us prove that the dynamical system $(Y, P, G \times S_\infty)$ is ergodic.

LEMMA 31.4. *Let (Y_k, σ_k), $k = 1, 2$, be Lebesgue spaces, $Y = Y_1 \times Y_2$, $\sigma = \sigma_1 \times \sigma_2$. On Y_1 let an action of a transformation group Γ be given such that the measure σ_1 is quasi-invariant and ergodic with respect to this action. Assume that a function $f(y_1, y_2)$, $y_k \in Y_k$, is measurable and $f(y_1\gamma, y_2) = f(y_1, y_2)$ for each $\gamma \in \Gamma$ almost everywhere on Y. Then there exists a measurable function $\varphi(y_2)$, $y_2 \in Y_2$, such that $f(y_1, y_2) = \varphi(y_2)$ almost everywhere (with respect to the measure σ) on Y.*

PROOF. We can assume that $Y_k = [0, 1]$ and $d\sigma_k = dy$, $k = 1, 2$. For each $y_2 \in Y_2$ the function $F(y_1, y_2) = \int_0^{y_2} f(y_1, s)\, ds$ is defined on $Y_1 \mod 0$. The mod 0 invariance of the function $f(y_1, y_2)$ implies that $F(y_1, y_2) = \text{const}$ almost everywhere on Y_1 (for a fixed y_2). Therefore, there exists a measurable set $K \subset Y_1$ such that $\operatorname{meas} K = 1$ and $F(y_1, r) = \text{const}$ for $y_1 \in K$ for each rational $r \in Y_2$. Since the function $F(y_1, y_2)$ is continuous in y_2, $F(y_1, y_2) = \text{const}$ on K for $y_2 \in Y_2$. Therefore, there exists an absolutely continuous function $\psi(y_2)$ such that $F(y_1, y_2) = \psi(y_2)$ for $y_1 \in K$, $y_2 \in Y_2 \mod 0$. Hence, $f(y_1, y_2) = \psi'(y_2)$ almost everywhere on Y_2. The lemma is proved.

Now let $f(y)$ be a measurable function on Y that is invariant mod 0 with respect to transformations from $G \times S_\infty$. Let us show that $f(y) = \text{const} \mod 0$ on Y.

Take an arc $\Delta \in S^1$ and a finite set I of natural numbers, and set $\Delta' = S^1 \setminus \Delta$, $Y(\Delta, I) = \{y \in Y \mid y_j \in \Delta \text{ for } j \in I,\ y_k \in \Delta' \text{ for } k \notin I\}$.

LEMMA 31.5. *There exists a measurable set $Y'(\Delta, I) \subset Y(\Delta, I)$ such that $P(Y'(\Delta, I)) = 0$ and $f(y) = f(y')$ for each pair of points $y = (y_1, y_2, \dots)$, $y' = (y_1', y_2', \dots)$ from $Y(\Delta, I) \setminus Y(\Delta', I)$ satisfying the condition $y_k = y_k'$ for $k \notin I$.*

PROOF. We start by representing the space Y in the form $Y = Y_1 \times Y_2$, where $Y_1 = \{y^{(1)} = \{y_j, j \in I\}\}$, $Y_2 = \{y^{(2)} = \{y_k, k \notin I\}\}$. Introduce the measure $\alpha = P_1 \times P_2$ on Y, where P_k is the projection of the measure P to Y_k, $k = 1, 2$. One can easily see that $\alpha \sim P$. We have $Y(\Delta, I) = Y_1^* \times Y_2^*$, where Y_1^* (resp. Y_2^*) is the set of all points from Y_1 (resp. Y_2) with all coordinates in Δ (resp. in

$\Delta' = S^1 \setminus \Delta$). Let Γ be the subgroup of all elements $(g, s) \in D(S^1) \times S_\infty$, where g acts identically on Δ' and s preserves all $k \notin I$. The group Γ acts transitively on Y_1^* and identically on Y_2^*. The mod 0 invariance of the function $f(y)$ with respect to Γ and Lemma 31.4 imply Lemma 31.5.

LEMMA 31.6. *For any natural m the function $f(y)$ is measurable with respect to $\{y_m, y_{m+1}, \dots\}$.*

PROOF. Consider the set $Y^* = Y \setminus (\bigcup Y'(\Delta, I) \cup Z)$, where Δ runs over the set of all arcs of the circle S^1 with ends in a fixed countable set $R \subset S^1$ that is dense in S^1, I is an arbitrary finite set of natural numbers, and Z is the set of all $y = (y_1, y_2, \dots) \in Y$ satisfying at least one of the following conditions: either $y_i = y_j$ for $i \neq j$ or $s(y) = \lim_{k\to\infty} y_k = y_i$ (recall that $s(y)$ exists almost everywhere on Y). It is clear that $P(Y^*) = 1$. Take two arbitrary points $y = (y_1, y_2, \dots), y' = (y_1', y_2', \dots) \in Y^*$ such that $y_k = y_k'$ for $k > m$. Let us show that $f(y) = f(y')$. Indeed, since $y_i \neq s(y) = s(y') \neq y_i'$ for all i, there exists an interval $\Delta \subset S^1$ such that its ends belong to the above countable set R and $s(y) \in \Delta' = S^1 \setminus \Delta$, $y_i \in \Delta$, $y_i' \in \Delta$ for $i < m$. Then there exists a finite set $I \supset \{1, 2, \dots, m\}$ such that $y_i, y_i' \in \Delta$ for $i \in I$, $y_k = y_k' \in \Delta'$ for $k \notin I$. Hence, both y and y' belong to $Y(\Delta, I) \setminus Y'(\Delta, I)$, so that $f(y) = f(y')$. This proves the lemma.

LEMMA 31.7. *Random points $(s, x_2, x_3, \dots)$ are independent, and the point s is distributed uniformly on S^1.*

PROOF. Let $q_m(x)$ be the density function of the distribution of the point $r_m = \sum_{k=m+1}^\infty x_k$. Then for any arcs $\Delta, \Delta_2, \dots, \Delta_m$ we have

$$\begin{aligned}\delta &= P\{s \in \Delta,\ x_k \in \Delta_k,\ 2 \leq k \leq m\} \\ &= \int_{S^1 \times D} p_1(x_1) \dots p_m(x_m) q_m(s - x_1 - \dots - x_m)\, dx_1 \dots dx_m\, ds,\end{aligned}$$

where $D = \{(x_2, \dots, x_m, s \mid x_k \in \Delta_k,\ 2 \leq k \leq m,\ s \in \Delta\}$. We have

$$\delta = \frac{1}{2}|\Delta| \prod_{k=2}^m P\{x_k \in \Delta_k\}.$$

Lemma 31.7 is proved.

LEMMA 31.8. *The function $f(y)$ is measurable with respect to s* mod 0.

PROOF. Let $\overline{Y}$ be the set of all sequences $\overline{y} = (s, x_2, x_3, \dots)$ with the measure $\overline{P}$ equal to the product of the Lebesgue measure ds and the measures $p_k(x_k)\, dx_k$. Lemma 31.7 implies that the correspondence $y = (y_1, y_2, \dots) \leftrightarrow \overline{y} = (s, x_2, x_3, \dots)$, where $s = s(y)$, gives a metric isomorphism between Y and $\overline{Y}$. By Lemma 31.6, the function $f(y)$ is mod 0 invariant under all transformations of the form $y \mapsto (y_1 + z_1, y_2 + z_2, \dots)$, with all but a finite number of the points z_k equal to zero. Therefore, the function $f_1(\overline{y}) = f(y)$ is mod 0 invariant under all transformations $\overline{y} \mapsto (s, x_2 + z_2', \dots)$, where $z_k' = 0$ for sufficiently large k. But the group of transformations $(x_2, x_3, \dots) \mapsto (x_2 + z_2', x_3 + z_3', \dots)$ acts ergodically on the space of sequences $(x_2, x_3, \dots)$ (the zero-one law). By Lemma 31.4, $f_1(\overline{y})$ coincides almost everywhere with a measurable function of the form $\varphi(s)$. The lemma is proved.

The function $\varphi(s)$, being invariant mod 0 under all transformations $s \mapsto sg$, is almost everywhere constant. We have thus proved that the dynamical system $(Y, P, D(S^1) \times S_\infty)$ is ergodic.

This implies the ergodicity of the dynamical system $(\Omega_*, \theta, D(S^1))$, proving Theorem 31.1. As we have already noted, Theorem 31.2 follows from the results of §28.

31.4. Generalization to compact manifolds of dimension > 1. For any compact manifold X we can construct a similar ergodic dynamical system of the form $(\Omega_*, \theta, D(X))$, where Ω_* is the space of countable subsets $\omega \subset X$ with a single limit point. The measure θ appears from a rank n Markov random walk on X, $n = \dim X$. This construction is rather long, and we do not present it here (see [**15**]).

§32. Representations of the group $D(\mathbb{R}^n)$ associated to the Poisson measure

Let us recall that $D(\mathbb{R}^n)$ is the connected component of the group of compactly supported diffeomorphisms of the space $\mathbb{R}^n$; everywhere in this section we assume that $n \geq 3$. Denote by $\overset{\circ}{D}$ the subgroup of elements $g \in D(\mathbb{R}^n)$ preserving the Lebesgue measure. We would like to describe all irreducible representations of the group $D = D(\mathbb{R}^n)$ that contain the trivial representation of the group $\overset{\circ}{D} = \overset{\circ}{D}(\mathbb{R}^n)$. We do not know a solution of this problem in such a general form. However, we give a solution under the additional assumption that the representation we consider is smooth (in the sense described in 30.5) and belongs to the class O. It turns out that such representations are related to the Poisson measure. The next subsection is devoted to the properties of the Poisson measure; in the description of these properties we omit some details.

32.1. The Poisson measure. For any unbounded set $E \subset \mathbb{R}^n$, denote by $\Omega^*(E)$ the set of all countable subsets (configurations) $\omega \subset E$ without limit points in $\mathbb{R}^n$. Denote $\Omega^* = \Omega^*(\mathbb{R}^n)$; this is a Borel subset of the space $\Omega = \Omega(\mathbb{R}^n)$ described in 28.1. There exists a unique probability measure μ_c in Ω^* (here c is a positive parameter) such that for any collection of disjoint Borel subsets F_i, $1 \leq i \leq m$, and any integers $l_i \geq 0$ we have

$$\mu_c\{\omega \in \Omega^* \mid |\omega \cap F_i| = l_i\} = \prod_i \frac{e^{-cm_i}}{(l_i)!}(cm_i)^{l_i}, \qquad m_i = \operatorname{meas} F_i;$$

recall that meas is the Lebesgue measure on $\mathbb{R}^n$. Let us list some properties of the measure μ_c that can be easily deduced from the definitions.

(α) To any function $f \in C_0^\infty(\mathbb{R}^n)$ associate the function $\omega \mapsto f^0(\omega)$, $\omega \in \Omega^*$, by the formula

$$f^0(\omega) = \prod_{x \in \omega}(1 + f(x)).$$

Then

$$\int_\Omega f^0(\omega)\, d\mu_c(\omega) = \exp\left\{c \int_{\mathbb{R}^n} f(x)\, dx\right\}.$$

This yields a useful isomorphism of Hilbert spaces:

$$\overline{S}(H) \simeq L_1(\Omega^*, \mu_c), \qquad H = L_2(\mathbb{R}^n); \tag{32.1}$$

this isomorphism is given by the formula

$$\operatorname{Exp} f \mapsto \exp\left\{-\sqrt{c}\int_{\mathbb{R}^n} f(x)\,dx\right\}\left(\frac{f}{\sqrt{c}}\right)^{0}.$$

(β) Take the balls $B_k = \{x \in \mathbb{R}^n \mid |x| \le d_k\}$ such that $\operatorname{meas} B_k = kc^{-1}$, $k = 1, 2, \ldots$.

LEMMA 32.1. *Almost every (with respect to the measure μ_c) point $\omega \in \Omega^*$ satisfies the condition*

$$|\omega \cap B_k| = k$$

for an infinite collection of numbers $k \in \{1, 2, \ldots\}$ (this collection might depend on ω).

PROOF. The random variables $\xi = |\omega \cap (B_k \setminus B_{k-1})| - 1$ are independent and identically distributed, with finite variance and zero expectation. The sequence $\xi_1 + \cdots + \xi_k$, $k \ge 1$, is a random walk on $\mathbb{Z}$. We know (see [**9,** Chapter II, §6]) that it is recurrent, i.e., the equality $\xi_1 + \cdots + \xi_k = 0$ holds with probability 1 for arbitrarily large k. It remains to notice that this equality is equivalent to the condition $|\omega \cap B_k| = k$. The lemma is proved.

The set of all configurations $\omega \in \Omega^*$ satisfying the conditions of the previous lemma will be denoted by Ω_1^*. For any $\omega \in \Omega_1^*$ write down the numbers

$$r_1(\omega) < r_2(\omega) < \ldots$$

of all balls for which $|\omega \cap B_k| = k$. This sequence of numbers will be important later on.

(γ) Let a measurable function $f \mapsto f(\omega)$, $\omega \in \Omega^*$, have the following property: for any ball $B_a = \{x \in \mathbb{R}^n \mid |x| < a\}$ there exists a function $\omega \mapsto g_a(\omega)$, $\omega \in \Omega^*(\mathbb{R}^n \setminus B_a)$, such that $f(\omega) = g_a(\omega \setminus B_a)$ for almost all $\omega \in \Omega^*$. Then $f(\omega) = \text{const}$, $\omega \in \Omega^* \bmod 0$. This is the zero-one law for Ω^*, μ_c); it can be easily deduced from the zero-one law for a sequence of independent random variables.

32.2. The dynamical system $(\Omega^*, \mu_c, D(\mathbb{R}^n))$; the principal cocycle. Let us consider the action $\omega \mapsto \omega g$ of the group $D = D(\mathbb{R}^n)$ on Ω^*. It is clear that the measure μ_c is quasi-invariant with respect to this action, and

$$\frac{d\mu_c(\omega g)}{d\mu_c(\omega)} = \prod_{x \in \omega} j(g, x), \qquad j(g, x) = \frac{d(xg)}{dx}.$$

We get a dynamical system $(\Omega^*, \mu_c, D(\mathbb{R}^n))$. In the next lemma we use the group $\overset{\circ}{D} = \overset{\circ}{D}(\mathbb{R}^n)$; let us recall that it consists of all diffeomorphisms $g \in D(\mathbb{R}^n)$ preserving the Lebesgue measure.

LEMMA 32.2. *The dynamical system $(\Omega^*, \mu_c, \overset{\circ}{D}(\mathbb{R}^n))$ is ergodic.*

PROOF. We must prove that $f_0(\omega) = 1$ is the unique (up to a factor) function that is invariant under the representation of the group $\overset{\circ}{D}(\mathbb{R}^n)$ acting by translations $f(\omega) \mapsto f(g\omega)$ in the space $L_2(\Omega^*, \mu_c)$. The isomorphism (32.1) carries this representation to the space $\overline{S}(L_2(\mathbb{R}^n, \text{meas}))$. This last representation decomposes into representations of the form ρ^k, $k = 0, 1, \ldots$, where ρ is the representation

$f(x) \mapsto f(xg)$ in $L_2(\mathbb{R}^n, \text{meas})$ and ρ^k is the kth symmetric power of ρ. This proves the lemma.

Choose a configuration $\omega_0 \in \Omega^*$ satisfying the condition

$$|\omega_0 \cap B_k| = k \text{ for } k = 1, 2, \dots,$$

where B_k are the balls from 32.1. Introduce the subgroup

$$D^{\omega_0} = \{g \in D(\mathbb{R}^n) \mid \omega_0 g = \omega_0\}$$

and construct a cocycle of the dynamical system (Ω^*, μ_c, D) with values in D^{ω_0} as follows. For any spherical layer $B_{km} = B_k \setminus B_m$, $k > m$, consider the set Ω_{km} of all configurations of $k - m$ points in B_{km}. Choose a piecewise continuous mapping $\varphi_{km} \colon \Omega_{km} \to D(B_{km})$ such that $(\omega_0 \cap B_{km})\varphi_{km}(\omega) = \omega$ for $\omega \in \Omega_{km}$. Take a configuration $\omega \in \Omega_1^*$. To this configuration corresponds a sequence $r_1(\omega) < r_2(\omega) < \dots$ such that $|\omega \cap B_{r_k(\omega)}| = r_k(\omega)$ for all $k \geq 1$. Therefore, each layer $B_{r_k(\omega), r_{k-1}(\omega)}$ contains exactly $r_k(\omega) - r_{k-1}(\omega)$ points from ω. This means that there is a diffeomorphism $\varphi_{r_k(\omega), r_{k-1}(\omega)}$ mapping the set $\omega_0 \cap B_{r_k(\omega), r_{k-1}(\omega)}$ to $\omega \cap B_{r_k(\omega), r_{k-1}(\omega)}$; for brevity, denote this isomorphism by $\theta_k(\omega)$. Set

$$\theta(\omega) = \prod_{k-1}^{\infty} \theta_k(\omega).$$

For any configuration $\omega \in \Omega_1^*$ this product is finite, with commuting factors. Finally, define the cocycle $a(g, \omega)$ by the formula

$$a(g, \omega) = \theta(\omega) g \big(\theta(\omega g)\big)^{-1}, \qquad \omega \in \Omega^* \bmod 0, \quad g \in D.$$

We call it the *principal cocycle.*

32.3. A representation of the group $D(\mathbb{R}^n)$. Take a unitary representation $g \mapsto V(g)$, $g \in D^{\omega_0}$, in a Hilbert space H^0. Then the operators

$$(T(g)f)(\omega) = \sqrt{j(g, \omega)} V(a(g, \omega)) f(\omega g), \qquad f \in H = L_2(\Omega^*, \mu_c, H^0)$$

form a unitary representation of the group D. This construction is close to that described in §30. We say that the representation T is induced by the representation V, denoting this by $T = \operatorname{Ind} V$. (Of course, the usual definition of the induced representation $\operatorname{Ind} V$ is not applicable here, so we allow ourselves a freedom of notation.) Similarly to 30.1, we formulate the following conditions that might be imposed on V:

(α) V is trivial on $D(\mathbb{R}^m \setminus \omega_0)$.

Therefore, T is an elementary representation (see 28.1).

THEOREM 32.1. *Let V_i, $i = 1, 2$, be two representations of the group D^{ω_0} satisfying the condition (α), and $T_i = \operatorname{Ind} V_i$. Then the natural mapping of linear spaces*

$$L(V_1, V_2) \to L(T_1, T_2) \tag{32.2}$$

is an isomorphism.

In particular, if V is irreducible, then $T = \operatorname{Ind} V$ is also irreducible; if $V_1 \perp V_2$, then $T_1 \perp T_2$.

In the next theorem we consider two Poisson measures μ_{c_i}, $i = 1, 2$, the corresponding configurations ω_0^i, the groups $D^{\omega_0^i}$, and, finally, the induced representations $T_i = \operatorname{Ind} V_i$.

THEOREM 32.2. *$T_1 \perp T_2$ for $c_1 \neq c_2$.*

PROF OF THEOREM 32.1. Let $M \in L(T_1, T_2)$. By Theorem 28.1, the operator M intertwines the cocycles $a_1(g, \omega)$ and $a_2(g, \omega)$, so that

$$(Mf)(\omega) = \mathcal{N}(\omega) f(\omega), \qquad \omega \in \Omega^* \bmod 0, \quad f \in H = L_2(\Omega^*, \mu_c, H_1^0),$$

where $\omega \mapsto \mathcal{N}(\omega)$ is a weakly measurable family of mappings from H_1^0 to H_2^0 such that $\operatorname{ess\,sup} \|\mathcal{N}(\omega\| < \infty$ and

$$\mathcal{N}(\omega) a_1(g, \omega) = a_2(g, \omega) \mathcal{N}(\omega g), \qquad g \in D, \quad \omega \in \Omega^* \bmod 0. \tag{32.3}$$

LEMMA 32.3. *The function $\omega \mapsto \mathcal{N}(\omega)$ is invariant* mod 0 *with respect to the action $\omega \mapsto \omega g$, $\omega \in \Omega^*$, $g \in D$.*

PROOF. Choose one of the balls B_k from 32.1 (property (β)) and define

$$\Omega(k) = \{\omega \in \Omega^* \mid |\omega \cap B_k| = k\}.$$

Represent an arbitrary configuration $\omega \in \Omega(k)$ in the form $\omega = (u, v)$, where $u \subset B_k$, $v \subset \mathbb{R}^n \setminus B_k$. In particular, $\omega_0 = (u_0, v_0)$. It follows from the definition of the cocycles $a_i(g, \omega)$, $i = 1, 2$, that

$$a_i(g, \omega) = a_i(g, u, v) = V_i(\theta_k(u) g (\theta_k(ug))^{-1}),$$
$$\omega - (u, v) \in \Omega_k, \quad i - 1, 2, \quad g \in D(B_k),$$

where $u \mapsto \theta_k(u)$ is a mapping from $\Omega_k(B_k)$ to $D(B_k)$ satisfying the condition $u_0 \theta_k(u) = u$. Formula (32.3) implies that

$$\mathcal{N}(u, v) V_1((\theta_k(u) g (\theta_k(ug))^{-1}) = V_2(\theta_k(u) g (\theta_k(ug))^{-1}) \mathcal{N}(ug, v),$$
$$u \in \Omega_k(B_k), \quad v \in \Omega^*(\mathbb{R}^n \setminus B_k), \quad g \in D(B_k).$$

Using Lemma 30.3, we come to the following conclusion: the function $\omega = (u, v) \mapsto \mathcal{N}(u, v)$, $\omega \in \Omega(k)$, coincides almost everywhere on $\Omega(k)$ with a measurable function depending only on $v \in \Omega^*(\mathbb{R}^n \setminus B_k)$. But then the function $\mathcal{N}(\omega)$, $\omega \in \Omega(k)$, is invariant with respect to the group $D(B_k)$.

Choose a diffeomorphism $g \in D(\mathbb{R}^n)$ supported is some ball B_{k^0}. The restriction of the function $\omega \mapsto \mathcal{N}(\omega)$ to each $\Omega(k)$, $k > k^0$, is invariant under g. But by Lemma 32.1, $\bigcup_{k > k^0} \Omega(k) = \Omega^* \bmod 0$. Therefore, the function $\omega \mapsto \mathcal{N}(\omega)$ is invariant under g, and the lemma is proved.

Lemmas 32.3 and 32.2 imply that $\mathcal{N}(\omega) = \mathcal{N}_0$ almost everywhere on Ω^*. The correspondence $M \mapsto \mathcal{N}_0$ yields the isomorphism (32.2). Theorem 32.1 is proved.

Theorem 32.2 follows from the fact that the measures μ_{c_1} and μ_{c_2} are mutually disjoint.

32.4. Spherical representations. Let us analyze the representations $T = \operatorname{Ind} V$ that have nonzero $\overset{\circ}{D}$-invariant vectors; such representations are said to be *spherical.* Below we will use the group $\overset{\circ}{D}{}^{\omega_0} = D^{\omega_0} \cap \overset{\circ}{D}$.

LEMMA 32.4. *The function $f \in H = L_2(\Omega^*, \mu_c, H^0)$ is invariant under all operators $T(g)$, $g \in \overset{\circ}{D}$, if and only if $f(\omega) = h_0$ for $\omega \in \Omega^* \bmod 0$, where $h_0 \in H^0$, $V(g)h_0 = h_0$ for $g \in \overset{\circ}{D}{}^{\omega_0}$.*

PROOF. It is clear that any function of the above form is invariant under $\overset{\circ}{D}$. Now let us take an arbitrary function $f \in H$ that is invariant under all operators $T(g)$, $g \in \overset{\circ}{D}$. Similarly to the proof of Lemma 32.3, let us take a ball B_k, $k \geq 1$, and consider the restriction of the function f to the set $\Omega(k) = \{\omega \in \Omega^* \mid |\omega \cap B_k| = k\}$. The invariance of the function f yields the formula

$$V(\theta_k(u)g(\theta_k(ug))^{-1})f(ug, v) = f(u, v), \qquad \omega = (u, v) \in \Omega(k), \quad g \in \overset{\circ}{D}.$$

Applying once again Lemma 30.3 (this time to the pair of representations $V_1 = V$, $V_2 = I$ of the group $\overset{\circ}{D}(B_k)$) and repeating arguments from the proof of Lemma 32.3, we see that $f(\omega) = h_0 \in H^0$, $\omega \in \Omega^* \bmod 0$, $V(g)h_0 = h_0$ for $g \in \overset{\circ}{D}{}^{\omega_0}$. The lemma is proved.

Hence, to distinguish among all representations $\operatorname{Ind} V$ those that are irreducible and spherical, it remains to describe irreducible representations of the group D^{ω_0} satisfying the condition (α) from 32.3 and containing a vector invariant under $\overset{\circ}{D}{}^{\omega_0}$. For that, we will need the group $A(\omega_0) \cdot S(\omega_0)$ that is described as follows: $A(\omega_0)$ is the additive group of all functions $f\colon \omega_0 \to \mathbb{R}$ vanishing outside a finite set, and $S(\omega_0)$ is the group of all bijections of the set ω_0 that coincide with the identity outside a finite subset. The group $S(\omega_0)$ acts on $A(\omega_0)$ by the formula $f(x) \mapsto f(xs)$, $x \in \omega_0$, $s \in S(\omega_0)$. The corresponding semidirect product is $A(\omega_0) \cdot S(\omega_0)$. There exists a homomorphism $\Delta\colon D^{\omega_0} \to A(\omega_0) \cdot S(\omega_0)$ mapping $g \in D^{\omega_0}$ to the element $(a, s) \in A(\omega_0) \cdot S(\omega_0)$ such that $xg = xs$, $a(x) = \log j(g, x)$, $x \in \omega_0$. (There exists also an obvious isomorphism of the group $A(\omega_0) \cdot S(\omega_0)$ to the semidirect product $\mathbb{R}_0^\infty \cdot S(\infty)$, where $\mathbb{R}_0^\infty$ is the group of functions $\varphi\colon \{1, 2, \dots\} \to \mathbb{R}$ with finite support, and $S(\infty)$ the group of permutations of the set $\{1, 2, \dots\}$ that coincide with the identity outside a finite set.)

Now we construct irreducible representations of the group $A(\omega_0) \cdot S(\omega_0)$ that are spherical with respect to $S(\omega_0)$. Take a probabilistic measure τ on $\mathbb{R}$ and the measure $\tau^\infty = \tau \times \tau \times \dots$ on the space $\mathbb{R}^{\omega_0}$ of all functions $\xi\colon \omega_0 \to \mathbb{R}$. The group $S(\omega_0)$ acts on $\mathbb{R}^{\omega_0}$ by the formula $s\xi(x) = \xi(xs)$, $s \in S(\omega_0)$, $x \in \omega_0$. Define a representation V_0 of the group $A(\omega_0) \cdot S(\omega_0)$ in the space $L_2(\mathbb{R}^{\omega_0}, \tau^\infty)$ by the formula

$$\begin{gathered}(V_0(a, s)f)(\xi) = e^{\langle \xi, a\rangle} f(s^{-1}\xi), \\ (a, s) \in A(\omega_0) \cdot S(\omega_0), \quad \langle \xi, a\rangle = \sum_{x \in \omega_0} \xi(x)a(x).\end{gathered} \tag{32.4}$$

The function $f_0(\xi) = 1$ is invariant under $S(\omega_0)$, and

$$(V_0(a, s)f_0, f_0) = \prod_{x \in \omega_0} \Phi(a(x)), \tag{32.5}$$

where the function $\Phi\colon \mathbb{R} \to \mathbb{C}$ is the Fourier transformation of the measure τ.

Let us return to the group D^{ω_0}. Consider a representation of this group of the form $V = V_0 \circ \Delta$ and the representation $T = \operatorname{Ind} V$ of the group D. The representation T has a unique (up to a factor) $\overset{\circ}{D}$-invariant vector F_0 such that $|F_0| = 1$. The corresponding spherical function can be easily obtained using (32.4), (32.5); it has the form

$$(T(g)F_0, F_0) = \exp\left\{c \int_{\mathbb{R}^n} (\sqrt{j(g,x)}\Phi(\log j(g,x)) - 1)\, dx\right\}. \tag{32.6}$$

Let us recall that

$$\Phi(t) = \int_{-\infty}^{\infty} e^{itu}\, d\tau(u). \tag{32.7}$$

Therefore, Φ is an arbitrary positive-definite function such that $\Phi(0) = 1$.

The representation T is determined by the number c in the definition of the Poisson measure μ_c, and the probability measure τ on the line $\mathbb{R}$; denote this representation by $T^{c,\tau}$.

32.5. Description of all smooth spherical class O representations of the group $D(\mathbb{R}^n)$, $n \geq 3$. The following theorem solves the problem mentioned at the beginning of this section.

THEOREM 32.3. *Any irreducible smooth class O representation S of the group $D(\mathbb{R}^n)$ containing a nonzero vector invariant under the group $\overset{\circ}{D}(\mathbb{R}^n)$ is equivalent to one of the representations $T^{c,\tau}$. The number c and the probability measure τ are determined uniquely by the representation.*

We omit the (rather long) proof of this theorem. However, we indicate the main steps of this proof. According to Theorem 28.2, the representation S is equivalent to an elementary representation. From the fact that the restriction of the representation S to $\overset{\circ}{D}$ contains an invariant vector F_0, $|F_0| = 1$, we can deduce that the corresponding measure μ on $\Omega = \Omega(\mathbb{R}^n)$ can be chosen to be $\overset{\circ}{D}$-invariant. Then we prove that any such measure is a mixture of Poisson measures μ_c, $c > 0$, and the ergodicity of the dynamical system (Ω, μ, D) implies that $\mu = \mu_c$ for some c. Then we use Theorem 30.4 to analyze the function $\psi_g(\omega) = (a(g,\omega)F_0(\omega g), F_0(\omega))$, where $a(g,\omega)$ is the cocycle defining the elementary representation. It follows that $\psi_g(\omega) = W(\log j(g,x_1), \log j(g,x_2), \dots)$, where $j(g,x) = d(xg)/dx$, $\omega = \{x_1, x_2, \dots\}$, and $W(t_1, t_2, \dots)$ is a symmetric positive-definite function (here $t_i = 0$ for sufficiently large i). The irreducibility of S implies that W is indecomposable. By the de Finetti theorem (see [**68**, Vol. 2, Chapter VII, §4]), $W(t_1, t_2, \dots) = \prod_k \Phi(t_k)$, where Φ is a function of the form (32.7). Therefore, $g \mapsto (S(g)F_0, F_0)$, $g \in D$, is given by (32.6), which proves the theorem.

32.6. Remarks. Representations associated to the space $\Omega^*(\mathbb{R}^n)$ were considered in [**8, 16**]. The paper [**8**] contains some general results about arbitrary measures on $\Omega^*(\mathbb{R}^n)$. Here we considered only the Poisson measure; its role in the description of $\overset{\circ}{D}$-spherical representations is shown in the paper [**16**] whose approach we partly followed in this section.

CHAPTER 6

Representations of Certain Group Extensions

This chapter starts with representations of the group $\Gamma^X \cdot M(X,\mu)$ (see §13). Then the group $C^\infty(\mathbb{R}^n,\mathbb{R}) \cdot D(\mathbb{R}^n)$ is considered, and a result of Goldin–Grodnik–Powers–Sharp describing a class of representations of $C^\infty(\mathbb{R}^n,\mathbb{R}) \cdot D(\mathbb{R}^n)$ is reproduced (with certain modifications).

§33. Representations of the group $\Gamma^X \cdot M(X,\mu)$

As in §13, (X,μ) is a space with measure; we assume that X does not contain atoms. The semidirect product $\Gamma^X \cdot M(X,\mu)$ was defined in §13.

33.1. Representations of the group $M(X,\mu)$. Recall that in 30.4 we constructed irreducible representations of the group $D(X,v^n)$ induced by representations π of the symmetric group, which were denoted T^π when $\mu(X)=\infty$ and T_0^π when $\mu(X)=1$ (here μ is the measure generated by the form v^n). These representations can be extended in the obvious way to the group $M(X,\mu)$; we keep the notation T^π and T_0^π for these extensions. The representations T^π and T_0^π are irreducible and pairwise nonequivalent. If $m=1$ and $\pi=I$, then we denote the representations T^π and T_0^π by ρ and ρ_0 respectively; they act by translations in the space $L_2(X,\mu)$ when $\mu(X)=\infty$ and in $L_2^0(X,\mu)=\{f\in L_2(X,\mu) \mid \int_X f\,d\mu=0\}$ when $\mu(X)=1$.

33.2. The representations $T^{U,r}$ and T^r. Let $\gamma\mapsto U(\gamma)$, $\gamma\in\Gamma$, be a unitary representation of the group Γ in a Hilbert space H^0 and $r\colon\Gamma\to H^0$ a cocycle of this representation. We restrict ourselves to the case when the 2-cocycle $\operatorname{Im}(r(\gamma_1^{-1}),r(\gamma_2))$ is trivial, i.e., can be written in the form $b(\gamma_1\gamma_2)-b(\gamma_1)-b(\gamma_2)$ for a function $b\colon\Gamma\to\mathbb{R}$. Let us extend U, r, and b to the group Γ^X as follows: $U(v)(x)=U(v(x))$, $r(v)(x)=r(v(x))$, $b(v)=\int_X b(v(x))\,d\mu$. According to 10.1, we obtain the following representation T_0 of the group Γ^X in the space $\overline{S}(H)$, $H=L_2(X,\mu,H^0)$:

$$T_0(v)\operatorname{Exp} f=\exp\left\{-\tfrac12|r(v)|^2-(f,r(v^{-1}))+ib(v)\right\}\operatorname{Exp}(U(v)f+r(v)).$$

Let us define a representation T_1 of the group $M(X,\mu)$ in the space $\overline{S}(H)$ by the formula

$$T_1(\sigma)\operatorname{Exp} f=\operatorname{Exp}(\rho(\sigma)f).$$

As a result, we obtain the representation $T=T^{U,r}$ of the group $\Gamma^X\cdot M(X,\mu)$ in $\overline{S}(H)$ given by the formula

$$T(v,\sigma)=T_0(v)T_1(\sigma),\qquad (v,\sigma)\in\Gamma^X\cdot M(X,\mu).$$

The above construction can be applied both when $\mu(X)=\infty$ and when $\mu(X)<\infty$. Now we consider separately the case when the measure of X is finite; we can assume that $\mu(X) = 1$. Let there be given a homomorphism $r\colon \Gamma \to \mathbb{R}$. Denote $H_0 = L_2^0(X,\mu) = \{f \in L_2(X,\mu) \mid \int_X f\,d\mu = 0\}$. Define representations of the groups Γ^X and $M(X,\mu)$ in the space $S(H_0)$ as follows:

$$T_0(v)\,\mathrm{Exp}\,f = \exp\left\{-(1/2)|r_0(v)|^2 - (f, r_0(v^{-1}))\right\}\mathrm{Exp}(f + r_0(v)),$$
$$T_1(\sigma)\,\mathrm{Exp}\,f = \mathrm{Exp}(\rho_0(\sigma)f),$$
$$(v,\sigma) \in \Gamma^X \cdot M(X,\mu), \quad r_0(v)(x) = r(v(x)) - \int_X r(v(x))\,d\mu.$$

We get a unitary representation $(v,\sigma) \mapsto T_0(v)T_1(\sigma)$, $(v,\sigma) \in \Gamma^X \cdot M(X,\mu)$. Denote it by T^r; this is in fact a representation of the group $\Gamma_0^X \cdot M(X,\mu)$, where $\Gamma_0 = \Gamma/[\Gamma,\Gamma]$.

Results obtained in §10 allow us to draw the following conclusions.

(1) If the representation U of the group Γ does not contain the trivial representation and the cocycle r is absolutely nontrivial (see 10.2), then the representation $T^{U,r}$ is irreducible.

(2) Representations T^{U_k,r_k}, $k = 1,2$, are equivalent if and only if there exists an isomorphism of Hilbert spaces H_1^0 and H_2^0 that maps U_1 to U_2 and r_1 to r_2.

(3) For $r \neq 0$ the representations T^r are irreducible, and T_1^r is equivalent to T_2^r if and only if $r_1 = \pm r_2$.

33.3. Spherical representations. To simplify the exposition we assume from now on that Γ is a countable discrete group. Let S be a unitary representation of the group $\Gamma^X \cdot M(X,\mu)$ in a Hilbert space H. Define

$$H_* = \{h \in H \mid S(\sigma)h = h \text{ for each } \sigma \in M(X,\mu)\}. \tag{33.1}$$

The representation S is called a *spherical representation* if H_* is a generating subspace, i.e., the vectors $S(v)h$, $h \in H_*$, $v \in \Gamma^X$, form a total subset in H. Denote by P_* the projection to H_* and define the operator-valued spherical function as follows:

$$A(v) = P_*S(v, e_\Gamma))P_*, \qquad v \in \Gamma^X \tag{33.2}$$

(here e_Γ is the unit element of the group Γ). The function (33.2) is two-side invariant with respect to $M(X,\mu)$, so in the case $\mu(X) = 1$ there exists a mapping $B\colon P(\Gamma) \to L(H_*)$ such that

$$A(v) = B(p^v), \qquad v \in \Gamma^X; \tag{33.3}$$

here we used the semigroup $P(\Gamma)$ and the mapping $v \mapsto p^v$, $v \in \Gamma^X$, introduced in §13. In the case $\mu(X) = \infty$ the equality (33.3) remains true for the mapping $B : P_1(\Gamma) \to L(H_*)$.

The representations $T^{U,r}$ and T^r from 33.2 are spherical ones. If $\mu(X) = \infty$, then the subspace (33.1) is one-dimensional. Therefore, the functions (33.2) and (33.3) take scalar values; these functions are given by the following formulas:

$$\begin{gathered} A(v) = \exp\left\{-\tfrac{1}{2}|r(v)|^2 + ib(v)\right\}, \\ B(p) = \exp\left\{-\frac{1}{2}\sum_{\gamma\neq e_\Gamma}|r(\gamma)|^2 p(\gamma) + i\sum_{\gamma\neq e_\Gamma} b(\gamma)p(\gamma)\right\}, \qquad p \in P_1(\Gamma). \end{gathered} \tag{33.4}$$

For the representation T^r (in the case $\mu(X)=1$) we also have $\dim H_* = 1$ and

$$A(v) = \exp\left\{-\tfrac{1}{2}|r_0(v)|^2\right\},$$

$$B(p) = \exp\left\{-\frac{1}{2}\sum_{\gamma\in\Gamma}|r(\gamma)|^2 p(\gamma) + \frac{1}{2}\left|\sum_{\gamma\in\Gamma} r(\gamma)p(\gamma)\right|\right\}, \qquad p \in P(\Gamma).$$

Finally, let us consider the representation $T^{U,r}$ in the case $\mu(X) = 1$. In this case the space (33.1) can be naturally identified with $\overline{S}(H^0)$. To describe the spherical function, let us extend U, r, and b to the semigroup $P(\Gamma)$ by the formulas $U(p) = \sum p(\gamma)U(\gamma)$, $r(p) = \sum p(\gamma)r(\gamma)$, $b(p) = \sum p(\gamma)b(\gamma)$. Let us also set $r_1(p) = \sum p(\gamma)|r(\gamma)|^2$. Then the operator $B(p)\colon \overline{S}(H^0) \to \overline{S}(H^0)$ is given by the formula

$$B(p)\operatorname{Exp} h = \exp\left\{-\tfrac{1}{2}r_1(p) + (h, r(p^*)) + ib(p)\right\}\operatorname{Exp}(U(p)h + r(p)).$$

33.4. The functional equation for spherical functions. Let $(v,\sigma) \mapsto S(v,\sigma)$, $(v,\sigma) \in \Gamma^X \cdot M(X,\mu)$, be a unitary spherical representation in a Hilbert space H. First, let us consider the case $\mu(X) = 1$. Consider the function $p \mapsto B(p)$, $p \in P(\Gamma)$, defining the spherical function of the representation S (see (33.3)). The next theorem contains the "functional equations" for the function B.

THEOREM 33.1. *The following equalities hold:*

$$\begin{aligned} &B(p_1 * p_2) = B(p_1)B(p_2), && p_1, p_2 \in P(\Gamma),\\ &B(p^*) = (B(p))^*, && p \in P(\Gamma),\\ &B(e_\Gamma) = I. && \end{aligned}$$

This means that the mapping $B : P(\Gamma) \to L(H_*)$ is a $*$-representation of the semigroup $P(\Gamma)$.

Now let $\mu(X) = \infty$. In this case the spherical function of the representation S is given by the formula (33.3).

THEOREM 33.2. *The following equalities hold:*

$$\begin{aligned} &B(p_1 + p_2) = B(p_1)B(p_2),\\ &B(p^*) = (B(p))^*,\\ &B(e_\Gamma) = I. \end{aligned} \tag{33.5}$$

The proof of Theorem 33.1 is based on the following lemma, in which it is assumed that $\mu(X) = 1$.

LEMMA 33.1. *Let there be given two finite decompositions $X = \bigcup_1^n E_i$, $X = \bigcup_1^m F_j$, such that $E_p \cap E_q = \varnothing$, $F_p \cap F_q = \varnothing$ for $p \neq q$ and the sets E_i and F_j are measurable. Then there exists a transformation $\sigma \in M(X,\mu)$ such that the decompositions $\{E_i\}$ and $\{F_j\sigma\}$ are independent (i.e., $\mu(E_i \cap F_j\sigma) = \mu(E_i)\mu(F_j)$ for all i, j).*

PROOF OF THE LEMMA. Decompose each F_j into disjoint subsets F_{ji} in such a way that $\mu(F_{ji}) = \mu(F_j)\mu(E_i)$. Then map each set $\bigcup_j F_{ji}$ onto E_i preserving the measure μ. This given the desired mapping σ.

PROOF OF THEOREM 33.1. Let us choose a vector $h \in H_*$, measures $p_1, p_2 \in P(\Gamma)$, and functions $f_1, f_2 \in \Gamma^X$ with distributions given by the measures p_1, p_2. The projection P_* can be represented as a strong limit

$$P_* = \lim_{n\to\infty} \sum_\sigma c_n(\sigma) S(e_\Gamma, \sigma), \tag{33.6}$$

where $\sigma \mapsto c_n(\sigma)$, $\sigma \in M(X,\mu)$, is a nonnegative function with finite support, $\sum_\sigma c_n(\sigma) = 1$. The equality (33.6) remains true if we replace each function $\sigma \mapsto c_n(\sigma)$ by $\sigma \mapsto c_n(\sigma_n\sigma)$ with arbitrary elements $\sigma_n \in M(X,\mu)$. Therefore,

$$\begin{aligned} B(p_1)B(p_2) &= P_* S(f_1, e_\Gamma) P_* S(f_2, e_\Gamma) P_* \\ &= \lim_{n\to\infty} \sum_\sigma c_n(\sigma) P_*(S(f_1, e_\Gamma) S(e_\Gamma, \sigma_n\sigma) S(f_2, e_\Gamma) P_* \\ &= \lim_{n\to\infty} \sum_\sigma c_n(\sigma) P_*(S(f_1 \cdot \sigma_n\sigma f_2, e_\Gamma) P_* \qquad (33.7) \\ &= \lim_{n\to\infty} \sum_\sigma c_n(\sigma) A(f_1 \cdot \sigma_n\sigma f_2); \end{aligned}$$

$$(\sigma f)(x) = f(x\sigma), \qquad f \in \Gamma^X, \quad \sigma \in M(X,\mu).$$

In this formula the function $f_1 \cdot \sigma_n\sigma f_2$ is defined by the formula

$$x \mapsto f_1(x) f_2(x\sigma_n\sigma), \qquad x \in X, \quad \sigma \in \operatorname{supp} c_n.$$

Since the set $\operatorname{supp} c_n$ is finite and each of the functions f_1, f_2 takes a finite number of values, Lemma 33.1 guarantees the existence of a transformation $\sigma_n \in M(X,\mu)$ such that the functions $x \mapsto f_1(x)$ and $x \mapsto f_2(x\sigma_n\sigma)$ are independent for each $\sigma \in \operatorname{supp} c_n$. Then the distribution of each of the functions $x \mapsto f_1(x)f_2(x\sigma_n\sigma)$ is given by the function $p_1 * p_2$. Therefore, continuing formula (33.7), we get

$$B(p_1)B(p_2) = \lim_{n\to\infty} \sum_\sigma c_n(\sigma) A(f_1 \cdot \sigma_n\sigma f_2) = \lim_{n\to\infty} \sum_\sigma c_n(\sigma) B(p_1 * p_2) = B(p_1 * p_2).$$

This proves the first statement of Theorem 33.1. The remaining two statements are obvious.

PROOF OF THEOREM 33.2. For the proof of (33.5) we again use (33.6). The transformations $\sigma_n \in M(X,\mu)$ we use in such a way that the supports of the functions $x \mapsto f_1(x)$, $x \mapsto (\sigma_n\sigma f_2)(x) = f_2(x\sigma_n\sigma)$ are disjoint (this can be done since $\mu(X) = \infty$). Then it is clear that the distribution of the function $f_1 \cdot \sigma_n\sigma f_2$ is given by the function $p_1 + p_2$. After that (33.7) gives (33.5). Theorem 33.2 is proved.

33.5. Description of irreducible spherical representations; the case $\mu(X) = \infty$. In the next theorem we use the representation ρ of the group $M(X,\mu)$ acting in the space $L_2(X,\mu)$ by the formula $(\rho(\sigma)f)(x) = f(x\sigma)$, $x \in X$; by ρ^k we denote its kth symmetric power.

THEOREM 33.3. *Let $\mu(X) = \infty$. Then each irreducible spherical representation of the group $\Gamma^X \cdot M(X,\mu)$ (nonequivalent to the trivial one-dimensional representation) is equivalent to one of the representations $T^{U,r}$. In particular, each representation whose restriction to the subgroup $M(X,\mu)$ is decomposed into the direct sum $\rho^0 \oplus \rho^1 \oplus \cdots$ is equivalent to one of the representations $T^{I,r}$.*

PROOF. Theorem 33.2 implies that the spherical function of a spherical representation can be found by the formula

$$B(p) = \exp\Big\{ -\sum_{\gamma\neq e_\Gamma} p(\gamma)F(\gamma)\Big\}, \qquad p \in P_1(\Gamma), \tag{33.8}$$

for some function $F\colon \Gamma \setminus \{e_\Gamma\} \to \mathbb{C}$. Taking $F(e_\Gamma) = 0$, we can assume that this function is defined everywhere on Γ. An explicit formula for the spherical function can be obtained from (33.8):

$$A(v) = \exp\Big\{ \int_X F(v(x))\, d\mu \Big\}, \qquad v \in \Gamma^X.$$

Since this function is positive definite on Γ^X, the function $\gamma \mapsto e^{tF(\gamma)}$, $\gamma \in \Gamma$, is positive definite for all $t > 0$. Then, using standard arguments, we get

$$\sum F(\gamma_k^{-1}\gamma_j)\xi_k\overline{\xi_j} \geq 0, \qquad \sum_k \xi_k = 0$$

(all sums are finite). Functions with this property are studied in the theory of spaces with indefinite metric (see [**27**]). Without going into details, we note that this theory results in formula (33.4), and Theorem 33.3 is proved.

33.6. Spherical representations in the case $\mu(X) = 1$. In this case the group $\Gamma^X \cdot M(X,\mu)$ has irreducible representations different from those described above. Any such representation is determined by a representation $\gamma \mapsto U(\gamma)$, $\gamma \in \Gamma$, in a Hilbert space H^0 and acts in $L_2(X,\mu,H^0)$. The subgroup $M(X,\mu)$ acts by translations $f(x) \mapsto f(x\sigma)$, $\sigma \in M(X,\mu)$, whereas the subgroup Γ^X acts by operators

$$f(x) \mapsto U(v(x))f(x), \qquad v \in \Gamma^X, \quad f \in L_2(X,\mu,H^0).$$

Tensor products of these representations yield some new representations.

However, we still do not have a complete description of spherical representations. The main difficulty is that the functional equations for a spherical function in Theorem 33.1 are much more complicated than the equations (33.5) we used in the case $\mu(X) = \infty$. The problem is not solved even in the case when Γ is an abelian group. In this case the problem is reduced to the description of indecomposable positive definite functions on Γ^X that are invariant with respect to $M(X,\mu)$. By Theorem 33.1, this problem is equivalent to the following one. Find all homomorphisms Φ of the semigroup $P(\Gamma)$ to the multiplicative group $\mathbb{C}^*$ that satisfy the following condition: for each family of numbers $\{\lambda_k\}_1^n$, $\lambda_k > 0$, $\sum \lambda_k = 1$, the function Φ_n on the group Γ^n defined by the formula

$$\Phi_n(\gamma_1,\dots,\gamma_n) = \Phi\Big(\sum_{k=1}^n \lambda_k\delta_{\gamma_k}\Big)$$

is positive definite (here δ_γ is the characteristic function of the set $\{\gamma\}$). A similar problem was solved by Vershik [**15**], who described indecomposable $M(X,\mu)$-invariant positive definite functions on $L_2(X,\mu)$.

We conclude this subsection (which does not contain any results) with the following problem. Describe all representations of the group $\Gamma^X \cdot M(X,\mu)$ whose restriction to $M(X,\mu)$ can be decomposed as $\rho_0^0 \oplus \rho_0^1 \oplus \cdots$, where ρ_0^k is the kth symmetric power of the representation ρ_0 acting by translations in $L_2^0(X,\mu)$. It seems plausible that such representations are exactly the representations T_0^r constructed in 33.2.

§34. Representations of the group $C^\infty(\mathbb{R}^n,\mathbb{R})\cdot D(\mathbb{R}^n)$; Goldin–Grodnik–Powers–Sharp uniqueness theorem

In this section we extend the simplest of the representations of the group $D(\mathbb{R}^n)$ constructed in §32 to the semidirect product $C^\infty(\mathbb{R}^n,\mathbb{R})\cdot D(\mathbb{R}^n)$. The resulting representation has a nonzero vector that is mapped to zero by a distinguished element of the complexification of the Lie algebra of this group. The theorem of four authors claims that such representations are completely characterized by this property. We will prove this claim, following mainly the paper [**52**]. After that (again following [**52**]) we present a realization of this representation using the non-Fock representations of CCR. This last realization serves as an example of a general construction from 4.3.

We mention also that in [**52**] a somewhat different group was considered. Namely, there the authors studied the semidirect product of the additive group of the Schwartz space $S(\mathbb{R}^n)$ with the group of all diffeomorphisms $x\mapsto xg$ such that the coordinates of the vector $x-xg$ belong to the space $S(\mathbb{R}^n)$.

34.1. The group $G=C^\infty(\mathbb{R}^n,\mathbb{R})\cdot D(\mathbb{R}^n)$. The group $D=D(\mathbb{R}^n)$ acts on $C_0^\infty=C_0^\infty(\mathbb{R}^n,\mathbb{R})$ by translations $(\rho(g)f)(x)=f(xg)$. This gives the above semidirect product G. Elements of the group G will be written as (φ,g), $\varphi\in C_0^\infty$, $g\in D$. The multiplication is given by the formula $(\varphi_1,g_1)(\varphi_2,g_2)=(\varphi_1+\rho(g_1)\varphi_2, g_1g_2)$. For an element $g\in D$ define the function $j_g\colon \mathbb{R}^n\to\mathbb{R}$ by the formula

$$j_g(x)=\frac{d(xg)}{dx},$$

where dx is the Lebesgue measure in $\mathbb{R}^n$.

The Lie algebra of the group G is the linear space $\mathfrak{g}=C_0^\infty\oplus V(\mathbb{R}^n)$ with the bracket $[(\varphi,\xi),(\varphi_1,\xi_1)]=(L_\xi\varphi_1-L_{\xi_1}\varphi,[\xi,\xi_1])$, where L_ξ is the derivative along the vector field ξ. The complexification $\mathfrak{g}_\mathbb{C}$ can be identified with the linear space $\mathfrak{g}_\mathbb{C}=C_0^\infty(\mathbb{R}^n,\mathbb{C})\oplus V_\mathbb{C}(\mathbb{R}^n)$ with the above bracket. For a vector field $\xi=\sum_{k=1}^n\xi_k\partial/\partial x_k$ define the function (divergence) by the formula

$$\nabla\xi=\sum_{k=1}^n\frac{\partial\xi_k}{\partial x_k}.$$

34.2. Representations of the group $G=C_0^\infty\cdot D$. As in §32, we consider the space Ω^* with the Poisson measure μ_c, $c>0$. Define the representation T^c of the group G in the space $L_2(\Omega^*,\mu_c)$ by the formula

$$(T^c(\varphi,g)f)(\omega)=\exp\left\{i\sum_{x\in\omega}\varphi(x)\right\}\sqrt{j(g,\omega)}f(\omega g),$$

$$f\in L_2(\Omega^*,\mu_c),\quad j(g,\omega)=\frac{d\mu_c(\omega g)}{d\mu_c(\omega)}=\prod_{x\in\omega}j_g(x).$$

Let $\mathcal{L}$ be the linear span of all functions $f\in L_2(\Omega^*,\mu_c)$ of the form $\omega\mapsto f^0(\omega)=\prod_{x\in\omega}(1+f(x))$, $f\in C_0^\infty(\mathbb{R}^n,\mathbb{C})$. The space $\mathcal{L}$ is invariant under the representation

T^c, and for any $f \in \mathcal{L}$ the vector-valued function $(\varphi, g) \mapsto T^c(\varphi, g)f$ is smooth. Hence, in the space $\mathcal{L}$ we get a representation of the Lie algebra $\mathfrak{g}$ and of its complexification $\mathfrak{g}_{\mathbb{C}}$; we denote this representation by t^c.

Consider the function $v_0 \in L_2(\Omega^*, \mu_c)$ defined by the formula $v_0(\omega) = 1$, $\omega \in \Omega^*$. One can easily see that

$$t^c(-\nabla\xi, 2i\xi)v_0 = 0, \tag{34.1}$$

$$(T^c(\varphi, g)v_0, v_0) = \exp\left\{c \int_{\mathbb{R}^n} \left(\sqrt{j_g(x)}e^{i\varphi(x)} - 1\right) dx\right\}. \tag{34.2}$$

(The proof of the last inequality uses formulas from 32.1.) The goal of the present section is to prove that relation (34.1) (together with certain technical conditions) completely determines the representation T^c.

34.3. The uniqueness theorem (Goldin, Grodnik, Powers, Sharp [**54**]). Let us consider an irreducible unitary representation $(\varphi, g) \mapsto T(\varphi, g)$, $(\varphi, g) \in G = C_0^\infty \cdot D$, in a Hilbert space H, $\dim H = \infty$. Let us assume that there exists a T-invariant dense linear subset $L \subset H$ such that for any $x \in L$ the vector-valued function $(\varphi, g) \mapsto T(\varphi, g)x$, $(\varphi, g) \in G$, is smooth. Therefore, we have a representation t of the Lie algebra $\mathfrak{g}$ in L; we denote the corresponding representation of the complexification $\mathfrak{g}_{\mathbb{C}}$ by t as well. Introduce the following notation:

$$\begin{aligned} A(\varphi) &= T(\varphi, e), \quad B(g) = T(0, g), \qquad \varphi \in C_0^\infty(\mathbb{R}^n, \mathbb{R}), \quad g \in D, \\ a(\varphi) &= t(\varphi, 0), \quad b(\xi) = t(0, \xi), \qquad \varphi \in C_0^\infty(\mathbb{R}^n, \mathbb{C}), \quad \xi \in V_{\mathbb{C}}(\mathbb{R}^n). \end{aligned}$$

For any $\varphi \in C_0^\infty(\mathbb{R}^n, \mathbb{C})$ define the operator $e^{a(\varphi)} \colon L \to L$ by the formula

$$e^{a(\varphi)}x = \sum_{k=0}^{\infty} \frac{(a(\varphi))^k}{k!} x, \qquad x \subset L.$$

It is assumed that the series converges in the strong topology and its sum lies in L. Finally, let us assume that there exists a vector $u_0 \in L$ satisfying the conditions $|u_0| = 1$ and

$$a(-\nabla\xi)u_0 = 2ib(\xi)u_0 \tag{34.3}$$

(this is equivalent to the condition (34.1) for the representation T^c).

THEOREM 34.1. *A representation T with the above properties is equivalent to one of the representations T^c, $c > 0$.*

LEMMA 34.1. *For any $g \in D$ we have*

$$B(g) = \exp\{-(i/2)a(\log j_g)\}\, u_0. \tag{34.4}$$

PROOF. Let us take a smooth (in s) family of diffeomorphisms $\{g(s), 0 \le s \le 1\}$, $g(0) = e$, $g(1) = g$; we assume that all $g(s)$ are supported in a fixed compact subset of $\mathbb{R}^n$. Consider the vector field

$$\xi(s) = \frac{d}{dw} (g(s))^{-1} g(s+w)\bigg|_{w=0}, \qquad 0 \le s \le 1.$$

One can easily see that

$$\frac{d}{ds}B(g(s)) = B(g(s))b(\xi(s)),$$

$$\frac{1}{j_{g(s)}}\frac{d}{ds}(j_{g(s)}) = \rho_{g(s)}\nabla\xi(s), \qquad \rho_g f(x) = f(xg).$$

Using (34.3), we get

$$\begin{aligned}
&\frac{d}{ds}\left(\exp\left\{(i/2)a(\log j_{g(s)})\right\}B(g(s))u_0\right)\\
&\quad = \exp\left\{(i/2)a(\log j_{g(s)})\right\}\left(\frac{i}{2}a\left(\frac{1}{j_{g(s)}}\frac{d}{ds}(j_{g(s)})\right)B(g(s)) + B(g(s))b(\xi(s))\right)u_0\\
&\quad = \exp\left\{(i/2)a(\log j_{g(s)})\right\}\left(\frac{i}{2}a\left(\rho_{g(s)}\nabla\xi(s)\right)B(g(s)) - \frac{i}{2}B(g(s))a(\nabla\xi)\right)u_0\\
&\quad = \frac{i}{2}\exp\left\{(i/2)a(\log j_{g(s)})\right\}B(g(s))\left(a(\nabla\xi(s)) - a(\nabla\xi(s))\right)u_0 = 0.
\end{aligned}$$

Therefore,

$$\exp\left\{(i/2)a(\log j_{g(s)})\right\}B(g(s))u_0 = \text{const} = u_0$$

and (34.4) follows.

Our next goal is to find an explicit formula for the functional $\varphi \mapsto (e^{a(\varphi)}u_0, u_0)$, $\varphi \in C_0^\infty(\mathbb{R}^n, \mathbb{C})$.

LEMMA 34.2. *We have*

$$(34.5)\qquad \begin{gathered}(e^{a(\varphi)}u_0, u_0) = (e^{a(\rho_g\varphi - i\log j_g)}u_0, u_0),\\ \varphi \in C_0^\infty(\mathbb{R}^n, \mathbb{C}), \quad g \in D = D(\mathbb{R}^n), \quad (\rho_g\varphi)(x) = \varphi(xg).\end{gathered}$$

PROOF. Lemma 34.1 implies that for $\varphi \in C_0^\infty(\mathbb{R}^n, \mathbb{C})$ and $g \in D$, we have

$$(e^{a(\varphi)}u_0, u_0) = (B(g^{-1})e^{a(\rho_g\varphi)}B(g)u_0, u_0) = (e^{a(\rho_g\varphi)}B(g)u_0, B(g)u_0).$$

To complete the proof, we continue this computation using (34.4) and taking into account that $ia(\log j_g)$ is an Hermitian operator. This implies (34.5).

LEMMA 34.3. *We have*

$$(34.6)\qquad (e^{a(f)}a(\varphi)u_0, u_0) = c(f)\int_{\mathbb{R}^n} e^{if(x)}\varphi(x)\,dx, \qquad f, \varphi \in C_0^\infty(\mathbb{R}^n, \mathbb{C});$$

here $c(f)$ depends on f but not on φ.

PROOF. First we prove that

$$(34.7)\qquad (e^{a(f)}a(L_\xi f - i\nabla\xi)u_0, u_0) = 0, \qquad f \in C_0^\infty(\mathbb{R}^n, \mathbb{C}), \quad \xi \in V_{\mathbb{C}}(\mathbb{R}^n).$$

It suffices to assume that ξ is a real vector field. Take a smooth family $\{g(s),\ 0 \le s \le 1\}$ such that $g(0) = e$, $(d/ds)g(s)\,|_{s=0} = \xi$. Writing down the equality (34.5) for $g = g(s)$ and taking the derivative of both sides at $s = 0$, we arrive at (34.7).

Choose a function $f \in C_0^\infty(\mathbb{R}^n, \mathbb{C})$ and consider the functional

$$S_f(\varphi) = (e^{a(f)}a(\varphi)u_0, u_0), \qquad \varphi \in C_0^\infty(\mathbb{R}^n, \mathbb{C}).$$

Write down this functional in the standard notation of distribution theory:

$$S_f(\varphi) = \int_{\mathbb{R}^n} s_f(x)\varphi(x)\,dx.$$

It follows from (34.7) that $S_f(e^{-if}\nabla(e^{if}\xi)) = 0$ for all $\xi \in V_{\mathbb{C}}(\mathbb{R}^n)$, whence $S_f(e^{-if}\nabla\xi) = 0$ for all $\xi \in V_{\mathbb{C}}(\mathbb{R}^n)$. Hence,

$$\int_{\mathbb{R}^n} s_f(x)e^{-if}\nabla\xi\,dx = 0, \qquad \xi \in V_{\mathbb{C}}(\mathbb{R}^n).$$

This implies that $s_f(x)e^{-if} = \text{const} = c(f)$. Therefore $s_f(x) = c(f)e^{if(x)}$, and Lemma 34.3 is proved.

Introduce the functional $K\colon C_0^\infty(\mathbb{R}^n, \mathbb{C}) \to \mathbb{C}$ by the formula

$$K(\varphi) = \int_{\mathbb{R}^n} (e^{i\varphi(x)} - 1)\,dx. \tag{34.8}$$

LEMMA 34.4. *Let $\{\varphi_t,\ 0 \le t \le 1\}$ be a continuous piecewise smooth family in the space $C_0^\infty(\mathbb{R}^n, \mathbb{C})$ such that $K(\varphi_t) = \text{const}$, $0 \le t \le 1$. Then*

$$(e^{a(\varphi_t)}u_0, u_0) = \text{const}, \qquad 0 \le t \le 1. \tag{34.9}$$

PROOF. We can assume that the family $\{\varphi_t\}$ is smooth in t. Using (34.6) we get (denoting $(d/dt)\varphi_t$ by $\dot\varphi_t$)

$$\begin{aligned}\frac{d}{dt}(e^{a(\varphi_t)}u_0, u_0) &= (e^{a(\varphi_t)}a(\dot\varphi_t)u_0, u_0)\\ &= c_{\varphi_t}\int_{\mathbb{R}^n} e^{i\varphi_t(x)}\dot\varphi_t(x)\,dx = \frac{1}{i}c_{\varphi_t}\frac{d}{dt}(K(\varphi_t)) = 0,\end{aligned}$$

implying (34.9). Lemma 34.4 is proved.

To formulate the next lemma we choose a ball $B_a = \{x \in \mathbb{R}^n \mid |x| < a\}$ and consider the space $C_0^\infty(B_a, \mathbb{R}^n)$ with the topology defined by the set of norms $\varphi \mapsto \max_x |D^\alpha\varphi(x)|$, where $D^\alpha = \prod(\partial/\partial x_i)^{\alpha_i}$, $\alpha_i \ge 0$, $\alpha_i \in \mathbb{Z}$.

LEMMA 34.5. *Let the functions $\varphi_1, \varphi_2 \in C_0^\infty(B_a, \mathbb{R})$ satisfy the conditions*

$$K(\varphi_1) = K(\varphi_2),$$
$$\operatorname{diam}\operatorname{supp}\varphi_k < a/8, \qquad k = 1, 2.$$

Then for any neighborhood of zero $U \subset C_0^\infty(B_a, \mathbb{R})$ there exists a continuous piecewise smooth family $\{\psi_t,\ 0 \le t \le 1\} \subset C_0^\infty(B_a, \mathbb{R})$ such that

$$\begin{gathered}K(\psi_t) = \text{const}, \quad 0 \le t \le 1,\\ \psi_0 - \varphi_1 \in U, \quad \psi_1 - \varphi_2 \in U.\end{gathered} \tag{34.10}$$

PROOF. We can assume that the sets $\operatorname{supp}\varphi_k$, $k = 1, 2$, are disjoint; this can always be achieved by applying a translation (more precisely, a one-parameter family of translations) to the function φ_1. Choose nonzero functions $\varphi_j \in C_0^\infty(B_a, \mathbb{R})$, $3 \le j \le 5$, satisfying the following conditions: (1) the sets $\operatorname{supp}\varphi_j$, $1 \le j \le 5$, are pairwise disjoint; (2) the function φ_5 is of the form $\varphi_5(x) = h(x) - h(x - x_0)$, where $h \in C_0^\infty(B_a, \mathbb{R})$ and the supports of the functions $h(x)$ and $h(x - x_0)$ are disjoint; (3) the function $h(x)$ equals 1 on a set $F \subset B_a$ such that

$$\operatorname{meas} F > \sum_{k=1}^{4} \operatorname{meas}\operatorname{supp}\varphi_k; \tag{34.11}$$

(4) we have

$$\int_{B_a} \varphi_3(x)\, dx = 0, \tag{34.12}$$

$$\int_{B_a} \varphi_4(x)\, dx = 1. \tag{34.13}$$

For any vector $s = (s_1, \dots, s_5) \in \mathbb{R}^5$ set

$$u_s(x) = \sum_{j=1}^{5} s_j \varphi_j(x),$$

$$H(s) = K(u_s) = \int_{\mathbb{R}^n} (e^{iu_s(x)} - 1)\, dx.$$

It is clear that $H(s) = \sum_{j=1}^{5} H_j(s_j)$, where

$$H_j(s_j) = \int_{\mathbb{R}^n} (e^{is_j\varphi_j(x)} - 1)\, dx.$$

It follows easily from (34.12), (34.13) that the numbers $\partial H/\partial s_3$, $\partial H/\partial s_4$ are linearly independent over $\mathbb{R}$ for $0 < s_3 < \delta$, $0 < s_4 < \delta$ with a sufficiently small δ. Choosing such a δ, consider the set

$$E_\delta = \{s \in \mathbb{R}^5 \mid 0 < s_3 < \delta,\ 0 < s_4 < \delta\}$$

and the mapping $s \mapsto H(s)$, $s \in E_\delta$. The following two properties of this mapping are essential: (A) it is nondegenerate (i.e., has rank 2 at all points $s \in E_\delta$); (B) the set $F_\delta = H(E_\delta)$ is contractible. Indeed, (A) follows from linear independence (over $\mathbb{R}^n$) of the numbers $\partial H/\partial s_3$, $\partial H/\partial s_4$ for $s \in E_\delta$. Let us verify (B). The equality

$$\operatorname{Re} H_k(\xi) = -2 \int_{\mathbb{R}^n} \sin^2 \frac{\xi\varphi_k(x)}{2}\, dx, \qquad 1 \le k \le 5,$$

implies that the set of values of the function $s \mapsto \sum_{k=1}^{4} H_k(s_k)$ is contained in the strip $-2\sum_{k=1}^{4} l_k < \operatorname{Re} z < 0$, where $l_k = \operatorname{meas}\operatorname{supp}\varphi_k$, and the set of values of the function $\xi \mapsto H_5(\xi)$, $\xi \in \mathbb{R}$, is the interval $[-M, 0]$, $M > 2 \operatorname{meas} F_5$. Together with the inequality (34.11), this implies that the set $F_\delta = H(E_\delta)$ is the union of intervals of length $> M$ parallel to the real line and intersecting the line $\operatorname{Re} z = -2\sum_{k=1}^{4} l_k$. Therefore, the set F_δ is contractible.

Hence, we have verified properties (A) and (B). These properties imply that for any point $z\in F_\delta$ the set $H^{-1}(z)$ is a connected smooth submanifold in E_δ. In particular, this is true for the set

$$\{s\in E_\delta \mid H(s)=H(1,0,\varepsilon,\varepsilon,0)=H(0,1,\varepsilon,\varepsilon,0)\} \tag{34.14}$$

for each $\varepsilon\in(0,\delta)$. Therefore, the functions $\varphi_1+\varepsilon(\varphi_3+\varphi_4)$ and $\varphi_2+\varepsilon(\varphi_3+\varphi_4)$ can be joined by a smooth family of functions $\{\psi_t,\ 0\le t\le 1\}$ with all ψ_t belonging to the set (34.14). This proves Lemma 34.5.

LEMMA 34.6. *The set of values of the functional* (34.8) *at* $\varphi\in C_0^\infty(\mathbb{R}^n,\mathbb{R})$ *is the entire half-plane* $\{z\mid \operatorname{Re} z\le 0\}$.

PROOF. For the proof, it suffices to consider the values of the functional (34.8) on the function $\varphi_1+\varphi_5$, where φ_1 and φ_5 are taken from the proof of Lemma 34.3.

Lemmas 34.4–34.6 imply that there exists a function $F(z)$, $\operatorname{Re} z\le 0$, satisfying the condition

$$(A(f)u_0,u_0)=(e^{a(f)}u_0,u_0)=F(K(f)),\qquad f\in C_0^\infty(\mathbb{R}^n,\mathbb{C}). \tag{34.15}$$

Furthermore, the function $F(z)=F(u+iv)$ is infinitely differentiable in u,v for $u\le 0$.

LEMMA 34.7. *The function* $z\mapsto F(z)$ *is holomorphic for* $\operatorname{Re} z\le 0$.

PROOF. Formulas (34.15) and (34.6) imply the equalities

$$\frac{d}{dt}(e^{a(f+th)}u_0,u_0)\Big|_{t=0}=\frac{\partial F}{\partial u}(K(f))\int_{\mathbb{R}^n}h\sin f\,dx+\frac{\partial F}{\partial v}(K(f))\int_{\mathbb{R}^n}h\cos f\,dx,$$

$$\frac{d}{dt}(e^{a(f+th)}u_0,u_0)\Big|_{t=0}=c_f\int_{\mathbb{R}^n}e^{if}h\,dx.$$

In these equalities, let us take $h=\nabla\xi+iL_\xi f$, $\xi\in V(\mathbb{R}^n)$. Then the last integral vanishes, so that

$$\begin{aligned}0&=-\frac{\partial F}{\partial u}(K(f))\int_{\mathbb{R}^n}(\nabla\xi+iL_\xi f)\sin f\,dx+\frac{\partial F}{\partial v}(K(f))\int_{\mathbb{R}^n}(\nabla\xi+iL_\xi f)\cos f\,dx\\&=\left(\frac{\partial F}{\partial u}(K(f))+i\frac{\partial F}{\partial v}(K(f))\right)\int_{\mathbb{R}^n}e^{-if}L_\xi f\,dx.\end{aligned}$$

Choosing a vector field ξ in such a way that the last integral does not vanish, we get the equality

$$\frac{\partial F}{\partial u}(z)+i\frac{\partial F}{\partial v}(z)=0,\qquad z=K(f).$$

Since the set of values of the functional K is the entire half-plane $\operatorname{Re} z\le 0$, the function F is holomorphic for $\operatorname{Re} z\le 0$.

LEMMA 34.8. *We have*

$$F(z_1+z_2)=F(z_1)F(z_2),\qquad \operatorname{Re} z_k\le 0. \tag{34.16}$$

PROOF. Take some functions $f_1, f_2 \in C_0^\infty(\mathbb{R}^n, \mathbb{R})$ and set $f_2^m(x) = f_2(x - ma)$, $m \in \mathbb{Z}$, $a \in \mathbb{R}^n$, $a \neq 0$. If m is sufficiently large, then $K(f_1 + f_2^m) = K(f_1) + K(f_2)$. Therefore,

$$(e^{a(f_1+f_2^m)}u_0, u_0) = F(K(f_1) + K(f_2)), \qquad m > m_0. \tag{34.17}$$

There exists a sequence $m_j \to \infty$ such that the sequence $e^{a(f_2^{m_j})}$ converges weakly to an operator U. Any element of the group G commutes with $f_2^{m_j}$ for a sufficiently large j. Therefore, the operator U commutes with each operator of the representation T. Since T is irreducible, we have $U = \lambda I$, where I is the identity operator. Moreover, $(e^{a(f_2^m)}u_0, u_0)$ does not depend on m (this follows from (34.15)), hence equals λ. Therefore,

$$\begin{aligned}(e^{a(f_1+f_2^{m_j})}u_0, u_0) &= (e^{a(f_1)}e^{a(f_2^{m_j})}u_0, u_0) \longrightarrow \lambda(e^{a(f_1)}u_0, u_0)\\ &= (e^{a(f_2)}u_0, u_0)(e^{a(f_1)}u_0, u_0) = F(K(f_1))F(K(f_2)).\end{aligned}$$

Comparing this equality with (34.17), we get (34.16).

Formula (34.16) and the holomorphicity of the function F imply that $F(z) = e^{cz}$ for some c. Hence,

$$(e^{a(f)}u_0, u_0) = e^{cK(f)}, \qquad f \in C_0^\infty(\mathbb{R}^n, \mathbb{R}), \tag{34.18}$$

where the functional K is defined by formula (34.8). Since $|(A(f)u_0, u_0)| \leq 1$ we have $c > 0$.

Formula (34.18) implies the equality

$$(e^{a(f)}u_0, u_0) = e^{cK(f)} = \exp\left\{c\int_{\mathbb{R}^n}(e^{if(x)} - 1)\,dx\right\}, \qquad f \in C_0^\infty(\mathbb{R}^n, \mathbb{C}), \tag{34.19}$$

which coincides with (34.18) for real f and is obtained by analytic continuation for complex f. Formulas (34.4) and (34.19) give the following formula for the matrix element of the representation T:

$$\begin{aligned}(T(\varphi, g)u_0, u_0) &= (e^{a(\varphi)}B(g)u_0, u_0) = (e^{a(\varphi-(i/2)\log j_g)}u_0, u_0)\\ &= \exp\left\{c\int_{\mathbb{R}^n}(e^{i\varphi+(1/2)\log j_g} - 1)\,dx\right\}\\ &= \exp\left\{c\int_{\mathbb{R}^n}(\sqrt{j_g(x)}e^{i\varphi(x)} - 1)\,dx\right\}.\end{aligned}$$

Comparing this with (34.2), we see that the representations T and T^c are equivalent.

34.4. Realization in the space $\overline{S}(H)$, $H = L_2(\mathbb{R}^n)$; relation with representations of CCR. Applying the natural isomorphism of Hilbert spaces $L_2(\Omega^*, \mu_c)$ and $\overline{S}(H)$, $H = L_2(\mathbb{R}^n)$ (this isomorphism was described in 32.1) to the representation T^c, we obtain a representation in the space $\overline{S}(H)$, for which we keep the notation T^c. This representation can be described as follows.

Define the representation U of the group $G = C_0^\infty \cdot D$ in the space $L_2(\mathbb{R}^n)$ (with the Lebesgue measure in $\mathbb{R}^n$) by the formula

$$(U(\varphi, g)f)(x) = \sqrt{j_g(x)}e^{-i\varphi(x)}f(xg).$$

One can easily see that the mapping $r\colon G \to L_2(\mathbb{R}^n)$ of the form

$$r(\varphi, g)(x) = c\big(\sqrt{j_g(x)}e^{-i\varphi(x)} - 1\big)$$

is a cocycle of the representation U. Following 5.2, we construct the corresponding projective representation $T^{U,r}$. It can be made an ordinary (nonprojective) representation of G by multiplication with an appropriate function $G \to \mathbb{T}^1$. The final formula for this representation is

$$T^c(\varphi, g)\,\mathrm{Exp}\, f = \exp\big\{\Phi(\varphi, g) + (f, r((\varphi, g)^{-1})\big\}\,\mathrm{Exp}(U(\varphi, g)f + r(\varphi, g)),$$

$$\Phi(\varphi, g) = c\int_{\mathbb{R}^n}\left(\sqrt{j_g(x)}e^{i\varphi(x)} - 1\right)dx.$$

Under the isomorphism $L_2(\Omega^*, \mu_c) \simeq \overline{S}(L_2(\mathbb{R}^n))$ the function $v_0(\omega) = 1$ becomes the vector $\mathrm{Exp}\, 0$.

Now we turn to the representation t^c of the Lie algebra $\mathfrak{g}$ in the space $\mathcal{L} \subset \overline{S}(L_2(\mathbb{R}^n))$. Following 4.2, we express this representation in terms of representations of CCR. Consider a standard (Fock) representation of CCR in the space $\overline{S}(L_2(\mathbb{R}^n))$. It is given by the operator-valued functions $a^+(x)$, $a(x)$, $x \in \mathbb{R}^n$, satisfying

$$[a^+(x), a(y)] = \delta(x - y)I, \quad [a^+(x), a^+(y)] = [a(x), a(y)] = 0.$$

Annihilation and creation operators are expressed in terms of a, a^+ as follows:

$$a(f) = \int_{\mathbb{R}^n} f(x)a(x)\,dx, \qquad a^+(f) = \int_{\mathbb{R}^n} f(x)a^+(x)\,dx$$

(here $f \in C_0^\infty(\mathbb{R}^n, \mathbb{C})$). Define a new (non-Fock) representation of CCR, replacing $a^+(x)$ and $a(x)$ in the above formulas by $a_1^+(x) = a^+(x) + \sqrt{c}I$ and $a_1(x) = a(x) + \sqrt{c}I$. Identify an element (φ, ξ) of the Lie algebra $\mathfrak{g}$ (here $\varphi \in C_0^\infty$, $\xi \in V(\mathbb{R}^n)$) with the operator in $L_2(\mathbb{R}^n)$ acting by the formula

$$f(x) \mapsto i\varphi(x)f(x) + (L_\xi - L_\xi^*)f(x).$$

Rewrite the above operators as integral operators with distribution kernels. As we have already explained in 4.2, under certain conditions to a kernel $K(x, y)$ we can associate the following operator in $\overline{S}(L_2(\mathbb{R}^n))$:

$$\int_{\mathbb{R}^n}\int_{\mathbb{R}^n} K(x, y)a_1^+(x)a_1(y)\,dx\,dy.$$

As a result, we get the following formulas that expresses the representation t^c in terms of non-Fock representations of CCR:

$$t^c(\varphi, \xi) = \int_{\mathbb{R}^n}\varphi(x)a_1^+(x)a_1(x)\,dx + \frac{i}{2}\sum_{k=1}^n\int_{\mathbb{R}^n}\xi_k(x)J_k(x)\,dx,$$

$$J_k(x) = \frac{1}{i}\left(\frac{\partial}{\partial x_k}a_1^+(x)a_1(x) - a_1^+(x)\frac{\partial}{\partial x_k}a_1(x)\right).$$

Of course, we obtained these formulas by purely formal computations; however, they can be rigorously proved.

CHAPTER 7

Representations of Diffeomorphism Groups in the Fock Space

In §30 we considered induced representations of diffeomorphism groups. In this section we introduce into the picture cocycles of these groups, and, following §6, we construct representations in the Fock space. The passage from §30 to the present chapter can be compared with the passage from quantization to second quantization.

§35. Description of representations; formulation of the problem

Let G be one of the groups $D(X)$, $D(X, v^n)$, $D(X, \omega^2)$, $D(X, \alpha^1)$ introduced in 20.2. Similarly to Chapter V, let us consider the space of configurations $\Omega_m = \Omega_m(X)$, the action $\omega \mapsto \omega g$, $\omega \in \Omega_m$, $g \in G$, the stabilizer C^{ω_0} of a point $\omega_0 \in \Omega_m$, a unitary representation $g \mapsto U_0(g)$, $g \in G^{\omega_0}$, in a Hilbert space H^0, and, finally, the induced representation $U = \operatorname{Ind} U_0$ in the space $H = L_2(\Omega_m, \mu_m, H^0)$ (we slightly changed the notation compared to 30.1). Similarly to Chapter 5, we restrict ourselves to representations U_0 that are trivial on G^{ω_0}, so that they are representations of the quotient group $K^{\omega_0} = G^\omega / G^{\omega_0}$ (see 26.3). We recall that the representation U is realized in the space $H = L_2(\Omega_m, \mu_m, H^0)$ by the formula

$$(U(g)f)(\omega) = \sqrt{j(g,\omega)}a(g,\omega)f(\omega g),$$
$$a(g,\omega) = U_0(\theta(\omega)g(\theta(\omega))^{-1}), \quad j(g,\omega) = \frac{d\mu_m(\omega g)}{d\mu_m(\omega)};$$

here $\theta\colon \Omega_m \to G$ is a Borel mapping satisfying the condition $\omega_0\theta(\omega) = \omega$ for all $\omega \in \Omega_m$ Of course, for the groups $D(X, v^n)$ and $D(X, \omega^2)$ the measure μ_m on Ω_m must be G-invariant; in this case $j(g,\omega) = 1$.

Let $r_0\colon G^{\omega_0} \to H^0$ be a cocycle of the representation $g \to \sqrt{j(g,\omega_0)}U_0(g)$, $g \in G^{\omega_0}$. Then the function

$$r(g)(\omega) = r_0(\theta(\omega)g(\theta(\omega g))^{-1}), \qquad g \in G,$$

is a cocycle of the representation U. Following 6.2, we consider the representation $T = T^{U,r}$ of the group G in the space $\overline{S}(H)$. This representation will be our primary object of investigation in the present chapter.

We consider the following problems.

(1) To describe representations (above all, irreducible ones) of the group K^{ω_0}.

(2) To describe one-dimensional cohomology of these representations.

(3) To analyze representations constructed (as described above) from representations of K^{ω_0} and their cocycles.

It would be useful also to have a more explicit description of the group K^{ω_0} (simple properties of this group were presented in 26.3).

The results of the present chapter can be considered as a small step towards the realization of this program.

§36. Representations of the groups $D(X, v^n)$ and $\widehat{D}(X, v^n)$ in the Fock space

In this section by X we denote a connected C^∞-smooth manifold (without boundary) of dimension $n < \infty$ with a volume form v^n. The groups $D(X, v^n)$ and $\widehat{D}(X, v^n)$ were defined in §20. We assume that $n \geq 3$, although this assumption is not always necessary. The results of this section are meaningful only when $\pi_1(X) \neq \{1\}$.

In 36.1–36.4 we assume that X is compact and $\int_X v^n = 1$. The case of a noncompact X is briefly discussed in 36.5.

36.1. The representation T_α, $\alpha \in H^1(X, \mathbb{R})$. Let μ be the probability measure on X generated by the form v^n and $H = L_2^0(X, \mu)$ the orthogonal complement to the space of constant functions in $L_2(X, \mu)$. Take an element $\alpha \in H^1(X, \mathbb{R})$ and construct a representation T_α of the group $G = D(X, v^n)$ in the space $\overline{S}(H)$, $H = L_2^0(X, \mathbb{R})$.

To do this consider the representation ρ_0 of the group G in the space $L_2^0(X, \mu)$ by the formula $(\rho_0(g)f)(x) = f(xg)$. Choose a closed 1-form a^1 representing the class α and, following 22.2, define the cocycle $g \to \psi_g$, $g \in G$, of the representation ρ_0 by the formula $d\psi_g = ga^1 - a^1$, $\psi_g \in H$. Finally, applying the general construction from §6, define the representation $g \mapsto T_\alpha(g)$, $g \in G$, in the space $\overline{S}(H)$ by the formula

$$T_\alpha(g) \operatorname{Exp} f = \exp\left\{-\tfrac{1}{2}|\psi_g|^2 + (f, \psi_{g^{-1}})\right\} \operatorname{Exp}(\rho_0(g)f + \psi_g),$$
$$f \in H = L_2^0(X, \mu), \quad g \in D = D(X, v^n).$$

(Strictly speaking, the representation T_α depends on the choice of the form a^1 from the class α; however, it is clear that the representations constructed from two closed forms are equivalent if the difference of the forms is exact.)

The analysis of representations T_α is based on the embedding of the group $G = D(X, v^n)$ into the group $\Gamma^X/\Gamma_0 \cdot M(X, \mu)$ as a dense subgroup (with respect to the convergence in $\Gamma^X/\Gamma_0 \cdot M(X, \mu)$ introduced in 23.1). Results of 23.2 show that the representation T_α can be extended to a representation of the group $\Gamma^X/\Gamma_0 \cdot M(X, \mu)$. This extension is of the form T^r, where $r\colon \Gamma \to \mathbb{R}$ is a homomorphism naturally related to the element $\alpha \in H^1(X, \mathbb{R})$ (this construction was described in more details in 23.2). Applying results from §33, we come to the following conclusion.

THEOREM 36.1. *For $\alpha \neq 0$ the representation T_α is irreducible. The representations T_α and T_β are equivalent if and only if either $\alpha = \beta$ or $\alpha = -\beta$.*

Now we determine the spherical functions of the representation T_α. For that, let us assume that we are given a region $Y \subset X$ satisfying the following conditions: (1) $Y \simeq \mathbb{R}^n$, (2) $\mu(X \setminus Y) = 0$, and (3) assuming that there is a Riemannian metric on X, there exists $M > 0$ such that any two points from Y can be joined by a curve of length $< M$ lying entirely in Y. For brevity, a region Y with the above properties will be called *suitable*. (One can easily see that Y can be represented in the form $Y = p(\widetilde{Y})$, where $p\colon \widetilde{X} \to X$ is the universal covering and $\widetilde{Y}$ is the region indicated in 23.1.) The above closed form a^1 is exact in Y. Hence, $a^1 = db_Y$, where $b_Y \in C^\infty(Y)$. It follows from the properties of region Y that the function b_Y is bounded. We can also assume that the integral of b_Y with respect to the measure μ vanishes, so that b_Y can be identified with an element of the space $H = L_2^0(X,\mu)$.

Let us return to the representation T_α and study its restriction to the subgroup $D(Y,v^n)$. Denote the restriction of ρ_0 to $D(Y,v^n)$ by ρ_Y.

THEOREM 36.2. *The restriction of the representation T_α to the subgroup $D(Y,v^n)$ is decomposed into irreducible components as $\rho_Y^0 \oplus \rho_Y^1 \oplus \cdots$. The unique (up to a factor of absolute value 1) normalized vector invariant with respect to $D(Y,v^n)$ is given by the formula*

$$f_Y = e^{-(1/2)|b_Y|^2}\operatorname{Exp} b_Y. \tag{36.1}$$

The corresponding spherical function is given by the formula

$$(T_\alpha f_Y, f_Y) = \exp\left\{\frac{1}{2}\int_X |\psi_g(x) - b_Y(x) + b_Y(xg)|^2\,d\mu\right\}, \quad g \in D(X,v^n). \tag{36.2}$$

Theorem 36.1 and the first statement of Theorem 36.2 follow from the results of §33. The proof of formulas (36.1) and (36.2) is easy.

Let us present another expression for the spherical function (36.2). Consider the homomorphism $r\colon \Gamma \to \mathbb{R}$ corresponding to the element $\alpha \in H^1(X,\mathbb{R})$, and the mapping $q\colon \widehat{D}(X,v^n) \to P(\Gamma)$ introduced in 23.3. We arrive at the formula

$$(T_\alpha f_Y, f_Y) = \exp\left\{-\frac{1}{2}\sum_\gamma |r(\gamma)|^2 p_{\widehat{g}}^Y + \frac{1}{2}\left|\sum_\gamma r(\gamma) p_{\widehat{g}}^Y\right|^2\right\},$$

$$g \in D(X,v^n), \quad \widehat{g} \in \widehat{D}(X,v^n), \quad p(\widehat{g}) = g, \quad p_{\widehat{g}}^Y = q(\widehat{g});$$

this formula follows from (36.2).

The answer to the following question remains unknown. Does the family $\{T_\alpha,\ \alpha \neq 0\}$ exhaust all irreducible representations of the group $D(X,v^n)$ whose restriction to $D(Y,v^n)$ (for any suitable region in X) can be decomposed as $\rho_Y^0 \oplus \rho_Y^1 \oplus \cdots$?

36.2. Representations associated to quadratic forms on $H_1(X,\mathbb{R})$. We can get new irreducible representations of the group $D(X,v^n)$ that are spherical with respect to the subgroup $D(Y,v^n)$ (where Y is a suitable region in X) by forming tensor products of the representations constructed above and separating an appropriate irreducible component. However, this construction can be replaced by the following simpler construction that gives the same result.

Let $H_1(X,\mathbb{R})$ be the one-dimensional real homology space of the manifold X, k a nonnegative quadratic form on $H_1(X,\mathbb{R})$. By the same letter k we denote the corresponding nonnegative definite inner product in the space $H_1(X,\mathbb{C}) = H_1(X,\mathbb{R}) \otimes \mathbb{C}$. Let K denote the Hilbert space of functions $f\colon X \to H_1(X,\mathbb{C})$ satisfying the condition

$$\int_X f(x)\,d\mu = 0,$$

with the inner product

$$(f_1, f_2) = \int_X k(f_1(x), f_2(x))\, d\mu.$$

Let ρ be the representation of the group $G = D(X, v^n)$ in K given by the formula $(\rho(g)f)(x) = f(xg)$. Construct a 1-cocycle of the representation ρ as follows. For any element $\alpha \in H^1(X, \mathbb{R})$ choose a closed 1-form a^1_α representing α, in such a way that a^1_α depends linearly on α. As in 22.2, for any $g \in G$ we obtain a function $\psi^\alpha_g \in C^\infty(X, \mathbb{R})$ such that its integral over X vanishes and $d\psi^\alpha_g = ga^1_\alpha - a^1_\alpha$. For any point $x \in X$ the linear functional $\alpha \mapsto \psi^1_\alpha(x)$ can be identified with an element $\theta_g(x) \in H_1(X, \mathbb{R})$. Hence, we get a mapping $g \mapsto \theta_g$ from G to the above space K; clearly, this mapping is the desired 1-cocycle of the representation ρ.

Finally, from the representation ρ and the cocycle $g \mapsto \theta_g$ we construct a representation of the group G in the space $\overline{S}(K)$. This representation is determined by the quadratic form k; we denote it by T_k.

THEOREM 36.3. *If $k \neq 0$, then the representation is irreducible. If $k_1 \neq k_2$, then the representations T_{k_1} and T_{k_2} are nonequivalent. If Y is a suitable region in X, then the restriction of the representation T_k to $D(Y, v^n)$ has a unique (up to a factor) nonzero invariant vector. The corresponding spherical function is given by the formula*

$$g \mapsto \exp\left\{\frac{1}{2}\sum_\gamma p^Y_{\widehat{g}}(\gamma)k(\mathring{\gamma}) + \frac{1}{2}k\Big(\sum_\gamma p^Y_{\widehat{g}}(\gamma)\mathring{\gamma}\Big)\right\}, \qquad g \in D(X, v^n),$$

where $\widehat{g}$ is an element of $\widehat{D}(X, v^n)$ satisfying the condition $p(\widehat{g}) = g$, and $\mathring{\gamma}$ is the image of the element $\gamma \in \Gamma$ under the natural homomorphism $\Gamma \to H_1(X, \mathbb{R})$.

Note that if the quadratic form k is of the form $k(\beta) = (\langle \alpha, \beta\rangle)^2$, where $\alpha \in H^1(X, \mathbb{R})$ and $\langle\cdot,\cdot\rangle$ is the natural pairing $H^1(X, \mathbb{R}) \times H_1(X, \mathbb{R}) \to \mathbb{R}$, then the representation T_k is equivalent to the representation T_α constructed in 36.1.

Since the proof of Theorem 36.3 does not require any new ideas, we omit it.

36.3. Representations $T^{\pi,\alpha}$. Here we use the construction from 22.3. Let $p\colon \widetilde{X} \to X$ be the universal covering space, Γ the group of deck transformations, π a unitary representation of the group Γ in a Hilbert space V. Recall that in 22.3 we constructed, starting from these data, a vector bundle $\xi\colon E \to X$, the space of its sections $S(X, \pi)$, and a representation ρ of the group $\widehat{D}(X, v^n)$ in $S(X, \pi)$. In the space $S(X, \pi)$ there is a natural inner product; by $L_2(X, \pi)$ we denote the completion of $S(X, \pi)$ with respect to the corresponding norm. The representation ρ can be extended to a unitary representation in the space $L_2(X, \pi)$. Taking an element $\alpha \in H^1(X, \pi)$ and a corresponding closed 1-form $a^1 \in Z^1(X, \pi)$, we define the cocycle $\sigma \mapsto f_\sigma$, $\sigma \in \widehat{D}(X, v^n)$, by formula (22.6). Given the representation ρ and the cocycle $\sigma \mapsto f_\sigma$, we construct a projective representation of the group $\widehat{D}(X, v^n)$ following the general construction from 6.2. According to 22.3, to the element $\alpha \in H^1(X, \mathbb{R})$ there corresponds $r \in H^1(\Gamma, \mathbb{R})$, determined by a cocycle $\overline{r}\colon \Gamma \to V$ of the representation V. Therefore, the natural notation for the constructed representation of the group $\widehat{D}(X, v^n)$ is $T^{\pi,r}$.

Results of 23.2 imply that $T^{\pi,r}$ can be extended to the group $\Gamma^X \cdot M(X, \mu)$. Using results from §33, we come to the following conclusions:

(1) If π does not contain a trivial subrepresentation and the cocycle r is completely nontrivial, then the representation $T^{\pi,r}$ is irreducible.

(2) The representations T^{π_k, r_k}, $k = 1, 2$, with π_k, r_k satisfying the conditions from (1), are equivalent if and only if there exists a unitary isomorphism of the spaces of representations π_1, π_2 that intertwines π_1 and π_2 and maps r_1 to r_2.

36.4. Locally weakly continuous representations. A unitary representation T of the group $D(X, v^n)$ is said to be *locally weakly continuous* if for any suitable region $Y \subset X$ (see 36.1) the restriction of T to $D(Y, v^n)$ can be extended to the group $M(Y, v^n)$. The last condition is equivalent to the requirement that the representation $g \mapsto T(g)$, $g \in D(Y, v^n)$, is continuous in the weak topology of the group $M(Y, v^n)$. One can easily see that representations of the group $D(X, v^n)$ obtained as restrictions to $D(X, v^n)$ of representations of $\Gamma^X/\Gamma_0 \cdot M(X, \mu)$ are locally weakly continuous. In particular, this is true for representations constructed in 36.1–36.5. It turns out that for $n \geq 5$ the converse is also true.

THEOREM 36.4. *Let $n = \dim X \geq 5$. Then any locally weak continuous representation of the group $D(X, v^n)$ can be continued to the group $\Gamma^X/\Gamma_0 \cdot M(X, \mu)$.*

We omit the (rather lengthy) proof of this theorem. Let us formulate a stronger statement using the language of inductive limits (in the category of metrizable groups with continuous homomorphisms as morphisms). For fixed X and v^n let us consider the group $D(X, v^n)$ and all subgroups $D(U, v^n)$, $M(U, v^n)$, where $U \subset X$; $U \simeq \mathbb{R}^n$ is such that any two points in U can be joined by a curve lying entirely in U with the length not exceeding a value that depends only on U. We have obvious inclusions $D(U_1, v^n) \hookrightarrow D(U_2, v^n)$ for $U_1 \subset U_2$, $D(U, v^n) \hookrightarrow D(X, v^n)$, $D(U, v^n) \hookrightarrow M(X, v^n)$. Then the group $\Gamma^X/\Gamma_0 \cdot M(X, v^n)$ together with the inclusions $D(U, v^n) \hookrightarrow M(U, v^n) \hookrightarrow \Gamma^X/\Gamma_0 \cdot M(X, v^n)$ is the inductive limit of the family of groups $\{D(U, v^n), M(U, v^n), D(X, v^n)\}$ with the above inclusions.

36.5. The case $\mu(X) = \infty$. Of course, in this case X is not compact. Choose a region $Y \subset X$ with the following properties: (1) $Y \simeq \mathbb{R}^n$; (2) $\mu(X \setminus Y) = 0$; (3) if we introduce a Riemannian metric in X, then for any compact subset $K \subset X$ there exists a number M such that any two points from $Y \cap K$ can be joined by a curve of length $< M$ lying entirely in Y. A region satisfying these conditions is said to be suitable.

Take an element $\alpha \in H^1(X, \mathbb{R})$ and construct the representation T_α of the group $D(X, v^n)$ that is spherical with respect to $D(Y, v^n)$ (i.e., contains a nonzero $D(Y, v^n)$-invariant vector) as follows. Let a^1 be a closed 1-form representing the class α. Similarly to 36.1, consider the natural representation ρ of the group $G = D_0(X, v^n)$ in the space $H = L_2(X, \mu)$ and the cocycle $g \mapsto \psi_g$ of this representation given by the formula $d\psi_g = ga^1 - a^1$, $\psi_g \in C_0^\infty$. In Y, we have $a^1 = db$ for some function $b \in L_2^{\mathrm{loc}}(X, \mu)$ that is smooth in Y. Take the cocycle $g \mapsto \psi_g - \rho(g)b + b = r_g$, $g \in G$, where $(\rho(g)b)(x) = b(xg)$. Using the representation ρ and the cocycle r_g, we construct the representation $T^{\alpha,Y}$ in the space $\overline{S}(H)$, $H = L_2(X, \mu)$, following the general procedure described in §6. Since the cocycle $g \mapsto r_g$ vanishes on $D(Y, v^n)$ and $\mu(X) = \infty$, we see that $\operatorname{Exp} 0$ is the unique (up to a factor) $D(Y, v^n)$-invariant vector. The corresponding spherical function is given by the formula similar to (36.1), (36.2). Two representations $T^{\alpha_k, Y}$, $k = 1, 2$, are equivalent if and only if $\alpha_1 = \alpha_2$ or $\alpha_1 = -\alpha_2$.

If Y, Z are two suitable regions, then the representations $T^{\alpha,Y}$ and $T^{\alpha,Z}$ are, in general, nonequivalent.

36.6. The case $\dim X = 2$. The example to be considered in this subsection first appeared in a different form in the paper by Neretin [**31**].

Let X be a connected manifold of dimension 2, v^2 an area form on X, μ the corresponding measure on X, $G = D(X, v^2)$. Choose a point $x_0 \in X$. There is an obvious homomorphism of the stabilizer G^{x_0} to the group $SL(2,\mathbb{R})$. Take one of the representations of $SL(2,\mathbb{R}) \simeq SU(1,1)$ mentioned in 11.1, and a cocycle of this representation. The composition of the representation with the above homomorphism gives a representation U_0 of the group G^{x_0} in the Hilbert space H^0 and a cocycle $r_0\colon G^{x_0} \to H^0$ of this representation. Following §35, we construct from these data first a representation U of the group $G = D(X, v^2)$ in the space $H = L^2(X, \mu, H^0)$ and a cocycle $r\colon G \to H$ of this representation, and then a representation (possibly projective) $T = T^{U,r}$ of the group G in the Fock space $\overline{S}(H)$.

THEOREM 36.5 (Neretin [**31**]). *The representation T is irreducible*

In the proof below some technical details will be omitted.

First let us consider the case when X is the disk $x_1^2 + x_2^2 < 1$, $v^2 = dx_1 \wedge dx_2$, μ is the Lebesgue measure, $x_0 = (0,0)$. Take a smooth mapping $\theta\colon X \to G$ such that $x_0\theta(x) = x$ for all $x \in X$ and at some neighborhood of the point x_0 the mapping $\theta(x)$ coincides with the translation by the vector $\overrightarrow{x_0 x}$ The representation T of the group G acts in $\overline{S}(H)$, $H = L_1(X, \mu, H^0)$, by the formula

$$T(g)\operatorname{Exp} f = \exp\left\{-\tfrac{1}{2}|r(g)|^2 + (f, r(g^{-1}))\right\}\operatorname{Exp}(U(g)f + r(g)), \tag{36.3}$$

where

$$(U(g)f)(x) = U_0(\theta(x)g(\theta(xg))^{-1})f(xg), \tag{36.4}$$

$$r(g)(x)x = r_0(\theta(x)g(\theta(xg))^{-1}). \tag{36.5}$$

Take a bijection $x \mapsto xg$, $x \in X$ that is smooth on a (possibly nonconnected) open subset $Y \subset X$ with $\mu(X \setminus Y) = 0$ and preserves the form v^2. Define the unitary operator $T(g)$ by formulas (36.3)–(36.5), where the right-hand sides of (36.4) and (36.5) are defined almost everywhere on X. Consider some examples of mappings g as above and prove that for these mappings the operators $T(g)$ belong to the strong closure $W = \overline{\{T(g),\ g \in G\}}$.

EXAMPLE 1. Take a rectangle $Q \subset X$; let g be the mapping that acts in A as the rotation by the angle π and is the identity outside S. It is easy to construct diffeomorphisms $g_\varepsilon \in D(X, v^2)$, $0 < \varepsilon < \varepsilon_0$, such that g_ε coincides with g outside the ε-neighborhood of ∂Q and in this neighborhood the partial derivatives of the vector-valued function $xg - x$ are bounded by $M_0\varepsilon^{-1}$, where $M_0 = \text{const}$. This gives the estimate $|r(g_\varepsilon)| = O(\log(1/\varepsilon))$. This easily implies that $T(g_\varepsilon)$ strongly converges to $T(g)$ as $\varepsilon \to 0$. Hence $T(g) \in W$.

EXAMPLE 2. Let Q_1, Q_2 be rectangles with sides of length a, b and with a common side of length a. Together, Q_1 and Q_2 form a rectangle Q_3 with sides of length a and $2b$. For each Q_i, $1 \le i \le 3$, take a mapping g_i as described in the previous example. Then the mapping $\overline{g} = g_1 g_2 g_3$ interchanges the rectangles Q_1 and Q_2, acting on each of them as a translation (without the rotation). The previous example shows that $\overline{g} \in W$.

Let us note that the choice of the mapping $\theta(x)$ ensures that $\theta(x)\overline{g}(\theta(x\overline{g}))^{-1} = e$ almost everywhere on X. Therefore, $T(\overline{g})$ acts by the formulas

$$T(\overline{g})\operatorname{Exp} f = \operatorname{Exp}(\rho(\overline{g})f), \tag{36.6}$$

$$(\rho(\overline{g})f(x) = f(x\overline{g}). \tag{36.7}$$

One can easily see that the transformations $\overline{g}$ constructed in Example 2 generate a subgroup that is dense in $M(X,\mu)$ in the weak topology. Therefore, for each $\overline{g} \in M(X,\mu)$ the unitary operator $T(\overline{g})$ defined by (36.6), (36.7) belongs to the group W. Now it is easy to prove that the group generated by all operators (36.3), $g \in G$, and all operators (36.6) acts irreducibly (we omit the easy details).

It is equally easy to pass from the disk to any two-dimensional manifold.

36.7. Remark (about the application of the braid group). We mention one more method to construct representations of the group $G = D(X, v^2)$ in the Fock space (in the case $\dim X = 2$). These representations arise from the natural homomorphism of the stabilizer G^{ω_0} (here ω_0 is a finite subset of X) to the braid group, a unitary representation of the braid group, and a corresponding cocycle. As in §30, we do not go into details.

§37. Representations of the group $D_0(X,\omega^2)$ in the Fock space

Here we consider a connected noncompact symplectic manifold (X,ω^2) with an exact form ω^2; let us recall that this group is generated by the subgroups $D(U,\omega^2)$, $U \subset X$, $U \simeq \mathbb{R}^n$.

37.1. Construction of representations. Let μ be a measure on X generated by the volume form $(\omega^2)^m$, $2m = \dim X$. By ρ we denote the representation of the group $D(X,\omega^2)$ in $L_2(X,\mu)$ given by the formula $(\rho(g)f)(x) = f(xg)$. If $\mu(X) < \infty$ (in this case we assume that $\mu(X) = 1$), we consider the space $L_2^0(X,\mu)$ consisting of all functions $f \in L_2(X,\mu)$ with zero integral over X; the restriction of the representation ρ to $L_2^0(X,\mu)$ will be denoted by ρ_0.

Following 24.2, we define certain cocycles of these representations. Choose a 1-form a^1 such that $da^1 = \omega^2$. Define a mapping $g \mapsto f_g$ from $D_0(X,\omega^2)$ to $C_0^\infty(X)$ by the formula $ga^1 - a^1 = df_g$, $f_g \in C_0^\infty(X)$. If $\mu(X) = 1$, we define additionally $f_g^0 = f_g - \int_X f_g\, d\mu$.

Finally, we construct the representation. If $\mu(X) = \infty$, then we take $\lambda \in \mathbb{R}$ and define (following §6) the representation $T = T^{a^1,\lambda}$ in the space $\overline{S}(H)$, $H = L_2(X,\mu)$, by the formula

$$T(g)\operatorname{Exp} f = \exp\left\{-\tfrac{1}{2}\lambda^2 |f_g|^2 + \lambda(f, f_{g^{-1}})\right\} \operatorname{Exp}(\rho(g)f + \lambda r_g).$$

If $\mu(X) = 1$, then we define the representation $T = T^{a^1,\lambda}$ in the space $\overline{S}(L_2^0(X,\mu))$ by the same formula with ρ replaced by ρ_0 and f_g replaced by f_g^0.

37.2. Irreducibility of representations $T^{a^1,\lambda}$. Following 24.2, define a homomorphism $K\colon D_0(X,\omega^2) \to \mathbb{R}$ by the formula

$$K(g) = \int_X f_g(x)\, d\mu, \qquad g \in D_0(X,\omega^2),$$

and set $D_*(X,\omega^2) = \ker K$. Let us note that $D_*(X,\omega^2)$ is the "small" subgroup defined in 26.1 for an arbitrary symplectic manifold.

THEOREM 37.1. *If $\lambda \neq 0$, then the restriction of the representation $T = T^{a^1,\lambda}$ to the subgroup $D_*(X, \omega^2)$ is irreducible. Here $\mu(X)$ can be either finite or infinite.*

PROOF. First we consider the case $\mu(X) = \infty$. We apply Theorem 8.1. For that, we must indicate a sufficiently large family of special subgroups in $D_*(X, \omega^2)$. This is done as follows.

Take a region $U \subset X$ such that in local coordinates x_k, y_k, $k = 1, \dots, m$, $2m = \dim X$, in U the form ω^2 is given by the formula

$$\omega^2 = \sum_{k=1}^{\infty} dx_k \wedge dy_k;$$

we will assume that the coordinates give a diffeomorphism $U \simeq \mathbb{R}^{2m}$. Introduce the function

$$r = \sum_{k=1}^{m} (x_k^2 + y_k^2). \tag{37.1}$$

Take the numbers a, b such that $0 < a < b < \infty$ and define regions U_{ab}, U_a, and U_b described by the inequalities $a < r < b$, $r < a$, and $r < b$ respectively. To a function $w \in C_0^\infty(a, b)$ we associate a diffeomorphism g_w that coincides with the identity outside U_{ab} and on U_{ab} is given by the formulas

$$\begin{aligned} x_k' &= x_k \cos w(r) - y_k \sin w(r), \\ y_k' &= x_k \sin w(r) + y_k \cos w(r). \end{aligned}$$

One can easily see that $g_w \in D(X, \omega^2)$. Let $A = A_U^{ab}$ be the group of all diffeomorphisms g_w satisfying the conditions

$$f_{g_w}(x) = 0 \quad \text{for all } x \in U_a, \tag{37.2}$$

$$K(g_w) = 0. \tag{37.3}$$

LEMMA 37.1. *A is a special subgroup (in the sense of* 8.3*).*

PROOF. In the region U, the form a^1 defined in 37.1 can be written as

$$a_U^1 = \frac{1}{2} \sum_{k=1}^{m} (x_k\, dy_k - y_k\, dx_k) + ds, \tag{37.4}$$

where $s \in C^\infty(U)$. A simple computation gives

$$g_w a_U^1 - a_U^1 = \frac{1}{2} r w'(r)\, dr + d(s(x g_w) - s(x)).$$

Therefore,

$$f_{g_w}(x) = \frac{1}{2} r w(r) + \frac{1}{2} \int_r^\infty w(t)\, dt + s(x g_w) - s(x). \tag{37.5}$$

Conditions (37.2) and (37.3) are equivalent to the conditions

$$\int_a^b w(r)\, dr = 0,$$

$$\int_a^b r^m w(r)\, dr = 0.$$

Denote by $C_0^0(a,b)$ the space of all real functions from $C_0^\infty(a,b)$ satisfying these conditions. The mapping $w \mapsto g_w$ establishes an isomorphism between the additive group of the space $C_0^0(a,b)$ and the group A. Below we will use this isomorphism to identify these two groups. The representation ρ of the group $D(X,\omega^2)$ in the space $H = L_2(X,\mu)$ yields a representation of the group $C_0^0(a,b)$, which we also denote by ρ.

Let H_0 be the subspace of functions from $H = L_2(X,\mu)$ vanishing outside the region U_{ab} and depending only on the function r defined by (37.1). Denote $H_1 = H \ominus H_0$. We must prove that the decomposition $H = H_0 \oplus H_1$ has all the properties listed in the definition of the special subgroup. First, the representation ρ of the group $C_0^0(a,b)$ is trivial in the subspace H_0 and the projection of the cocycle $w \mapsto f_{g_w}$ to H_0 is given by the formula

$$w \mapsto \frac{1}{2} r w(r) + \frac{1}{2}\int_r^\infty w(t)\,dt = f_w^0.$$

Second, the projection of the cocycle $w \mapsto f_{g_w}$ to H_1 is a trivial cocycle; this can be seen from (37.5). It remains to describe the measures ν_0 and ν_1 entering the definition of the special subgroup, and to prove that they are singularly related.

Let us recall that the space $\Omega = (C_0^0(a,b))'$ can be embedded into the group of characters of the group $A = C_0^0(a,b)$ as follows: to any functional $\omega \in \Omega$ there corresponds the character $f \mapsto \exp(i\langle f,\omega\rangle)$, $f \in A$. The required measures ν_0 and ν_1 turn out to be measures on Ω.

Let us recall that ν_0 is a Gaussian measure on Ω generated by the quadratic form

$$w \mapsto \lambda^2 (f_w^0, f_w^0)_{L_2} = \gamma_m \lambda^2 \int_a^b \left(w(r) + \frac{1}{r}\int_r^b w(t)\,dt \right)^2 r^{m+1}\,dr, \tag{37.6}$$

where the number γ_m depends on m only. Together with ν_0, consider the Gaussian measure ν_0^1 on Ω generated by the quadratic form

$$w \mapsto \gamma_m \lambda^2 \int_a^b (w(r))^2 r^{m+1}\,dr. \tag{37.7}$$

The quadratic forms (37.6) and (37.7) are related by the transformation $w \mapsto (I+K)w$, where

$$(Kw)(r) = \frac{1}{r}\int_r^b w(t)\,dt.$$

Consider K as an operator in the real Hilbert space V obtained from $C_0^\infty(a,b)$ with the quadratic form (37.7). One can see that the operator $I+K$ is invertible and the operators $K+K^*$ and KK^* are nuclear. Therefore, $I+K = S\mathcal{O}$, where $\mathcal{O}$ is an invertible orthogonal operator in V, and $S = \sqrt{(I+K)(I+K^*)} = I+L$, where L is a selfadjoint nuclear operator. This implies that the measures ν_0 and ν_0^1 are equivalent.

Let us return to the measure ν_1. Recall (8.3) that it can be written in the form $\nu_1 = \sum_{k\geq 0} 2^{-k}\tau^{*k}$, where τ is a probability measure that determines the maximal spectral type of the representation $w \mapsto \rho_{g_w}$, $w \in C_0^0(a,b)$, in the space H_1. To describe τ, we introduce in the region U_{ab} polar coordinates r_k, φ_k by the formulas

$x_k = r_k \cos\varphi_k$, $y_k = r_k \sin\varphi_k$. Then

$$H_1 = H_1^* \oplus \left(\bigoplus_k H_{1k}\right), \quad k = (k_1, \dots, k_m) \in \mathbb{Z}^m, \qquad H_1^* = L_2(X \setminus U_{ab}, \mu),$$

where for $k \neq 0$ the space H_{1k} consists of the functions of the form

$$x \mapsto f(r_1, \dots, r_m) e^{i(k_1\varphi_1 + \dots + k_m\varphi_m)}, \qquad x \in U_{ab},$$

and the space H_{10} consists of the functions of the form $f(r_1, \dots, r_m)$ that are orthogonal to H_0. In each space H_{1k} the representation $w \mapsto \rho_{g_w}$, $w \in A$, acts as the multiplication by $\exp\{i(k_1 + \dots + k_m)w(r)\}$. This means that the desired measure τ is supported on the set of functionals $k\delta_r \in \Omega$, where $k \in \mathbb{Z}$, $a < r < b$; here we used the above-mentioned embedding of the space $\Omega = (C_0^0(a,b))'$ into the group of characters of the group $C_0^0(a,b)$. Therefore, the measure ν_1 is supported on the set of functionals of the form $\sum k_j \delta_{r_j}$, $k_j \in \mathbb{Z}$, $r_j \in (a,b)$.

Therefore, we completely described the measures ν_0 and ν_1. Results of 15.5 imply that they are singularly related. Lemma 37.1 is proved.

Lemma 37.1 gives us a family of special subgroups A_U^{ab}. From the irreducibility of the representation ρ it follows that condition (1) of Theorem 8.1 is satisfied. Condition (2) is satisfied for obvious reasons. This proves Theorem 37.1 in the case $\mu(X) = \infty$. The case $\mu(X) = 1$ requires certain simple modifications of the arguments, which we do not present here.

37.3. Equivalence of representations $T^{a^1,\lambda}$. Again, we consider the case $\mu(X) = \infty$.

THEOREM 37.2. *Representations $T^{a_k^1,\lambda_k}$, $k = 1, 2$, $\lambda_k \neq 0$, are equivalent if and only if $\lambda_1 = \pm\lambda_2$, the form $a_2^1 - a_1^1$ is exact, and, moreover, $a_2^1 - a_1^1 = db$ for a function $b \in C^\infty(X) \cap L_2(X,\mu)$.*

PROOF. The representations $T^{a^1,\lambda}$ and $T^{a^1,-\lambda}$ are equivalent (the intertwining operator is $\operatorname{Exp} f \mapsto \operatorname{Exp}(-f)$). Next, the representations $T^{a^1,\lambda}$ and $T^{a^1+db,\lambda}$, where $b \in C^\infty(X) \cap L_2(X,\mu)$, are equivalent (the intertwining operator is $\operatorname{Exp} f \mapsto \exp\{-(|b|^2/2) + (f,b)\} \operatorname{Exp}(f+b)$). Therefore we have proved that if the pairs (a_1^1, λ_1) and (a_2^1, λ_2) satisfy the conditions of the theorem, then the representations $T^{a_1^1,\lambda_1}$ and $T^{a_2^1,-\lambda_2}$ are equivalent. It remains to prove that these conditions are also necessary for the equivalence of representations $T^{a_1^k,\lambda_k}$, $k = 1, 2$.

Let $T_1 \simeq T_2$, $T_k = T^{a_1^k,\lambda_k}$. First, we prove that $|\lambda_1| = |\lambda_2|$. Take a region U_{ab} as in 37.2. Consider the restrictions of representations T_k to the subgroup $A = A_U^{ab}$; as in 37.2, we identify this subgroup with the additive group of the space $C_0^0(a,b)$. Maximal spectral types of the resulting representations of the group $C_0^0(a,b)$ are given by the measures $\nu_0^{\lambda_k} * \nu_1$ (the measures $\nu_0^{\lambda_k}$, $k = 1, 2$, and ν_1 are described in the proof of Lemma 37.1). Since $T_1 \simeq T_2$, these measures are equivalent. According to Corollary 15.3, we have $\lambda_1 = \pm\lambda_2$.

Now we can assume that $\lambda_1 = \lambda_2 = \lambda$ and $T^{a_1^1,\lambda} \simeq T^{a_2^1,\lambda}$. We must prove that $a_2^1 - a_1^1 = db$ for a function $b \in C^\infty(X) \cap L_2(X,\mu)$. Take a region $Y \subset X$ such that $T \simeq \mathbb{R}^{2m}$, $\mu(X \setminus Y) = 0$. Then $a_2^1 - a_1^1 = db^Y$, $b^Y \in C^\infty(Y)$. Restrictions of the representations T_1 and T_2 to $D(Y,\omega^2)$ act by the formula

$$(37.8)\qquad \begin{gathered} T_k(g)\operatorname{Exp} f = \exp\{-(\lambda^2/2)|f_g^k|^2 + \lambda(f, f_{g^{-1}}^k)\}\operatorname{Exp}(\rho(g)f + \lambda f_g^k), \\ f_g^2(x) - f_g^1(x) = b^Y(xg) - b^Y(x). \end{gathered}$$

Let U be a unitary operator satisfying the condition $UT_1 = T_2U$. We analyze this operator in more detail.

Take a region $B \subset Y$ such that $B \simeq \mathbb{R}^n$, $\overline{B} \subset Y$, and $\overline{B}$ is compact. Define the function $h_B^Y\colon Y \to \mathbb{R}$ by the formula

$$h_B^Y = \begin{cases} 0, & x \notin B, \\ b^Y(x) - \dfrac{1}{\mu(B)} \int_B b^Y(x)\,d\mu, & x \in B. \end{cases}$$

Then $h_B^Y \in L_2^0(B,\mu)$. We have an obvious isomorphism of Hilbert spaces

$$(37.9)\qquad \begin{gathered} \overline{S}(L_2(X,\mu)) \simeq \overline{S}(L_2^0(B,\mu)) \otimes \overline{S}(\widetilde{H}), \\ \widetilde{H} = L_2(X,\mu) \ominus L_2^0(B,\mu). \end{gathered}$$

Restrict the representations T_1 and T_2 to the subgroup $D_*(B,\omega^2)$. Under the isomorphism (37.9) these restrictions act only on the first factor $\overline{S}(L_2^0(B,\mu))$ of the tensor product on the right-hand side of (37.9). These actions are irreducible (by Theorem 37.1) and are given by formula (37.8), where the functions f should be taken from $L_2^0(B,\mu)$. Therefore, the intertwining operator has the form $U = V_B \otimes K_B$, where V_B is an operator in the space $\overline{S}(L_2^0(B,\mu))$ given by the formula

$$V_B \operatorname{Exp} f = \exp\{-(\lambda^2/2)|h_B^Y|^2 + \lambda(f, h_B^Y)\}\operatorname{Exp}(f + h_B^Y), \qquad f \in L_2^0(B,\mu),$$

and K_B is an arbitrary unitary operator in $\overline{S}(\widetilde{H})$.

Choose functions $f_k \in C_0^\infty(Y) \cap L_2^0(Y,\mu)$ such that $(U\operatorname{Exp} f_1, \operatorname{Exp} f_2) = h \neq 0$. Take a region B satisfying the above conditions and such that $\operatorname{supp} f_k \subset B$, $k = 1, 2$. Denote by f_k^B the restriction of the function f_k to B and by 0^B the zero vector in the space H^B. Then

$$\begin{aligned} h &= (U\operatorname{Exp} f_1, \operatorname{Exp} f_2) = (V_B \operatorname{Exp} f_1^B, \operatorname{Exp} f_2^B)(K_B \operatorname{Exp} 0^B, \operatorname{Exp} 0^B) \\ &= \exp\{-\tfrac{1}{2}\lambda^2|h_B^Y|^2 + \lambda(f_1, h_B^Y) + (f_1 + \lambda h_B^Y, f_2)\}(K_B \operatorname{Exp} 0^B, \operatorname{Exp} 0^B). \end{aligned}$$

Since $(K_B \operatorname{Exp} 0^B, \operatorname{Exp} 0^B) \leq 1$, we have

$$\exp\{\tfrac{1}{2}\lambda^2|h_B^Y|^2 - \lambda(f_1 + f_2, h_B^Y) - (f_1, f_2)\} \leq 1/h.$$

Therefore, $|h_B^Y| \leq M$ for all regions B satisfying the above conditions, where M does not depend on B. This implies that $b^Y \in c + L_2(Y,\mu)$. Hence, we can assume that $b^Y \in L_2(Y,\mu)$.

The irreducibility of the representations T_k imply that the intertwining operator U is given by the formula

$$U\operatorname{Exp} f = \theta \exp\{-\tfrac{1}{2}|b^Y|^2 + (f, b^Y)\}\operatorname{Exp}(f + b^Y), \qquad \theta = \text{const}.$$

Therefore, the functions b^Y constructed for different regions Y coincide with a certain function $b \in C^\infty(X) \cap L_2(X,\mu)$. This proves the theorem.

In the case $\mu(X) < \infty$ the formulation (and the proof) of the theorem requires simple modifications.

§38. Concluding remarks about the representations of diffeomorphism groups

The constructions of representations of "large" groups presented in §30 and in this chapter are based, first, on representations of the group $K_r^{\omega_0}$, and second, on the one-dimensional cohomology of these representations. In this book we do not give a systematic presentation of these two themes related to $K_r^{\omega_0}$. Zaitsev (see [**13**]) studied representations of the group of r-jets of diffeomorphisms stabilizing a point; this group is a special case of the group $K_r^{\omega_0}$.

CHAPTER 8

Representations of Groups of Foliation Preserving Diffeomorphisms

Let X be a connected manifold, F a foliation on X, $D(X,F)$ the subgroup of all elements $g \in D(X)$ such that there exists a smooth family

$$\{g(t),\, 0 \le t \le 1\} \subset D(X), \qquad g(0) = e, \quad g(1) = g,$$

such that each $g(t)$ preserves each leaf of the foliation F and is supported on a compact set that does not depend on t. A systematic theory of representations of the group $D(X,F)$ does not exist yet. Without making any attempts to decide what problems such a theory might solve, we consider here only one example related to the foliation of the two-dimensional torus with the irrational rotation number. The chapter concludes with simple remarks about the construction of unitary representations of the group $D(X,F)$ and "large" subgroups of this group.

§39. Brief description of the results

It is known that the group $D(S^1)$ of orientation preserving diffeomorphisms of the circle can be embedded into the infinite-dimensional symplectic group. Using the Shale–Weil representation, we obtain from this embedding projective representations of the group $D(S^1)$. Among the extensive literature on this subject we mention [**30**] and [**64**]. In this chapter we present a modification of this subject related to II_∞ factors and diffeomorphisms of the torus $\mathbb{T}^2$ preserving the flow with irrational rotation number (the one-dimensional foliation whose leaves have constant angle with the equator of the torus, with irrational tangent). More precisely, for each factor M of type II_∞ we define the group, denoted Sp_M, which can be considered as one more version of the infinite symplectic group. Then we consider the group $D(\mathbb{T}^2,\theta)$ of diffeomorphisms of the torus $\mathbb{T}^2$ preserving leaves of the above foliation (here θ is the irrational number defining this foliation). We construct an embedding of the group $D(\mathbb{T}^2,\theta)$ into Sp_M, where the factor M is associated (in the standard way) with the group of rotations of the circle of the form

$$x \mapsto x + m\theta, \qquad x \in \mathbb{T}^1 = \mathbb{R}/\mathbb{Z}, \quad m \in \mathbb{Z}.$$

This provides us with a family of projective representations of the group $D(\mathbb{T}^2,\theta)$. To these representations there correspond projective representations of the Lie algebra $V(\mathbb{T}^2,\theta)$ of all vector fields on $\mathbb{T}^2$ tangential to the leaves of the foliation. We obtain representations of a central extension V_Γ, which can be considered as an analog of the Virasoro algebra; generators of this extensions are indexed not by integers (as in the Virasoro algebra), but by numbers $\gamma = m\theta + n$, $m,n \in \mathbb{Z}$. Let us note that the representations are not unitarizable.

§40. Groups and homogeneous domains associated to type II factors

In this section we start with a type II_∞ factor and define an analog of the symplectic group and a family of projective representations of this group.

40.1. The group Sp_M and the domain D_M. Let H be a Hilbert space of countable dimension, endowed with an involution $h \mapsto \overline{h}$, i.e., with a one-to-one antilinear mapping satisfying the conditions $\overline{\overline{h}} = h$, $(\overline{h_1}, \overline{h_2}) = (h_2, h_1)$. We extend the involution to linear operators by the formula $\overline{A}h = \overline{A\overline{h}}$. One can easily see that $\overline{A^*} = (\overline{A})^*$. Define the transposition of operators by the formula $A' = \overline{A^*}$.

Let M be a type II_∞ factor acting in H and invariant under the involution. Introduce the ideal

$$J_2 = \{A \in M \mid \operatorname{tr} A^*A < \infty\}$$

and define

$$D_M = \{X \in J_2 \mid X = X',\ \|X\| < 1\},$$

$$Sp_M = \left\{ g = \begin{pmatrix} A & B \\ \overline{B} & \overline{A} \end{pmatrix} \middle| A \in M,\ B \in J_2,\ g^*Jg = gJg^* = J \right\}, \quad J = \begin{pmatrix} -I & 0 \\ 0 & I \end{pmatrix}.$$

It is clear that Sp_M is a group. The domain D_M is an analog of one of the classical domains (see [**34,** Chapter II, §6]). The equality $g^*Jg = J$ is equivalent to the fact that the real-linear transformation $h \mapsto AH + B\overline{h}$ preserves the antisymmetric form $\operatorname{Im}(h_1, h_2)$. The formula

$$Xg = (XB + \overline{A})^{-1}(XA + \overline{B}), \qquad X \in D_M, \quad g \in Sp_M, \tag{40.1}$$

define a transitive action of the group Sp_M on the domain D_M.

This construction is completely analogous to the construction from §3. Below we will see how far this analogy can be extended.

40.2. The group Sp_M^0 and the domain D_M^0. In our further considerations it will be convenient to present the factor M in matrix form, and also make the group Sp_M and the domain D_M slightly smaller. Take a factor M^0 of type II_1 acting in a Hilbert space H^0. In the Hilbert space $H = H^0 \oplus H^0 \oplus \cdots$, whose elements are written in the form $h = (h_0, h_1, \dots)$, consider the algebra M of all bounded operators that can be represented by a matrix $A = (A_{pq})$, where $p, q \in \{0, 1, \dots\}$, $A_{pq} \in M^0$. Then M is a factor of type II_∞; it is the factor we will use in this chapter. The involution in H is given by the formula $\overline{h} = (\overline{h_1}, \overline{h_2}, \dots)$.

To define the objects mentioned in the title of this subsection, we introduce certain subsets of M. For any $\alpha \in (0, 1]$ denote by $M(\alpha)$ the set of all operators $A = (A_{pq}) \in M$ satisfying the condition

$$\|A_{pq}\| \le L_r(1 + p + q)^{-r} \tag{40.2}$$

for each $r > 0$ in each of the region $p \le \alpha q$, $q \le \alpha p$; here L_r does not depend on p and q. In particular, the set $M(1)$ consists of all matrices satisfying the condition (40.2) for all $p, q \ge 0$ and all $r > 0$. It is easy to verify that

$$M(\alpha_1)M(\alpha_2) \subset M(\alpha_1\alpha_2), \quad M(\alpha)M(1) \subset M(1). \tag{40.3}$$

Define

$$D^0_M = D_M \cap M(1),$$
$$Sp^0_M = \left\{ g = \begin{pmatrix} A & B \\ \bar{B} & \bar{A} \end{pmatrix} \in Sp_M \,\middle|\, B \in M(1),\ A \in M(\alpha) \right\}$$

(for some $\alpha \in (0,1]$). Using (40.3), one can easily see that Sp^0_M is a subgroup of Sp_M, and formula (40.1) defines a transitive action of Sp^0_M on D^0_M. The group Sp^0_M will be used in §41 to study the group $D(\mathbb{T}^2, \theta)$.

Now we turn to projective representations of the group Sp_M. We need the notion of the determinant in a factor.

40.3. The function det **on a factor.** In [**50**], Fuglede and Kadison introduced the determinant function on a factor of type II_1 by the formula $\Delta(A) = \exp \operatorname{tr} \log A^*A$. In particular, they proved multiplicativity: $\Delta(AB) = \Delta(A)\Delta(B)$. We need a modification of this definition. Define the ideal $J_1 = J_2 \cdot J_2$, so that J_1 consists of sums of products of elements from J_2. For an element $A \in M$ of the form $A = I + K$, $K \in J_1$, $\|K\| < 1$, define

$$\det A = \exp \operatorname{tr} \log A, \quad \text{where } \log A = \sum_{n=1}^{\infty} \frac{(-1)^{n-1} K^n}{n}.$$

LEMMA 40.1. $\det AB = \det A \cdot \det B$ *whenever* A, B, AB *satisfy the conditions in the definition of* det.

PROOF. Our argument is close to the proof of a similar result about the function Δ from [**50**]. Let there be given a smooth family $X(t) \in M$, $t \in [0,1]$, and a holomorphic function $f(z)$, $z \in E$, where E is a simply connected region in $\mathbb{C}$ containing the spectra of all operators $X(t)$. Under these assumptions the following equality is proved in [**50**]:

$$(\operatorname{tr} f(X(t)))' = \operatorname{tr}(f'(X(t))X'(t), \qquad t \in [0,1]. \tag{40.4}$$

Let us apply this equality to the function $f(z) = \log z$ (the principal branch of the logarithm) and to the family $X(t) = A(t)B$, $0 \le t \le 1$, where $A(t) = tI + (1-t)A$. First we show that $\|X(t) - I\| < 1$ for $0 \le t \le 1$. Indeed, by the conditions of the lemma this equality holds for $X(0) = AB$ and for $X(1) = B$. Since the ball $\{Y \mid \|I - Y\| < 1\}$ is a convex set, this equality holds for all $t \in [0,1]$ Therefore, the spectrum of the operator $X(t)$ is contained in the disk $|1 - z| < 1$, and we can apply (40.4). Hence,

$$\begin{aligned}(\operatorname{tr} \log X(t))' &= \operatorname{tr}((X(t))^{-1}X'(t)) = \operatorname{tr}(B^{-1}(A(t))^{-1}(I - A)B) \\ &= \operatorname{tr}((A(t))^{-1}(I - A)) = (\operatorname{tr} \log A(t))'.\end{aligned}$$

Therefore, $\operatorname{tr} \log X(t) - \operatorname{tr} \log A(t) = h = \text{const}$. Taking $t = 1$ we get $h = \operatorname{tr} \log B$. Taking $t = 0$ we get $\operatorname{tr} \log B = \operatorname{tr} \log AB - \operatorname{tr} \log A$. This implies the lemma.

Let us note also that for any $\lambda \in \mathbb{C}$ we have the number $\det A^\lambda$ defined by the formula $\det A^\lambda = \exp(\lambda \operatorname{tr} \log A)$.

40.4. Representations T^λ. By $F = F(D_M)$ we denote the space of holomorphic functions $f\colon D_M \to \mathbb{C}$ that can be extended to the region $(1+\varepsilon)D_M$ (where ε depends on f). For $\lambda \in \mathbb{C}$ define the operators $T^\lambda(g)$ in the space F by the formula

$$(T^\lambda(g)f)(X) = \det(I + XB\overline{A}^{-1})^\lambda f(Xg), \qquad g = \begin{pmatrix} A & B \\ \overline{B} & \overline{A} \end{pmatrix} \in Sp_M, \quad f \in F.$$

Using Lemma 40.1, we can easily verify that

$$T^\lambda(g_1g_2) = \tau^\lambda(g_1, g_2)T^\lambda(g_1)T^\lambda(g_2),$$

where the function $\tau^\lambda(g_1, g_2)$ is given by the formula

$$\tau^\lambda(g_1, g_2) = \det(I + \overline{A_1^{-1}B_1B_2A_2^{-1}})^\lambda$$

(see the last sentence of 40.3). Note that τ^λ can be rewritten in the form

$$\tau^\lambda(g_1, g_2) = \det((g_1)_{22}^{-1}(g_1g_2)_{22}(g_2)_{22}^{-1})^\lambda, \tag{40.5}$$

where g_{22} is the right lower element of g. We get a projective representation of the group Sp_M; the corresponding 2-cocycle is given by formula (40.5) (compare with 3.10).

For the group Sp_M^0, a natural space of the representation (also denoted T^λ) is the space of functions $f\colon D_M^0 \to \mathbb{C}$ admitting a holomorphic extension to the region $(1+\varepsilon)D_M^0$, where $\varepsilon > 0$ depends on f.

40.5. The representation r^λ of the Lie algebra sp_M. How should we define the Lie algebras of the above groups? We have chosen the following definition. By sp_M we denote the Lie algebra of all matrices

$$l = \begin{pmatrix} a & b \\ \overline{b} & \overline{a} \end{pmatrix}, \qquad a = (a_{pq}), \quad b = (b_{pq}), \quad a_{pq}, b_{pq} \in M^0, \quad p, q \geq 0,$$

satisfying the following conditions:

(1) $a_{pq} = 0$ for $|p - q| > N$, $b_{pq} = 0$ for $p + q > N$, where N depends on l;
(2) $\|a_{pq}\| \leq \mathrm{const}(1+p)^c$, where c depends on l;
(3) $a = -a^*$, $b = b'$.

We will consider sp_M as the Lie algebra of the group Sp_M (and of the group Sp_M^0).

Now we pass on to representations. In the space F consider the subspace F^0 consisting of polynomials, i.e., sums of products of functions of the form

$$X \mapsto \mathrm{tr}(K_1X \cdots K_nX), \tag{40.6}$$

where K_j are finite matrices from M (a matrix $K = (K_{pq})$ is said to be finite if for some N we have $K_{pq} = 0$ whenever $p + q > N$). For any function from F^0 there exists an obvious extension from D_M to the space of arbitrary matrices of the form $X = (X_{pq})$, $X_{pq} \in M^0$ (so that X does not necessarily correspond to a bounded operator). For any such matrix X_0 define an operator in F^0 by the formula

$$(D_{X_0}f)(X) = \frac{d}{dt}f(X + tX_0)\big|_{t=0};$$

it is clear that this operator maps F^0 into itself. Now we define operators $r^\lambda(l)$, $l \in sp_M$, acting in F^0 by the formula

$$(r^\lambda(l)f)(X) = (D_Y f)(X) + \lambda \operatorname{tr}(bX) f(X), \qquad l = \begin{pmatrix} a & b \\ \bar{b} & \bar{a} \end{pmatrix}, \tag{40.7}$$

where $Y = a'X + Xa + \bar{b} - XbX$.

LEMMA 40.2. *We have*

$$[r^\lambda(l_1), r^\lambda(l_2)] = r^\lambda([l_1, l_2]) + \lambda \operatorname{tr}(b_2\bar{b}_1 - b_1\bar{b}_2), \tag{40.8}$$

where

$$l_k = \begin{pmatrix} a_k & b_k \\ \bar{b}_k & \bar{a}_k \end{pmatrix}, \qquad k = 1, 2.$$

The proof follows from the direct computation of the action of operators $r^\lambda(l)$ on functions of the form (40.6).

The next lemma somewhat justifies our intention to declare operators (40.7) a representation of the algebra sp_M corresponding to the representation T^λ of the group Sp_M.

LEMMA 40.3. *Let there be given elements*

$$l = \begin{pmatrix} a & b \\ \bar{b} & \bar{a} \end{pmatrix} \in sp_M, \qquad g_t = \begin{pmatrix} A_t & B_t \\ \bar{B}_t & \bar{A}_t \end{pmatrix} \in Sp_M, \quad 0 \le t \le 1,$$

satisfying the following conditions (*as* $t \to 0$):

$$\sum_k \|(t^{-1}(A_t - I) - a)_{kn}\|^2 \to 0, \tag{40.9}$$

$$\sum_k \|(t^{-1}(A_t - I) - a)_{nk}\|^2 \to 0, \tag{40.10}$$

for each n, *and the condition*

$$\sum_{m,n} \|(t^{-1}B_t - b)_{mn}\|^2 \to 0. \tag{40.11}$$

Then

$$t^{-1}(T^\lambda(g_t)f - f)(X) \to (r^\lambda(l)f)(X) \quad \text{as } t \to 0 \tag{40.12}$$

for any function $f \in F^0$.

PROOF. It suffices to prove that

$$t^{-1}(f(Xg_t) - f(X)) \to D_Y f(X), \qquad Y = a'X + Xa + \bar{b} - XbX, \tag{40.13}$$

$$t^{-1} \operatorname{tr} \log(I + XB_t\overline{A_t^{-1}}) \to \operatorname{tr} Xb. \tag{40.14}$$

Let us prove (40.13). Condition (40.11) shows that

$$B_t = tb + tC_t, \qquad \sum_{p,q} \|(C_t)_{pq}\| \to 0. \tag{40.15}$$

Therefore, $B_t = tb + o(t)$; here and later $o(t)$ is understood in the sense of the operator norm. Hence, $A_t^{-1} = A_t^*(I + B_t B_t^*)^{-1} = A_t^* + o(t)$. Therefore, for each $X \in D_M$ we have

$$\begin{aligned} Xg_t &= (\overline{A_t} + XB_t)^{-1}(XA_t + \overline{B_t}) \\ &= \overline{A_t^{-1}}(I + XB_t\overline{A_t^{-1}})(XA_t + \overline{B_t}) \\ &= A_t'XA_t + tA_t'b - tA_t'XBA_t'XA_t + o(t) \end{aligned}$$

(recall that $A' = \overline{A^*}$). Using conditions (40.9) and (40.10) we see that for any finite matrices $K_1, K_2 \in M$

$$t^{-1}K_1(Xg_t - X)K_2 \to K_1(a'X + Xa + \bar{b} - XbX)K_2 \quad \text{as } t \to 0$$

(in the operator norm). This implies (40.13).

Let us prove (40.14). It is clear that

$$\log(I + XB_t\overline{A_t^{-1}}) = XB_t\overline{A_t^{-1}} + (XB_t\overline{A_t^{-1}})^2 V_t,$$

where $\|V_t\| \le \text{const}$ for $0 \le t \le 1$. Further estimates use the norm $|Y|_2 = \sqrt{\operatorname{tr} Y^*Y}$ and its obvious properties. Using (40.15), we get

$$|\operatorname{tr}(XB_t\overline{A_t^{-1}})V_t| \le \text{const}\, \|B_t\|^2 = o(t). \tag{40.16}$$

Further,

$$\begin{aligned} |t^{-1}\operatorname{tr}(XB_t\overline{A_t^{-1}}) - \operatorname{tr} Xb| &\le |\operatorname{tr}(Xb(\overline{A_t^{-1}} - I))| + |\operatorname{tr}(XC_t\overline{A_t^{-1}})| \\ &\le |X|_2(\|b(I - A_t)\| + |C_t\|_2) \\ &\le \text{const}\left(\left(\sum_{p \le N} \|(I - A_t)_{pq}\|^2\right)^{1/2} + \left(\sum_{p,q} \|(C_t)_{pq}\|^2\right)^{1/2}\right), \end{aligned}$$

where N is defined by the condition $b_{pq} = 0$ for $p + q > N$. Conditions (40.10) and (40.15) show that the last sum tends to zero. This, together with (40.16), implies (40.14). Lemma 40.3 is proved.

40.6. Representations of the Lie algebra $sp_M^{\mathbb{C}}$. By $sp_M^{\mathbb{C}}$ we denote the complexification of the Lie algebra sp_M. It can be identified with the complex Lie algebra formed by matrices of the form

$$l = \begin{pmatrix} a & b \\ c & -a' \end{pmatrix},$$

where a, b, c are matrices of the form (a_{pq}), etc., with elements from M^0 such that $a_{pq} = 0$ for $|p - q| > N$, $\|a_{pq}\| \le \text{const}(1 + p)^\alpha$ with α depending on l, $b_{pq} = c_{pq} = 0$ for $p + q > N$, $b = b'$, $c = c'$. The representation r^λ is extended to $sp_M^{\mathbb{C}}$ by the formula

$$(r^\lambda(l)f)(X) = D_Y f(X) + \lambda \operatorname{tr}(Xb)f(X), \quad Y = a'X + Xa - XbX + c. \tag{40.17}$$

We have

$$[r^\lambda(l_1), r^\lambda(l_2)] = r^\lambda([l_1, l_2]) + \lambda \operatorname{tr}(b_2c_1 - b_1c_2), \tag{40.18}$$

where

$$l_k = \begin{pmatrix} a_k & b_k \\ c_k & -a_k' \end{pmatrix} \in sp_M^{\mathbb{C}}, \qquad k = 1, 2.$$

§41. The group $D(\mathbb{T}^2, \theta)$, its embedding into Sp_M^0, and its projective representations

41.1. The group $D(\mathbb{T}, \theta)$. We realize the flow with the irrational rotation number on the torus as follows. Consider the circle $\mathbb{T}^1 = \mathbb{R}/\mathbb{Z}$, then the cylinder $\mathbb{T}^1 \times \mathbb{R}$, and identify in the cylinder the points (x, y) and $(x - \theta, y + 1)$. We obtain the torus $\mathbb{T}^2$. The images of the lines $x = x_0$ give the flow on the torus with the irrational rotation number.

By $D(\mathbb{T}^2, \theta)$ we denote the connected component of the identity in the group of all diffeomorphisms of the torus that preserve each leaf of the foliation and can be joined with the identity diffeomorphism by a smooth curve consisting of diffeomorphisms with the same property. It is convenient to identify elements $g \in D(\mathbb{T}^2, \theta)$ with diffeomorphisms of the cylinder $\mathbb{T}^1 \times \mathbb{R}$ that commute with the above transformation $(x, y) \mapsto (x - \theta, y + 1)$. Therefore, g is given by the formula $(x, y)g = (x, y + h(x, y))$, where the C^∞-function $h(x, y)$, $(x, y) \in \mathbb{T}^1 \times \mathbb{R}$, satisfies the conditions $h(x, y) = h(x - \theta, y + 1)$ and $h_y'(x, y) > -1$. We endow the group $D(\mathbb{T}^2, \theta)$ with the uniform convergence of the corresponding functions $h(x, y)$ together with all partial derivatives. Let us remark that this topology differs from that induced from $D(\mathbb{T}^2)$; moreover, the closure of the subgroup $D(\mathbb{T}^2, \theta)$ in the latter topology coincides with the subgroup of fiber preserving diffeomorphisms.

The goal of further exposition is to introduce an appropriate factor of type II_∞ and then to construct an embedding $D(\mathbb{T}^2, \theta) \hookrightarrow Sp_M$. This is done in subsections 41.2 and 41.3.

41.2. Description of the factor M. The number θ determines a subgroup in the group of rotations of the circle $\mathbb{T}^1 = \mathbb{R}/\mathbb{Z}$ consisting of the rotations $x \mapsto x + m\theta$, $\theta \in \mathbb{Z}$. Starting with this subgroup, the classical theory (see [**29,** Chap. 3]) provides us with a factor of type II_1. Let us recall that this factor acts in the Hilbert space $H^0 = L_2(\mathbb{T}^1 \times \mathbb{Z})$ and consists of bounded operators commuting with the operators $(V_1 f)(x, n) = f(x - \theta, n + 1)$, $f \in H^0$, and with the operators $M_\varphi f(x, n) = \varphi(x) f(x, n)$, where $\varphi(x)$ is an arbitrary bounded measurable function on $\mathbb{T}^1$. Denote this factor by M_0. The trace in M_0 is given by the formula $\operatorname{tr} A = (Ag_0, g_0)$, where $g_0(x, n) = 0$ for $n \neq 0$, $g_0(x, 0) = 1$. The involution $f(x, n) \mapsto \overline{f(x, n)}$ in H^0 can be extended to M_0. Now we introduce the factor M acting in the Hilbert space $H = H^0 \oplus H^0 \oplus \cdots$ and consisting of the bounded operators $A = (A_{pq})$, $p, q = \{0, 1, 2, \dots\}$, $A_{pq} \in M_0$. The factor M is of type II_∞, with the trace given by the formula $\operatorname{tr} A = \sum_{k=0}^{\infty} \operatorname{tr} A_{kk}$.

Below it will more convenient to use another realization of the factor M. Consider the Hilbert space $L_2(\mathbb{T}^1 \times [0, \infty))$ of functions $f(x, s)$ with the inner product

$$(f_1, f_2) = \int_0^1 \int_0^\infty f_1(x, s) \overline{f_2(x, s)} \, dx \, ds.$$

The isomorphism between the space $L_2(\mathbb{T}^1 \times [0, \infty))$ and the above space $H = H^0 \oplus H^0 \oplus \cdots$ is constructed as follows. Represent the space $L_2(\mathbb{T}^1 \times [0, \infty))$ as follows:

$$L_2(\mathbb{T}^1 \times [0,\infty)) = \bigoplus_0^\infty L_2(\mathbb{T}^1 \times [2\pi k, 2\pi(k+1)])$$

and then identify each space $L_2(\mathbb{T}^1 \times [2\pi k, 2\pi(k+1)])$ with $H^0 = L_2(\mathbb{T}^1 \times \mathbb{Z})$ by associating to a function $f(x,s)$ from $L_2(\mathbb{T}^1 \times [2\pi k, 2\pi(k+1)])$ the function

$$\widehat{f}(x,m) = \frac{1}{2\pi}\int_{2\pi k}^{2\pi(k+1)} f(x,s)e^{-ims}ds$$

from H^0. We obtain an isomorphism $L_2(\mathbb{T}^1 \times [0,\infty)) \to H$, which maps the above operators M_φ and V_1 to the operators $M_\varphi f(x,s) = \varphi(x)f(x,s)$, $(V_1 f)(x,s) = f(x-\theta,s)e^{is}$. The factor M becomes the factor (which we still denote M) consisting of all bounded operators commuting with these operators M_φ and V_1. The trace on M is given by the formula

$$\text{(41.1)} \qquad \operatorname{tr} A = \sum_{k=0}^{\infty}(Ag_k, g_k), \qquad g_k(x,s) = \begin{cases} 0 & \text{for } s \neq [2\pi k, 2\pi(k+1)], \\ 1 & \text{for } s \in [2\pi k, 2\pi(k+1)]. \end{cases}$$

Finally, the above involution in H becomes the involution $f \to \overline{f}$ in $L_2(\mathbb{T}^1 \times [0,\infty))$ given by the formula $\overline{f}(x,s) = \overline{f(x,\sigma(s))}$, where $s \mapsto \sigma(s)$ acts in each interval $(2\pi k, 2\pi(k+1))$ as the symmetry with respect to the center; in other words, $\sigma(s) = 4\pi k + 2\pi - s$ for $s \in (2\pi k, 2\pi(k+1))$.

41.3. The embedding $D(\mathbb{T}^2,\theta) \hookrightarrow Sp^0_M$. Denote by S the space of all real C^∞ functions $f(x,y)$, $(x,y) \in \mathbb{T}^1 \times \mathbb{R}$, that decrease rapidly as $|y| \to \infty$. The group $D(\mathbb{T}^2,\theta)$ acts on S by the formula $f(x,y) \mapsto f(x, y + h(x,y))$, where $h(x,y) = h(x-\theta, y+1)$, $h_y(x,y) > -1$. This action preserves the antisymmetric form

$$\text{(41.2)} \qquad (f_1, f_2) = \int_0^1 \int_{-\infty}^{\infty} f_1(x,y)\frac{\partial}{\partial y} f_2(x,y)\,dx\,dy.$$

Our goal is to map S to a complex Hilbert space with involution. To achieve it, we associate to a function $f(x,y)$ from S the function $\varphi(x,s)$, $(x,s) \in \mathbb{T}^1 \times [0.\infty)$, by the formula

$$\varphi(x,s) = \sqrt{s}\int_{-\infty}^{\infty} f(x,y)e^{-iys}\,dy.$$

We obtain an $\mathbb{R}$-linear embedding of the space S into the complex Hilbert space $L_2(\mathbb{T}^1 \times [0,\infty))$. Denote by $\widehat{S}$ the image of the space S under this embedding. The form (41.2) becomes the form $\operatorname{Im}(\varphi_1, \varphi_2)$, $\varphi_k \in \widehat{S}$, $k = 1,2$. The action of the group $D(\mathbb{T}^2,\theta)$ in the space S becomes the action in $\widehat{S}$ given by the formula

$$\text{(41.3)} \qquad (R_g)(\varphi)(x,s) = \int_0^\infty N_g(x,s,t)\varphi(x,t)\,dt + \int_0^\infty L_g(x,s,t)\overline{\varphi(x,t)}\,dt,$$

where

$$\text{(41.4)} \qquad N_g(x,s,t) = \frac{1}{2\pi}\sqrt{\frac{s}{t}}\int_{-\infty}^{\infty} \exp\{i(y + h(x,y))t - iys\}\,dy,$$

$$\text{(41.5)} \qquad L_g(x,s,t) = \frac{1}{2\pi}\sqrt{\frac{s}{t}}\int_{-\infty}^{\infty} \exp\{-i(y + h(x,y))t - iys\}\,dy$$

(the integrals should be understood in the sense of distribution theory). Let us recall that we have defined the involution $f \mapsto \overline{f}$ in $L_2(\mathbb{T}^1 \times [0,\infty))$ by the formula $\overline{f}(x,s) = \overline{f(x,\sigma(s))}$ (see 41.2). Using this involution, we can rewrite the operator (41.3) in the form $R_g\varphi = A_g\varphi + B_g\overline{\varphi}$, where A_g, B_g are the integral operators with the kernels $N_g(x,s,t)$ and $L_g(x,\sigma(x),\sigma(t))$. The operator R_g preserves the form $\operatorname{Im}(\varphi_1,\varphi_2)$.

THEOREM 41.1. *The operators A_g and B_g can be extended to bounded operators in the space $L_2(\mathbb{T}^1 \times [0,\infty))$ (we continue to denote these extensions by A_g and B_g). For each $g \in D(\mathbb{T}^2,\theta)$ the matrix*

$$K_g = \begin{pmatrix} A_g & B_g \\ \overline{B_g} & \overline{A_g} \end{pmatrix} \tag{41.6}$$

belongs to the group Sp^0_M.

PROOF. First we consider the case when the function $h(x,y)$ defining the element $g \in D(\mathbb{T}^2,\theta)$ satisfies the following condition: there exists $\varepsilon > 0$ such that

$$h(x,y) = 0 \quad \text{whenever } |y-n| < \varepsilon \text{ for some } n \in \mathbb{Z}. \tag{41.7}$$

We transform formula (41.4) taking into account that $h(x-\theta, y+1) = h(x,y)$; everywhere below we have $t > 0$, $s > 0$. We have

$$\begin{aligned}
N_g(x,s,t) &= \delta(s-t) + \frac{1}{2\pi}\sqrt{\frac{s}{t}}\int_{-\infty}^{\infty} \exp\{iy(t-s)\}\exp\{ith(x,y)) - 1\}\,dy \\
&= \delta(s-t) + \frac{1}{2\pi}\sqrt{\frac{s}{t}}\sum_{n=-\infty}^{\infty}\int_0^1 \exp\{i(y+n)(t-s)\}(\exp\{ith(x,y+n)\} - 1)\,dy \\
&= \delta(s-t) + \frac{1}{2\pi}\sqrt{\frac{s}{t}}\sum_{n=-\infty}^{\infty}\int_0^1 \exp\{i(y+n)(t-s)\}(\exp\{ith(x,y+n)\} - 1)\,dy \\
&= \delta(s-t) + \sum_{n=-\infty}^{\infty} u(x+n\theta,s,t)\exp\{in(t-s)\},
\end{aligned}$$

where

$$u(x,s,t) = \frac{1}{2\pi}\sqrt{\frac{s}{t}}\int_0^1 \exp\{iy(t-s)\}(\exp\{ith(x,y)\} - 1)\,dy.$$

Taking the Fourier series of the latter function in the variable x, we obtain

$$u(x,s,t) = \sum_k u_k(s,t)\exp\{2\pi xki\},$$

where

$$u_k(s,t) = \frac{1}{2\pi}\sqrt{\frac{s}{t}}\int_0^1\int_0^1 \exp\{iy(t-s) - 2\pi kxi\}(\exp\{ith(x,y)\} - 1)\,dx\,dy. \tag{41.8}$$

Substitute the Fourier series in the formula for $N_g(x,s,t)$ and apply the Poisson summation formula $\sum_k \exp(ikx) = 2\pi\sum_r \delta(x+2\pi r)$. We obtain

$$N_g(x,s,t) = \delta(s-t) + 2\pi\sum_{k,r} u_k(s,t)\delta(t-s+2\pi k\theta+2\pi r)\exp(2\pi kxi). \tag{41.9}$$

Similar computations give the formula

$$L_g(x,s,t) = 2\pi \sum_{k,r} v_k(s,t)\delta(-t-s+2\pi k\theta+2\pi r)\exp(2\pi kxi), \tag{41.10}$$

where

(41.11)

$$v_k(s,t) = \frac{1}{2\pi}\sqrt{\frac{s}{t}}\int_0^1\int_0^1 \exp\{-iy(t+s)-2\pi kxi\}(\exp\{-ith(x,y)\}-1)\,dx\,dy.$$

Now we estimate the functions $u_k(s,t)$ and $v_k(s,t)$. First, consider the function

$$\psi(s,t) = \int_0^1 \exp\{i(sy+t\varphi(y))\}f(y)\,dy$$

constructed for arbitrary functions $\varphi \in C^\infty(0,1)$ and $f \in C_0^\infty(0,1)$ with $\varphi'(y) > 0$.

LEMMA 41.1. *For any $m > 0$ we have*

$$|\psi(s,t)| \le A_m(1+|s|+|t|)^{-m}, \quad A_m = \text{const},$$

whenever $st > 0$. A similar inequality holds in each of the regions $|s| \le \alpha|t|$ and $|t| \le \beta|s|$, where α and β satisfy the conditions $0 < \alpha < \min\varphi'(y)$, $0 < \beta < (\max\varphi'(y))^{-1}$.

Let us prove the lemma (although it may be considered known). First we prove that in each of the regions indicated in the lemma the function $\psi(x,s)$ can be represented, for each $m > 0$, in the form

$$\psi(x,s) = \int_0^1 \exp\{i(sy+t\varphi(y))\}f_m(y,s,t)\,dy, \tag{41.12}$$

where

$$f_m(y,s,t) = \sum_{k=0}^{m} \frac{t^k q_k(y)}{(s+t\varphi'(y))^{m+k}},$$

and the functions $q_k(y)$ are polynomials in $f(y)$, $\varphi(y)$, and their derivatives. We use induction on m. If (41.12) holds for some m, then, integrating by parts, we get

$$\psi(x,s) = -\int_0^1 \exp\{i(sy+t\varphi(y))\}\left(\frac{f_m(y,s,t)}{s+t\varphi'(y)}\right)'_y dy.$$

It is clear that the function

$$f_{m+1}(y,s,t) = -\left(\frac{f_m(y,s,t)}{s+t\varphi'(y)}\right)'_y$$

satisfies the required conditions. This proves (41.12). It is clear that $|f_m(y,s,t)| \le \text{const}\,|s+t\varphi'(y)|^{-m}$ in the regions indicated in the lemma, and the proof of Lemma 41.1 is complete.

Now we estimate the functions $u_k(s,t)$ and $v_k(s,t)$.

LEMMA 41.2. *In each region of the form* $0 < s < \alpha t$, $0 < t < \beta s$, *where* $0 < \alpha < \min(1 + h_y)$, $0 < \beta < (\max(1 + h'_y))^{-1}$, $(x, y) \in \mathbb{T}^1 \times \mathbb{R}$, *for any* r *we have*

$$|u_k(s, t)| \le A_r (k^2 + 1)^{-1} (1 + s + t)^{-r}, \quad A_r = \text{const}. \tag{41.13}$$

In addition, in the region $r > 0$, $s > 0$ *we have*

$$|v_k(s, t)| \le A_r (k^2 + 1)^{-1} (1 + s + t)^{-r}. \tag{41.14}$$

PROOF OF THE LEMMA. First let $k \neq 0$. We rewrite (41.8) and (41.11) as follows. Integrating by parts twice we get the factor k^{-2} in front of the integral and get rid of one factor $\sqrt{1/t}$. Then we apply Lemma 41.1, proving our lemma for $k \neq 0$. If $k = 0$, then, integrating (41.11) by parts in y, we get

$$\begin{aligned} v_0(s, t) &= \frac{1}{2\pi} \sqrt{\frac{s}{t}} \int_0^1 \int_0^1 \exp\{-iy(t + s)\}(\exp\{-ith(x, y)\} - 1)\, dx\, dy \\ &= i \frac{1}{2\pi} \frac{1}{s + t} \sqrt{\frac{s}{t}} \int_0^1 \int_0^1 \exp\{-iy(t + s) - ith(x, y)\} h'_y(x, y)\, dx\, dy, \end{aligned}$$

and Lemma 41.1 yields (41.14) for $k = 0$.

The estimate (41.13) of $u_0(s, t)$ in the region $0 < s < \alpha t$ immediately follows from Lemma 41.1. If $0 < t < \beta s$, we integrate twice by parts in y to obtain

$$u_0(s, t) = i \frac{1}{2\pi} \frac{1}{s - t} \sqrt{st} \int_0^1 \int_0^1 \exp\{iy(t - s) + ith(x, y)\} h'_y(x, y)\, dx\, dy.$$

For $s > 1$ the estimate (41.13) follows from Lemma 41.1; for $0 < s < 1$ we have $0 < t < \beta$ and (41.13) follows from the continuity of the function $u_0(s, t)$. Lemma 41.2 is proved.

Now we analyze the operators A_g and B_g. Introduce the projection operators $P_m f(x, s) = \tau_m(s) f(x, s)$ in the space $L_2(\mathbb{T}^1 \times [0, \infty))$, where $\tau_m(s)$ is the characteristic function of the interval $[2\pi m, 2\pi(m + 1)]$.

LEMMA 41.3. *Let* α, β *be the same as in Lemma* 41.2. *Then*

$$\|P_m A_g P_n\| \le A_r (1 + m + n)^{-r}, \quad A_r = \text{const},$$

for any $r > 0$, *provided that the numbers* m *and* n *satisfy one of the inequalities* $0 \le m \le \alpha n$ *or* $0 \le n \le \beta m$. *Furthermore,*

$$\|P_m B_g P_n\| \le A_r (1 + m + n)^{-r}, \quad A_r = \text{const},$$

for all $r > 0$, $m \ge 0$, $n \ge 0$. *Here* A_r *does not depend on* m *and* n.

PROOF. Take a function $f \in L_2(\mathbb{T}^1 \times [0, \infty))$ vanishing for $s \notin [2\pi n, 2\pi(n+1)]$. Then formula (41.9) shows that for $m \neq n$ we have

$$\begin{aligned} &(P_m A_g f)(x, s) \\ &\qquad = 2\pi \tau_m(s) \sum_{k, r} u_k(s, s - 2\pi k\theta - 2\pi r) e^{2\pi kxi} f(x, s - 2\pi k\theta - 2\pi r), \end{aligned} \tag{41.15}$$

where the sum is taken over those k, r for which $s - 2\pi k\theta - 2\pi kr \in [2\pi n, 2\pi(n+1)]$, $s \in [2\pi m, 2\pi(m+1)]$. This implies that $k\theta + r \in [m - n - 1, m - n + 1]$. Therefore, in the sum (41.15) for each $k \in \mathbb{Z}$ the parameter r can take at most two values. Therefore (41.15) implies that for $m \neq n$ we have

$$\|P_m A_g P_n\| \\ \leq \text{const} \sum_k \max \{|u_k(s,t)|, s \in [2\pi m, 2\pi(m+1)], t \in [2\pi n, 2\pi(n+1)]\}.$$

From Lemma 41.2 it follows that if m and n satisfy one of the conditions of Lemma 41.3, we have

$$\|P_m A_g P_n\| \leq \text{const} \sum_k (k^2+1)^{-1}(m+n+1|^{-r} = \text{const}(m+n+1)^{-r}.$$

This proves the first statement of the lemma. The proof of the second statement is similar.

Lemma 41.3 implies that the operator B_g is bounded. Since the operator $\varphi \mapsto A_g\varphi + B_g\overline{\varphi}$ preserves the form $\operatorname{Im}(\varphi_1, \varphi_2)$, we have $|A_g\varphi|^2 - |B_g\overline{\varphi}|^2 = |\varphi|^2$. Therefore, the operator A_g is also bounded. Hence, the operators A_g and B_g can be extended to the entire space $L_2(\mathbb{T}^1 \times [0,\infty))$; we continue to denote these extensions by A_g and B_g.

LEMMA 41.4. *The operators A_g and B_g belong to M, and $B_g \in J_2$.*

PROOF. Let us recall that M consists of all operators in $L_2(\mathbb{T}^1 \times [0,\infty))$ that commute with the operators $(M_\varphi f)(x,s) = \varphi(x)f(x,s))$ and $(V_1 f)(x,s)) = f(x-\theta,s)e^{is}$, where the function φ is measurable and bounded. The operator $f \mapsto R_g f = A_g f + B_g\overline{f}$ commutes with the operators M_φ for real φ and with the operator V_1. Let T be one of the operators M_φ or V_1. Then $R_g T f = T R_g f$; that is, $A_g T f + B_g \overline{Tf} = T A_g f + T B_g \overline{f}$. Therefore, $A_g T = T A_g$, $B_g \overline{T} = T B_g$. But the operator T is real with respect to the involution in M, i.e., $\overline{T} = T$ (this easily follows from the definition of the involution in M). Therefore, $B_g T = T B_g$, i.e., the operators A_g and B_g commute with T, hence belong to M.

The results proved above imply that the matrix (41.6) lies in the group Sp_M^0. Recall that up to now we have considered only those diffeomorphisms $(x,y) \mapsto (x, y+h(x,y))$ from $D(\mathbb{T}^2,\theta)$ for which the function $h(x,y)$ satisfies the condition (41.7). Now we consider diffeomorphisms of the form $(x,y) \mapsto (x, y+c)$, $c \in \mathbb{R}$. For such a diffeomorphism the operator A_g is multiplication by e^{cis}, and $B_g = 0$. It is clear that the corresponding matrix belongs to Sp_M^0. One can easily see that all these diffeomorphisms together generate the group $D(\mathbb{T}^2,\theta)$. This proves Theorem 41.1.

41.4. Projective representations of the group $D(\mathbb{T}^2,\theta)$. The embedding $D(\mathbb{T}^2,\theta) \hookrightarrow Sp_M^0$ and the projective representation T^λ of the group Sp_M^0 constructed in 40.4 yield a projective representation of the group $D(\mathbb{T}^2,\theta)$; we denote it by the same letter T^λ. The fact that this is a genuine projective representation (i.e., it cannot be reduced to an ordinary representation) will be shown later.

§42. The Lie algebra $V_0^{\mathbb{C}}(\mathbb{T}^2,\theta)$, its embedding into sp_M^0, and projective representations

42.1. The embedding $V_0^{\mathbb{C}}(\mathbb{T}^2,\theta) \hookrightarrow sp_M^0$. The Lie algebra of the group $D(\mathbb{T}^2,\theta)$, denoted below by $V(\mathbb{T}^2,\theta)$, consists of all vector fields on the torus $\mathbb{T}^2$ that are tangent to the leaves of the foliation defined by the irrational number θ. Any such field is of the form $p(x,y)\partial/\partial y$, where $p(x,y) \in C^\infty(\mathbb{T}^1 \times \mathbb{R})$ satisfies the condition $p(x-\theta, y+1) = p(x,y)$. Denote by $V_0(\mathbb{T}^2,\theta)$ the subalgebra of fields $p(x,y)\partial/\partial y$, where $p(x,y)$ is a trigonometric polynomial, and by $V_0^{\mathbb{C}}(\mathbb{T}^2,\theta)$ the complexification of $V_0(\mathbb{T}^2,\theta)$. In $V_0^{\mathbb{C}}(\mathbb{T}^2,\theta)$ we have a natural basis indexed by elements of the group $\Gamma = \{\gamma = m\theta + n \mid (m,n) \in \mathbb{Z}^2\}$ and consisting of the fields

$$f_\gamma = (2\pi i)^{-1}\exp\{2\pi m i(x+\theta y) + 2\pi n y i\}(\partial/\partial y), \quad \gamma = m\theta + n \in \Gamma. \tag{42.1}$$

It is clear that

$$[f_{\gamma_1}, f_{\gamma_2}] = (\gamma_2 - \gamma_1) f_{\gamma_1+\gamma_2}. \tag{42.2}$$

Theorem 41.1 gives an embedding $V_0(\mathbb{T}^2,\theta) \hookrightarrow sp_M$, which maps a vector field $f = p(x,y)(\partial/\partial y)$ the matrix

$$l_f = \begin{pmatrix} a_f & b_f \\ \overline{b_f} & \overline{a_f} \end{pmatrix} \tag{42.3}$$

defined by the formula $l_f = (d/dt)g_t\,|_{t=0}$, where $g_t \in Sp_M^0$ is the image of the diffeomorphism $\exp\{tf\} \in D(\mathbb{T}^2,\theta)$ under the embedding introduced in Theorem 41.1. Passing to complexifications, we get the embedding $V_0^{\mathbb{C}}(\mathbb{T}^2,\theta) \hookrightarrow sp_M^{\mathbb{C}}$. Let us give an explicit formula for the matrix (42.3); instead of the integral operators a_f and b_f we write the corresponding kernels (these kernels are distributions). Carrying out computations similar to those leading to formulas for the kernels (41.4) and (41.5) (see 41.3), we get

$$l_f = \frac{i}{2\pi}\begin{pmatrix} \sqrt{st}\,\widehat{p}(x, s-t) & -\sqrt{s\sigma(t)}\,\widehat{p}(x, s+\sigma(t)) \\ \sqrt{\sigma(s)t}\,\widehat{p}(x, -\sigma(s)-t) & -\sqrt{\sigma(s)\sigma(t)}\,\widehat{p}(x, \sigma(t)-\sigma(s)) \end{pmatrix},$$

where $\widehat{p}(x,s)$ is the Fourier transform of the function $p(x,y)$ in the variable y. This formula implies that the matrix corresponding to the basic vector fields f_γ, $\gamma \in \Gamma$, is given by the formula

$$h_f = \frac{\exp\{2\pi m x i\}}{2\pi}\begin{pmatrix} \sqrt{st}\,\delta(s-t-2\pi\gamma) & -\sqrt{s\sigma(t)}\,\delta(s+\sigma(t)-2\pi\gamma) \\ \sqrt{\sigma(s)t}\,\delta(\sigma(s)+t+2\pi\gamma) & -\sqrt{\sigma(s)\sigma(t)}\,\delta(\sigma(t)-\sigma(s)-2\pi\gamma) \end{pmatrix}. \tag{42.4}$$

(Recall that the transformation $t \mapsto \sigma(t)$, $t \geq 0$, was defined in 41.2.)

42.2. Projective representations of the Lie algebra $V_0^{\mathbb{C}}(\mathbb{T}^2,\theta)$ corresponding to the representation T^λ of the group $D(\mathbb{T}^2,\theta)$. Restricting the representation r^λ constructed in 40.6 to the subalgebra $V_0(\mathbb{T}^2,\theta)$, we get a representation of the latter subalgebra; denote it also by r^λ. Let us show that it is naturally related to the representation T^λ of the group $D(\mathbb{T}^2,\theta)$.

LEMMA 42.1. *For $f \in V_0(\mathbb{T}^2,\theta)$ and $\varphi \in F^0$, setting $g_t = \exp\{tf\}$, we have*

$$(r^\lambda(f)\varphi)(X) = \frac{d}{dt}(T^\lambda(g_t)\varphi)(X)\Big|_{t=0},$$

Let us recall that the relation between the representations r^λ and T^λ of the group Sp_M and the Lie algebra sp_M (see Lemma 40.3) was not that direct.

PROOF OF LEMMA 42.1. Under the indicated embeddings the element g_t goes to the matrix

$$K(t) = \begin{pmatrix} Ag_t & Bg_t \\ \overline{Bg_t} & \overline{Ag_t} \end{pmatrix}$$

(see (41.6)) and the field $f = \sum a_\gamma f_\gamma$, $a_{-\gamma} = \overline{a_\gamma}$, to the matrix $h = \sum a_\gamma h_\gamma$, where h_γ is given by formula (42.4). We have $K'(t) = hK(t)$. Using the estimates from Lemma 41.3, we obtain, for any $t \geq 0$, the inequalities

$$\left\| \left(\frac{d}{dt} Bg_t \right)_{pq} \right\| \leq \text{const}(1+p+q)^{-r}$$

for all $p \geq 0$, $q \geq 0$, $r > 0$, and

$$\left\| \left(\frac{d}{dt} Ag_t \right)_{pq} \right\| \leq \text{const}(1+p+q)^{-r}$$

for $m \leq \alpha n$ and $n \leq \beta m$. Therefore, the family of matrices $K(t)$ satisfies the conditions of Lemma 40.3, and Lemma 42.1 is proved.

Now we derive the commutation relations for the operators $L_\gamma = r^\lambda(f_\gamma)$.

THEOREM 42.1.

$$[L_{\gamma_1}, L_{\gamma_2}] = (\gamma_3 - \gamma_2) L_{\gamma_1+\gamma_2} - \frac{\lambda}{6}\gamma_2^3 \delta(\gamma_1+\gamma_2), \qquad \gamma_1, \gamma_2 \in \Gamma.$$

(*As usual,* $\delta(\gamma) = 0$ *for* $\gamma \neq 0$, $\gamma \in \Gamma$, *and* $\delta(0) = 1$.)

PROOF. Formulas (40.8) and (42.4) imply that

$$[L_{\gamma_1}, L_{\gamma_2}] = (\gamma_2 - \gamma_1) L_{\gamma_1+\gamma_2} + \lambda \operatorname{tr}\left((h_{\gamma_2})_{12}(h_{\gamma_1})_{21} - (h_{\gamma_2})_{21}(h_{\gamma_1})_{12}\right),$$

where h_{ij} are the entries of the matrix h. Using (42.2) one easily sees that the operator K_{γ_1,γ_2} under the trace sign in the last formula is given by the kernel

$$(4\pi^2)^{-1} e^{2\pi i(m_1+m_2)x} \big(\sqrt{\sigma(s)\sigma(t)}(2\pi\gamma_1 - \sigma(t))_+ \delta(\sigma(s) - \sigma(t) + 2\pi(\gamma_1+\gamma_2)) \\ - \sqrt{st}(2\pi\gamma_2 - s)_+ \delta(t - s + 2\pi(\gamma_1+\gamma_2))\big),$$

where $a_+ = \max(a, 0)$. We compute $\operatorname{tr} K_{\gamma_1,\gamma_2}$ using formula (41.1). If $m_1+m_2 \neq 0$, then clearly $\operatorname{tr} K_{\gamma_1,\gamma_2} = 0$ (because of the factor $\exp\{2\pi i(m_1+m_2)x\}$). If $m_1+m_2 = 0$, $n_1+n_2 \neq 0$, then $\gamma_1+\gamma_2 = n_1+n_2$ is an integer. Therefore, the operator K_{γ_1,γ_2} maps the function g_n (see (41.1)) to a function whose support is disjoint from the support of g_n. Therefore, $\operatorname{tr} K_{\gamma_1,\gamma_2} = 0$. Finally, if $m_1 + m_2 = 0$, $n_1 + n_2 = 0$, i.e., $\gamma_1 + \gamma_2 = 0$, and, for example, $\gamma_2 \geq 0$, then K_{γ_1,γ_2} is given by the kernel $-(4\pi^2)^{-1} s(2\pi\gamma_2 - s)_+ \delta(t-s)$. In other words,

$$(K_{\gamma_1,\gamma_2}\varphi)(x,s) = -(4\pi^2)^{-1}(2\pi\gamma_2 - s)_+ \varphi(x,s).$$

Therefore,

$$\begin{aligned}\operatorname{tr} K_{\gamma_1,\gamma_2} &= \sum_{n=0}^{\infty}(K_{\gamma_1,\gamma_2} g_n, g_n) = -\frac{1}{4\pi^2}\frac{1}{2\pi}\sum_{n=0}^{\infty}\int_{2\pi n}^{2\pi(n+1)} s(2\pi\gamma_2 - s)_+\, ds \\ &= -\frac{1}{8\pi^3}\int_0^{\infty} s(2\pi\gamma_2 - s)_+\, ds = -\frac{1}{8\pi^3}\int_0^{2\pi\gamma_2} s(2\pi\gamma_2 - s)\, ds = \frac{1}{6}\gamma_2^3.\end{aligned}$$

Theorem 42.1 is proved.

42.3. The Lie algebra V_Γ (an analog of the Virasoro algebra). Theorem 42.1 indicates that we can consider the Lie algebra V_Γ with generators $\{\psi_\gamma,\ \gamma \in \Gamma\}$ and e and relations

$$\begin{aligned}[\psi_{\gamma_1}, \psi_{\gamma_2}] &= (\gamma_2 - \gamma_1)\psi_{\gamma_1+\gamma_2} + \gamma_2^3\delta(\gamma_1+\gamma_2)e, \\ [\psi_\gamma, e] &= 0.\end{aligned}$$

This algebra is a one-dimensional central extension of the algebra $V_0^{\mathbb{C}}(\mathbb{T}^2, \theta)$. Theorem 42.1 describes the representation of this algebra that maps the central element e to $-\frac{1}{6}\lambda$. The function $f_0(X) = 1$ is a weight vector of weight 0. The author does not claim that the algebra V_Γ deserves to be analyzed as closely as the ordinary Virasoro algebra (Verma modules, etc.). We confine ourselves to one simple observation.

Consider a representation π of the algebra V_Γ in a space E satisfying the following conditions:

(1) $\pi(e) = cI$, $c \in \mathbb{C}$;

(2) there exists a cyclic vector $v_0 \in E$ such that $L_\gamma v_0 = 0$, $\gamma \le 0$, $L_\gamma = \pi(f_\gamma)$.

An example of such a representation is the restriction of the representation r^λ to the subspace generated by the function $f_0(X) = 1$. Denote $\Gamma_+ = \Gamma \cap [0, \infty)$. One can easily see that $E = \bigoplus_{\gamma\in\Gamma_+} E_\gamma$, where $E_\gamma = \{v \mid L_0 v = \gamma v\}$. The subspace E_γ, $\gamma > 0$, is the linear span of the vectors $L_{\gamma_n}\cdots L_{\gamma_1}v_0$, $\gamma_i \in \Gamma_+$, $\sum \gamma_i = \gamma$, $\gamma_i > 0$.

LEMMA 42.2. *$\dim E_\gamma = \infty$ for $\gamma > 0$, and $\dim E_0 = 1$.*

PROOF. Consider the vectors of the form $L_{y-\gamma}L_{-y}L_x L_{\gamma-x}v_0$, where $0 < x < \gamma$, $0 < y < \gamma$. The weight of each such vector is zero; hence it has the form $A(x, y)v_0$, where $A(x, y)$ is a function. One can easily see that if x and y are sufficiently close to γ, then

$$A(x,y) = \begin{cases} (x+y)(y+1-2x)(1-y)^3 & \text{if } x < y, \\ (x+y)(x+1-2y)(1-x)^3 & \text{if } x > y, \\ 2x(1-x)^4 & \text{if } x = y. \end{cases}$$

One can easily see that $A(x, y)$ cannot be written as a finite sum $\sum a_k(x)b_k(y)$, i.e., $A(x, y)$ is not a "degenerate kernel." This implies that the linear span of the vectors $L_x L_{\gamma-x}v_0$ is infinite-dimensional, and Lemma 42.2 is proved.

Returning to the representation r^λ in the space generated by the function $f_0(X)$, we see that the set of eigenvalues of L_0 is Γ_+, with the eigenvalue 0 having multiplicity 1, and all eigenvalues $\gamma > 0$ multiplicity ∞.

§43. Concluding remarks

43.1. Let us outline the following generalization of the previous construction. Consider a compact manifold with an ergodic flow determined by a smooth vector field. We obtain a one-dimensional foliation F. Denote by $D(X, F)$ the group of leaf-preserving diffeomorphisms. How to generalize the previous construction to $D(X, F)$? It is reasonable to consider the following more general group G. Let (Y, μ) be a space with measure, S a measure preserving transformation. Identifying points (y, s) and $(Sy, s + 1)$ in $Y \times \mathbb{R}$, we obtain a space Z. This is a well-known construction of the special flow. The images of lines $y = y_0$ in Z are called leaves. Let G be the group of all measurable transformation of the space Z that map each leaf diffeomorphically to itself. As before, we can use (Y, μ, S) to construct a factor M (assuming S to be ergodic) and the group Sp_M. If S has a point spectrum, then G can be embedded into Sp_M.

Returning to the group $D(X, F)$, let us note that it can often be embedded into G; for Y one can take a disk on the manifold that is transversal to leaves, and for S the time shift transformation (it remains to require the existence of an S-invariant measure on Y).

43.2. Similarly to Sp_M, one can construct the orthogonal group O_M. Into this group one can embed the current group $C^\infty(\mathbb{T}^2, U_n)$. However, the construction of projective representations cannot be extended to O_M. The reason is that although one can construct a local (near the identity) 2-cocycle similarly to (40.5), this cocycle cannot be extended to the entire group O_M. However, the embedding of the current group into O_M might be of interest. In particular, this embedding gives an example of the representation of the current group by operators from a type II factor.

43.3. Let us return to an arbitrary manifold X with a foliation F and to the group $D(X, F)$. Similarly to the group $D(X)$, one can consider in $D(X, F)$ the subgroups $D(X, F, v^k)$, $D(X, F, \omega^2)$, $D(X, F, \alpha^1)$. For example, $D(X, F, v^n)$ is the connected component of the unit element in the group of all diffeomorphisms $g \in D(X, F)$ preserving the volume form v^k defined on leaves (and depending smoothly on $x \in X$); here k is the dimension of a leaf.

Unitary (possibly projective) representations of the group $D(X, F)$ were not considered before. Even the simplest constructions require the consideration of transversal measures on the set of leaves (in particular, ergodic measures are of interest) and 1-cohomology of the action of $D(X, F)$ on groups and on spaces of functions on X.

References

1. V. I. Arnold, *The one-dimensional cohomology of the Lie algebra of divergence-free vector fields, and the winding numbers of dynamical systems*, Funktsional. Anal. i Prilozhen. **3** (1969), no. 4, 77–78; English transl. in Functional Anal. Appl. **3** (1969).
2. F. A. Berezin, *The method of second quantization*, "Nauka", Moscow, 1965; English transl., Academic Press, New York, 1966.
3. ———, *Representations of the continuous direct product of universal coverings of the group of motions of a complex ball*, Trudy Moskov. Mat. Obshch. **36** (1978), 275–293; English transl. in Trans. Moscow Math. Soc. **1979**, no. 2 (36).
4. N. Bourbaki, *Éléments de mathématique*. Fasc. XXXIII, XXXVI: *Variétés différentielles et analytiques*, Hermann, Paris, 1967, 1971.
5. A. M. Vershik, *A description of invariant measures for actions of certain infinite-dimensional groups*, Dokl. Akad. Nauk SSSR **218** (1974), 749–752; English transl. in Soviet Math. Dokl. **15** (1974).
6. A. M. Vershik, I. M. Gelfand, and M. I. Graev, *Representations of the group* SL(2, R)*, where R is a ring of functions*, Uspekhi Mat. Nauk **28** (1973), no. 5, 83–128; English transl. in Russian Math. Surveys **28** (1973).
7. ———, *Irreducible representations of the group G^X and cohomology*, Funktsional. Anal. i Prilozhen. **8** (1974), no. 2, 67–69; English transl. in Functional Anal. Appl. **8** (1974).
8. ———, *Representations of the group of diffeomorphisms*, Uspekhi Mat. Nauk **30** (1975), no. 6, 1–50; English transl. in Russian Math. Surveys **30** (1975).
9. I. I. Gikhman and A. V. Skorokhod, *Theory of stochastic processes*. Vol. 1, "Nauka", Moscow, 1971; English transl., Springer-Verlag, Berlin, 1974.
10. A. Guichardet, *Cohomologie des groupes algébriques et des algèbres de Lie*, CEDIC, Paris, 1980.
11. C. Godbillon, *Géométrie différentielle et mécanique analytique*, Hermann, Paris, 1969.
12. M. I. Golenishcheva-Kutuzova, *Irreducible unitary representations of the group* $\mathrm{Diff}(S_2, \omega)$ *of functional dimension two*, Funktsional. Anal. i Prilozhen. **25** (1991), no. 3, 85–88; English transl. in Functional Anal. Appl. **25** (1991).
13. A. A. Zaĭtsev, *Unitary representations of a special group of jets*, Funktsional. Anal. i Prilozhen. **6** (1972), no. 4, 85–86; English transl. in Functional Anal. Appl. **6** (1972).
14. R. S. Ismagilov, *The unitary representations of the group of diffeomorphisms of a circle*, Funktsional. Anal. i Prilozhen. **5** (1971), no. 3, 45–53; English transl. in Functional Anal. Appl. **5** (1971).
15. ———, *The unitary representations of the diffeomorphism group of a compact manifold*, Izv. Akad. Nauk SSSR Ser. Mat. **36** (1972), 180–208; English transl. in Math. USSR-Izv. **6** (1972).
16. ———, *The unitary representations of the group of diffeomorphisms of the space R^n, $n \geq 2$*, Mat. Sb. **98** (1975), no. 1, 55–71; English transl. in Math. USSR-Sb. **27** (1975).
17. ———, *Unitary representations of the group $C_0^\infty(X, G)$, $G = \mathrm{SU}_2$*, Mat. Sb. **100** (1976), no. 1, 117–131; English transl. in Math. USSR-Sb. **29** (1976).
18. ———, *On the group of volume-preserving diffeomorphisms*, Izv. Akad. Nauk SSSR Ser. Mat. **44** (1980), 831–867; English transl. in Math. USSR-Izv. **17** (1981).
19. ———, *Representations of the group of smooth mappings of a segment in a compact Lie group*, Funktsional. Anal. i Prilozhen. **15** (1981), no. 2, 73–74; English transl. in Functional Anal. Appl. **15** (1981).

20. ______, *Groups and homogeneous domains connected with factors of type* II, *and representations of the group of diffeomorphisms of a torus that preserve the irrational windings*, Algebra i Analiz **5** (1993), no. 1, 215–231; English transl. in St. Peterburg Math. J. **5** (1994).
21. ______, *On extensions of current groups and their representations*, Algebra i Analiz **6** (1994), no. 1, 159–171; English transl. in St. Petersburg Math. J. **6** (1995).
22. S. I. Karpushev, *On the irreducibility of the multiplicative integral of representations*, Vestnik Leningrad. Univ. **1981**, no. 7 (Ser. Mat. Mekh. Astr. vyp. 2), 37–42; English transl. in Vestnik Leningrad Univ. Math. **14** (1982).
23. J.-P. Kahane, *Some random series of functions*, D. C. Heath, Lexington, MA, 1968.
24. A. A. Kirillov, *Elements of representation theory*, "Nauka", Moscow, 1972; English transl., Springer-Verlag, Berlin, 1976.
25. B. Kostant, *Quantization and unitary representations*, Lectures in Modern Analysis and Applications. III, Lecture Notes in Math., vol. 170, Springer-Verlag, Berlin, 1970, pp. 87–280.
26. M. A. Naimark, *Normed rings*, 2nd rev. ed., "Nauka", Moscow, 1968; English transl., *Normed algebras*, Noordhoff, Groningen, 1972.
27. ______, *Structure of unitary representations of locally bicompact groups and symmetric representations of algebras in the Pontrjagin spaces* Π_k, Izv. Akad. Nauk SSSR Ser. Mat. **30** (1966), 1111–1132; English transl. in Amer. Math. Soc. Transl. (2) **73** (1968).
28. ______, *Degenerate operator algebras in the Pontrjagin space* Π_k, Izv. Akad. Nauk SSSR Ser. Mat. **30** (1966), 1229–1256; English transl. in Amer. Math. Soc. Transl. (2) **68** (1968).
29. F. J. Murray and J. von Neumann, *On rings of operators.* I, II, III, IV, Ann. of Math. (2) **37** (1936), 116–229; Trans. Amer. Math. Soc. **41** (1937), 208–248; Ann. of Math. (2) **41** (1940), 94–161; Ann. of Math. (2) **44** (1943), 716–808.
30. Yu. A. Neretin, *Representations of Virasoro algebra and affine algebras*, Itogi Nauki i Tekhniki: Sovremennye Problemy Mat.: Fundametal′nye Napravleniya, vol. 22, VINITI, Moscow, 1988, pp. 163–244; English transl. in Representation Theory and Noncommutative Harmonic Analysis. I, Encyclopaedia of Math. Sci., vol. 22, Springer-Verlag, Berlin, 1994.
31. ______, *Categories of bistochastic measures and representations of some infinite-dimensional groups*, Mat. Sb. **183** (1992), no. 2, 52–76; English transl. in Russian Acad. Sci. Sb. Math. **75** (1993).
32. G. I. Olshanskiĭ, *Spherical functions and characters of the group* G^X, Uspekhi Mat. Nauk **37** (1982), no. 2, 217–218; English transl. in Russian Math. Surveys **37** (1982).
33. A. Pressley and G. Segal, *Loop groups*, Oxford Univ. Press, Oxford, 1986.
34. I. I. Pyatetskii-Shapiro, *Automorphic functions and the geometry of classical domains*, Fizmatgiz, Moscow, 1961; English transl., Gordon and Breach, New York, 1969.
35. M. Reed and B. Simon, *Methods of modern mathematical physics.* Vols. I, II, III, IV, Academic Press, New York and London, 1972, 1975, 1979, 1978.
36. D. Fuchs, *Cohomology of infinite-dimensional Lie algebras*, "Nauka", Moscow, 1984; English transl., Plenum Press, New York, 1986.
37. E. T. Shavgulidze, *A measure that is quasi-invariant with respect to the action of a group of diffeomorphisms of a finite-dimensional manifold*, Dokl. Akad. Nauk SSSR **303** (1988), no. 4, 811-814; English transl. in Soviet Math. Dokl. **38** (1989).
38. V. S. Shulman, *Symmetric Banach algebras of operators in a space of type* Π_1, Mat. Sb. **89** (1972), 264–279; English transl. in Math. USSR-Sb. **18** (1972).
39. S. Twareque Ali and G. A. Goldin, *Quantization, coherent states and diffeomorphisms groups*, Lecture Notes in Physics, vol. 379, Springer-Verlag, Berlin, 1991, pp. 379–401.
40. S. Albeverio, R. Høegh-Krohn, and D. Testard, *Irreducibility and reducibility for the energy representation of the group of mappings of a Riemanian manifold into a compact semisimple Lie group*, J. Funct. Anal. **41** (1981), no. 3, 378–396.
41. S. Albeverio, R. Høegh-Krohn, D. Testard, and A. Vershik, *Factorial representations of path groups*, J. Funct. Anal. **51** (1983), no. 1, 115–131.
42. H. Araki, *Factorisable representations of current algebra*, Publ. Res. Inst. Math. Sci. Kyoto Univ. **5** (1970), 311–422.
43. H. Araki and J. Words, *Complete Boolean algebras of type* I *factors*, Publ. Res. Inst. Math. Sci. Kyoto Univ. **2** (1966), 157–242.
44. A. Banyaga, *Sur le groupe des difféomorphisms symplectiques*, Lecture Notes in Math., vol. 484, Springer-Verlag, Berlin, 1975, pp. 50–56.

45. E. Calabi, *On the group of automorphisms of a symplectic manifold*, Problems in Analysis, Symposium in Honor of S. Bochner, Princeton Univ. Press., Princeton, NJ, 1977, pp. 1–26.
46. P. R. Chernoff, *Irreducible representations of infinite dimensional trasformation groups and Lie algebras*, Bull. Amer. Math. Soc. (N.S.) **13** (1985), 46–49.
47. P. Delorme, *Irreducibilité de certaines représentations de* G^X, J. Funct. Anal. **30** (1978), 36–47.
48. A. Fathi and Yves-Marie Visetti, *Structure du groupe des homéomorphisms préservant une mesure sur une variété compacte*, C. R. Acad. Sci. Paris **284** (1977), no. 15, 848–852.
49. R. H. Fox and L. P. Newirth, *The braid groups*, Math. Scand. **10** (1962), 119–126.
50. B. Fuglede and R. Kadison, *Determinant theory in finite factors*, Ann. of Math. (2) **55** (1952), 510–530.
51. I. Gelfand, M. Graev, and A. Vershik, *Remarks on representations of the group of functions with values in compact Lie groups*, Compositio Math. **42** (1981), 217–243.
52. G. A. Goldin, J. Grodnik, R. T. Powers, and D. H. Sharp, *Nonrelativistic current algebra in the* N/V *limit*, J. Math. Phys. **15** (1974), 88–100.
53. G. A. Goldin and D. H. Sharp, *Particle statistics and representations of diffeomorphism groups*, Comm. Math. Phys. **92** (1983), 217–241.
54. A. Guichardet, *Symmetric Hilbert spaces and related topics*, Lecture Notes in Math., vol. 261, Springer-Verlag, Berlin, 1972.
55. J. P. Kahane, *Sur le chaos multiplicatif*, Ann. Sci. Math. Québec **9** (1985), 105–150; correction, ibid. **10** (1986), 117-118.
56. A. A. Kirillov, *Infinite dimensional Lie groups, their orbits, invariants and representations*, Lecture Notes in Math., vol. 970, Springer-Verlag, Berlin, 1982, pp. 101–123.
57. S. Kusuoka, *Høegh-Krohn's model of quantum fields and the absolute continnity of measures*, Ideas and Methods in Quantum and Statistical Physics, vol. 2, Cambridge Univ. Press, Cambridge, 1992, pp. 405–424.
58. G. W. Mackey, *Imprimitivity for representations of locally compact groups*, Proc. Nat. Acad. Sci. USA **35** (1939), 123–125.
59. J. Mickelsson, *Current algebras and groups*, Plenum Press, New York, 1988.
60. J. Mickelsson and S. Rajeev, *Current algebras in* $d+1$ *dimensions and determinant bundles over infinite-dimensional Grassmanians*, Comm. Math. Phys. **116** (1988), 365–400.
61. J. Moser, *On the volume elements on manifolds*, Trans. Amer. Math. Soc. **120** (1965), 280–296.
62. K. R. Parthasarathy and K. Schmidt, *Factorisable representations of current groups and the Araki-Woods embedding theorems*, Acta Math. **128** (1972), 53–71.
63. R. T. Seeley, *Complex powers of an elliptic operator*, Proc. Sympos. Pure Math., vol. 10, Amer. Math. Soc., Providence, RI, 1967, pp. 288–307.
64. G. B. Segal, *Unitary representations of some infinite dimensional groups*, Comm. Math. Phys. **88** (1981), 301–342.
65. A. Weinstein, *Cohomology of symplectomorphism groups and critical values of Hamiltonians*, Math. Z. **201** (1989), 75–82.
66. F. Wilczek, *On braid statistics*, Phys. Rev. Lett. **49** (1982), 957–960.
67. V. Guillemin and S. Sternberg, *Geometric asymptotics*, Amer. Math. Soc., Providence, RI, 1977.
68. W. Feller, *An introduction to probability theory and its applications.* Vol. 1, Wiley, New York, 1966.

Selected Titles in This Series

(Continued from the front of this publication)

115 **Nikolai V. Ivanov,** Subgroups of Teichmüller modular groups, 1992
114 **Seizô Itô,** Diffusion equations, 1992
113 **Michail Zhitomirskiĭ,** Typical singularities of differential 1-forms and Pfaffian equations, 1992
112 **S. A. Lomov,** Introduction to the general theory of singular perturbations, 1992
111 **Simon Gindikin,** Tube domains and the Cauchy problem, 1992
110 **B. V. Shabat,** Introduction to complex analysis Part II. Functions of several variables, 1992
109 **Isao Miyadera,** Nonlinear semigroups, 1992
108 **Takeo Yokonuma,** Tensor spaces and exterior algebra, 1992
107 **B. M. Makarov, M. G. Goluzina, A. A. Lodkin, and A. N. Podkorytov,** Selected problems in real analysis, 1992
106 **G.-C. Wen,** Conformal mappings and boundary value problems, 1992
105 **D. R. Yafaev,** Mathematical scattering theory: General theory, 1992
104 **R. L. Dobrushin, R. Kotecký, and S. Shlosman,** Wulff construction: A global shape from local interaction, 1992
103 **A. K. Tsikh,** Multidimensional residues and their applications, 1992
102 **A. M. Il′in,** Matching of asymptotic expansions of solutions of boundary value problems, 1992
101 **Zhang Zhi-fen, Ding Tong-ren, Huang Wen-zao, and Dong Zhen-xi,** Qualitative theory of differential equations, 1992
100 **V. L. Popov,** Groups, generators, syzygies, and orbits in invariant theory, 1992
99 **Norio Shimakura,** Partial differential operators of elliptic type, 1992
98 **V. A. Vassiliev,** Complements of discriminants of smooth maps: Topology and applications, 1992 (revised edition, 1994)
97 **Itiro Tamura,** Topology of foliations: An introduction, 1992
96 **A. I. Markushevich,** Introduction to the classical theory of Abelian functions, 1992
95 **Guangchang Dong,** Nonlinear partial differential equations of second order, 1991
94 **Yu. S. Il′yashenko,** Finiteness theorems for limit cycles, 1991
93 **A. T. Fomenko and A. A. Tuzhilin,** Elements of the geometry and topology of minimal surfaces in three-dimensional space, 1991
92 **E. M. Nikishin and V. N. Sorokin,** Rational approximations and orthogonality, 1991
91 **Mamoru Mimura and Hirosi Toda,** Topology of Lie groups, I and II, 1991
90 **S. L. Sobolev,** Some applications of functional analysis in mathematical physics, third edition, 1991
89 **Valeriĭ V. Kozlov and Dmitriĭ V. Treshchëv,** Billiards: A genetic introduction to the dynamics of systems with impacts, 1991
88 **A. G. Khovanskiĭ,** Fewnomials, 1991
87 **Aleksandr Robertovich Kemer,** Ideals of identities of associative algebras, 1991
86 **V. M. Kadets and M. I. Kadets,** Rearrangements of series in Banach spaces, 1991
85 **Mikio Ise and Masaru Takeuchi,** Lie groups I, II, 1991
84 **Đào Trọng Thi and A. T. Fomenko,** Minimal surfaces, stratified multivarifolds, and the Plateau problem, 1991
83 **N. I. Portenko,** Generalized diffusion processes, 1990
82 **Yasutaka Sibuya,** Linear differential equations in the complex domain: Problems of analytic continuation, 1990

(See the AMS catalog for earlier titles)